Ajith Abraham, Aboul-Ella Hassanien,
André Ponce de Leon F. de Carvalho, and Václav Snášel (Eds.)

Foundations of Computational Intelligence Volume 6

T0206147

Studies in Computational Intelligence, Volume 206

Editor-in-Chief

Prof. Janusz Kacprzyk
Systems Research Institute
Polish Academy of Sciences
ul. Newelska 6
01-447 Warsaw
Poland
E-mail: kacprzyk@ibspan.waw.pl

Ajith Abraham, Aboul-Ella Hassanien,
André Ponce de Leon F. de Carvalho, and
Václav Snášel (Eds.)

Foundations of Computational Intelligence Volume 6

Data Mining

Springer

Dr. Ajith Abraham
Machine Intelligence Research Labs
(MIR Labs)
Scientific Network for Innovation
and Research Excellence
P.O. Box 2259 Auburn,
Washington 98071-2259
USA
E-mail: ajith.abraham@ieee.org
http://www.mirlabs.org
http://www.softcomputing.net

Prof. Aboul-Ella Hassanien
College of Business Administration
Quantitative and Information System
Department
Kuwait University
P.O. Box 5486
Safat, 13055
Kuwait
E-mail: abo@cba.edu.kw

Prof. André Ponce de Leon F. de
Carvalho
Department of Computer Science
University of São Paulo
SCE - ICMSC - USP
Caixa Postal 668
13560-970 Sao Carlos, SP
Brazil
E-mail: andre@icmc.usp.br

Václav Snášel
Technical University Ostrava
Dept. Computer Science
Tr. 17. Listopadu 15
708 33 Ostrava
Czech Republic
E-mail: vaclav.snasel@vsb.cz

ISBN 978-3-642-10167-0   e-ISBN 978-3-642-01091-0

DOI 10.1007/978-3-642-01091-0

Studies in Computational Intelligence   ISSN 1860949X

Typeset & Cover Design: Scientific Publishing Services Pvt. Ltd., Chennai, India.

Printed in acid-free paper

9 8 7 6 5 4 3 2 1

springer.com

Preface

Foundations of Computational Intelligence

Volume 6: Data Mining: Theoretical Foundations and Applications

Finding information hidden in data is as theoretically difficult as it is practically important. With the objective of discovering unknown patterns from data, the methodologies of data mining were derived from statistics, machine learning, and artificial intelligence, and are being used successfully in application areas such as bioinformatics, business, health care, banking, retail, and many others. Advanced representation schemes and computational intelligence techniques such as rough sets, neural networks; decision trees; fuzzy logic; evolutionary algorithms; artificial immune systems; swarm intelligence; reinforcement learning, association rule mining, Web intelligence paradigms etc. have proved valuable when they are applied to Data Mining problems. Computational tools or solutions based on intelligent systems are being used with great success in Data Mining applications. It is also observed that strong scientific advances have been made when issues from different research areas are integrated.

This Volume comprises of 15 chapters including an overview chapter providing an up-to-date and state-of-the research on the applications of Computational Intelligence techniques for Data Mining.

The book is divided into 3 parts:

Part-I: Data Click Streams and Temporal Data Mining
Part-II: Text and Rule Mining
Part-III: Applications

Part I on Data Click Streams and Temporal Data Mining contains four chapters that describe several approaches in Data Click Streams and Temporal Data Mining.

Hannah and Thangavel in Chapter 1, "Mining and Analysis of Clickstream Patterns" propose a Multi-Pass CSD-Means algorithm for clustering web clickstream patterns from web access logs of Microsoft Web site, which is available in the

UCI repository. Using these algorithms can make a meaningful contribution for the clustering analysis of web logs. This algorithm estimates the optimum number of clusters automatically and is capable of manipulating efficiently very large data sets.

In Chapter 2, "An Overview on Mining Data Streams" Gama and Rodrigues discuss the main challenges and issues when learning from data streams. Authors illustrate the most relevant issues in knowledge discovery from data streams: incremental learning, cost-performance management, change detection, and novelty detection. Illustrative algorithms are presented for these learning tasks, and a real-world application illustrating the advantages of stream processing.

Chapter 3, "Data Stream Mining Using Granularity-based Approach" by Gaber presents novel approach to solving the problem of mining data streams in resource constrained environments that are the typical representatives for data stream sources and processing units in many applications. The proposed Algorithm Output Granularity (AOG) approach adapts the output rate of a data-mining algorithm according to available resources and data rate. AOG has been formalized and the main concepts and definitions have been introduced followed by a rigorous discussion.

In Chapter 4, "Time Granularity in Temporal Data Mining" Cotofrei1 and Stoffel, illustrate formalism for a specific temporal data mining task: the discovery of knowledge, represented in the form of general Horn clauses, inferred from databases with a temporal dimension. The theoretical framework proposed by the authors is based on first-order temporal logic, which permits to define the main notions (event, temporal rule, and constraint) in a formal way. The concept of a consistent linear time structure allows us to introduce the notions of general interpretation, of support and of confidence, the lasts two measure being the expression of the two similar concepts used in data mining.

In Chapter 5, "Mining User Preference Model from Utterances" Takama1 and Muto, introduce a method for mining user preference model from user's behavior including utterances. The authors also analyze utterance of user. The proposed approach is applied to TV program recommendation, in which the log of watched TV programs as well as utterances while watching TV is collected. First, user's interest in a TV program is estimated based on fuzzy inference, of which inputs are watching time, utterance frequency, and contents of utterances obtained by sentiment analysis. Then, user profile is generated by identifying features common to user's favorite TV programs.

Part II on Text and Rule Mining contains six chapters discussing many approaches in text and rule mining problem.

One of the most relevant today's problems called information overloading has increased the necessity of more sophisticated and powerful information compression methods or summarizers. Chapter 6, "Text Summarization: an Old Challenge and New Approaches" by Steinberger and Karel, firstly introduce taxonomy of summarization methods and an overview of their principles from classical ones, over corpus based, to knowledge rich approaches. Authors considered various aspects, which can affect the categorization. A special attention is devoted to application of recent information reduction methods, based on algebraic transformations including Latent Semantic Analysis. Evaluation measures for assessing quality of a summary and taxonomy of evaluation measures is also presented.

Chapter 7, "From Faceted Classification to Knowledge Discovery of Semi-Structured Text Records", by Yee et al. implement a faceted classification approach to enhance the retrieval and knowledge discovery from extensive aerospace in-service records. The retrieval mechanism afforded by faceted classification can expedite responses to urgent in-service issues as well as enable knowledge discovery that could potentially lead to root-cause findings and continuous improvement.

Lui and Chiu in Chapter 8, "Multi-Value Association Patterns and Data Mining" discuss three related multi-value association patterns and their relationships. All of them have shown to be very important for data mining involving discrete valued data. Furthermore, authors generalized sequential data and have a more specific interpretation to multi-variable patterns. Further evaluations with respect to their conceptual mathematical properties are also presented in the Chapter.

Chapter 9, "Clustering Time Series Data: An Evolutionary Approach" by Monica et.al. discusses the state-of-the-art methodology for some mining time series databases and presents a new evolutionary algorithm for times series clustering an input time series data set. The data mining methods presented include techniques for efficient segmentation, indexing, and clustering time series.

Chapter 10, "Support Vector Clustering: From Local Constraint to Global Stability" by Khanloo et. al. Study the unsupervised support vector method for clustering and establishing a reliable framework for automating the clustering procedure and regularizing the complexity of the decision boundaries. The studied method takes advantage of the information obtained from a Mixture of Factor Analyzers (MFA) assuming that lower dimensional non-linear manifolds are locally linearly related and smoothly changing.

Chapter 11, "New algorithms for generation decision trees: Ant-Miner and its modifications" by Boryczka and Kozak propose new modifications in the original Ant-Miner rule producer. Compared to the previous implementations and settings of Ant-Miner, authors experimental studies illustrate how these extensions improve, or sometimes deteriorate, the performance of Ant-Miner.

The final Part of the book deals with the data mining applications. It contains five chapters, which discusses Adaptive Path Planning on Large Road Networks and provide a Framework for Composing Knowledge Discovery Workflows in Grids

Chapter 12, "Automated Incremental Building of Weighted Semantic Web Repository" by Martin and Roman, introduce an incremental algorithm creating a self-organizing repository and it describes the processes needed for updates and inserts into the repository, especially the processes updating estimated structure driving data storage in the repository. The process of building repository is foremost aimed at allowing the well-known Semantic web tools to query data presented by the current web sources.

Chapter 13, "A Data Mining Approach for Adaptive Path Planning on Large Road Networks" by Awasthi et al. presents a statistical approach for approximating fastest paths under stepwise constant input flows and initial states of the arcs on urban networks. Hybrid clustering and canonical correlation analysis have been used to find arc states and input flows that govern the fastest paths on the network.

In Chapter 14,"Linear Models for Visual Data Mining in Medical Images" by Machado proposes an analysis of available methods for data mining for very high-dimensional sets of data obtained from medical imaging modalities. When applied to imaging studies, data reduction methods may be able to minimize data redundancy and reveal subtle or hidden patterns. Author analysis is concentrated on linear transformation models based on unsupervised learning that explores the relationships among morphologic variables, in order to find clusters with strong correlation.

In Chapter 15, "A Framework for Composing Knowledge Discovery Workflows in Grids" by Lackovic et.al., present a framework to support the execution of knowledge discovery workflows in computational grid environments by executing data mining and computation intelligence algorithms on a set of grid nodes. The framework is an extension of Weka, an open-source toolkit for data mining and knowledge discovery, and makes use of Web Service technologies to access Grid resources and distribute the computation. We present the implementation of the framework and show through some applications how it supports the design of knowledge discovery workflows and their execution on a Grid.

In Chapter 16, "Distributed Data Clustering: A Comparative Analysis" Karthikeyani and Thangavel, compare the performance of two distributed clustering algorithms namely, Improved Distributed Combining Algorithm and Distributed K-Means algorithm against traditional Centralized Clustering Algorithms. Both algorithms use cluster centroid to form a cluster ensemble, which is required to perform global clustering. The centroid based partitioned clustering algorithms namely K-Means, Fuzzy K-Means and Rough K-Means are used with each distributed clustering algorithm, in order to analyze the performance of both hard and soft clusters in distributed environment.

We are very much grateful to the authors of this volume and to the reviewers for their great efforts by reviewing and providing interesting feedback to authors of the chapter. The editors would like to thank Dr. Thomas Ditzinger (Springer Engineering Inhouse Editor, Studies in Computational Intelligence Series), Professor Janusz Kacprzyk (Editor-in-Chief, Springer Studies in Computational Intelligence Series) and Ms. Heather King (Editorial Assistant, Springer Verlag, Heidelberg) for the editorial assistance and excellent cooperative collaboration to produce this important scientific work. We hope that the reader will share our joy and will find it useful!

December 2008   Ajith Abraham, Trondheim, Norway
                                                    Aboul Ella Hassanien, Cairo, Egypt
                                                    Václav Snášel, Ostrava, Czech Republic

Contents

Part III: Data Mining Applications

Part I
Data Click Streams and Temporal Data Mining

Mining and Analysis of Clickstream Patterns

H. Hannah Inbarani and K. Thangavel

Abstract. The explosive growth of the web has drastically changed the way in which information is managed and accessed. The large-scale of web data sources and the wide availability of services over the internet have increased the need for effective web data mining techniques and mechanisms . A sophisticated method to organize the layout of the information and assist user navigation is therefore particularly important. In this work, we focus on web usage mining, applying data mining techniques to web server logs. Web usage mining is the non-trivial process of distinguishing implicit, previously unknown but potentially useful clickstream patterns that may exist in any collection of web access logs. The required abstraction can be generated by clustering the web access logs based on some sort of similarity measure. Clustering is done such that the web access logs within the same group or cluster are more similar than data points from different clusters. In this chapter, we propose a partitional algorithm namely Multi Pass Combined Standard Deviation(CSD) Means algorithm which automatically generates the optimum number of clusters from the web clickstream patterns. The quality of clusters obtained using these algorithms are compared using K-Means algorithm, Rough K-Means algorithm and model based algorithms ANTCLUST and ACCANTCLUST. The experimental analysis of mined clickstream patterns shows the effectiveness of the proposed algorithm.

Keywords: Clickstreams, Clustering, K-Means, Ant-Clustering, Rough K-Means.

1 Introduction

In the highly competitive world and with the broad use of the web in E-commerce, E-learning, and E-news, finding users' needs and providing useful information are the primary goals of web site owners. Therefore, analyzing clickstream patterns of web users becomes increasingly important[42]. This increase stems from the realization that added value for web site visitors is not gained merely through larger quantities of data on a site, but through easier access to the required information at the right time and in the most suitable form. With competitors being 'one-click away', the requirement for adding value to E-services on the web

H. Hannah Inbarani and K. Thangavel
Department of Computer Science, Periyar University, Salem-636 011
e-mail: hhinba@yahoo.co.in, indrktvelu@yahoo.com

A. Abraham et al. (Eds.): Foundations of Comput. Intel. Vol. 6, SCI 206, pp. 3–27.
springerlink.com   © Springer-Verlag Berlin Heidelberg 2009

has become a necessity towards the creation of loyal visitors for a web site. This added value can be realized by focusing on specific individual needs and providing tailored products and services. Web currently constitutes one of the largest dynamic data repositories. So, it becomes necessary to extract useful knowledge from this raw and dynamic data by knowledge discovery called as web mining on the Internet.

Recently many researchers have proposed a new unifying area for all methods that apply data mining to web data, named web mining [9]. Web mining tools aim to extract knowledge from the web, rather than retrieving information. Commonly, web mining work is classified into the following three categories [5],[34] : web content mining, web usage mining and web structure mining. Web content mining is concerned with the extraction of useful knowledge from the content of web pages, by using data mining. Web structure mining is a new area, concerned with the application of data mining to the structure of the web graph.

Most of the researches in web usage mining techniques are used to discover user's browsing behaviour. This method processes the web clickstream data directly to find 'an interesting pattern' [41]. A click stream is then the sequence of page views that are accessed by the user. A user session is the click-stream of page views for a single user across the entire web [35],[31]. Originally, the aim of web usage mining has been to support the human decision making process and thus, the outcome of the process is typically a set of data models that reveal implicit knowledge about data items, like web pages, or products available at a particular web site. These models are evaluated and exploited by human experts, such as the market analyst who seeks business intelligence, or the site administrator who wants to optimize the structure of the site and enhance the browsing experience of visitors.

During the process of web usage mining, the rules and patterns in web log records are explored and analyzed mainly by means of the techniques relating to artificial intelligence, data mining, database theory and so on. A variety of machine learning methods have been used for pattern discovery in web usage mining. These methods represent the four approaches that most often appear in the data mining literature: clustering, classification, association discovery and sequential pattern discovery. Similar to most of the work in data mining, classification methods were the first to be applied to web usage mining tasks[9]. However, the difficulty of labeling large quantities of data for supervised learning has led to the adoption of unsupervised methods, especially clustering. The large majority of methods that have been used for pattern discovery from web data are clustering methods. Clustering aims to divide a data set into groups that are very different from each other and whose members are very similar to each other. The purpose of clustering users based on their web clickstreams in a particular web site is to find groups of users with similar interests and motivations for visiting that web site[9]. If the site is well designed there will be strong correlation between the similarity among user clickstreams and the similarity among the users' interests or intentions. Therefore, clustering of the former could be used to predict groupings for the latter.

One important constraint imposed by web usage mining is the choice of clustering method. In practical applications of clustering algorithms, several problems must be solved, including determination of the number of clusters and evaluation of the quality of the partitions. In this research work, we explore the problem of clickstream clustering based on users navigation behavior. There are many methods and algorithms for clustering based on crisp [11], fuzzy [4], probabilistic [38] and possibilistic approaches [22].

Clusters can be hard or soft in nature. In conventional clustering, objects that are similar are allocated to the same cluster while objects that differ significantly are put in different clusters. These clusters are disjoint and are called hard clusters. In soft clustering, an object may be a member of two or more clusters. Soft clusters may have fuzzy or rough boundaries [24]. In fuzzy clustering, each object is characterized by partial membership whereas in rough clustering objects are characterized using the concept of a boundary region. A rough cluster is defined in a similar manner to a rough set. The lower approximation of a rough cluster contains objects that only belong to that cluster. The upper approximation of a rough cluster contains objects in the cluster which are also members of other clusters [38]. The advantage of using rough sets is that, unlike other techniques, rough set theory does not require any prior information about the data such as apriori probability in statistics and a membership function in fuzzy set theory. In this chapter, both hard and soft clustering techniques are used for clustering clickstream patterns.

In this chapter, we propose Multi Pass CSD Means algorithm [37] to cluster clickstream patterns. This intelligent algorithm automatically determines the expected number of clusters from the given clickstream patterns. A comparative analysis is made with other intelligent clustering algorithms such as ANTCLUST [28], ACCANTCLUST[18], K-Means algorithm[25] and Rough-K-Means algorithm[17]. Empirical results clearly show that the proposed Multi Pass CSD Means algorithm performs well and provides stable results compared with other clustering algorithms.

The rest of the chapter is organized as follows. In section 2, an overview of web usage mining process is described. In section 3, the work related to clickstream clustering is summarized. Section 4, discusses the work related to clickstream analysis. In section 5, the partitional and artificial intelligence based algorithms applied for clustering clickstream patterns are presented. In section 6, the experimental results and comparative analysis of the proposed algorithms are discussed. Section 7 concludes this chapter with directions for further research.

2   Web Usage Mining – An Overview

In general, web usage mining consists of three stages, namely data preprocessing, pattern discovery and pattern analysis. Web data collected in the first stage of data mining are usually diverse and voluminous. These data must be assembled into a consistent, integrated and comprehensive view, in order to be used for pattern discovery. As in most applications of data mining, data preprocessing involves removing and filtering redundant and irrelevant data, predicting and filling in missing values, removing noise, transforming and encoding data, as well as resolving any inconsistencies. The task of data transformation and encoding is particularly important for the success of data mining. In Web usage mining, this stage includes the identification of users and user sessions, which are to be used as the basic building blocks for pattern discovery.

2.1 Data Preprocessing

2.1.1 Data Cleaning

The purpose of data cleaning is to eliminate irrelevant items, and these kinds of techniques are of importance for any type of web log analysis not only data mining.

Data Filtering

Most web log records are irrelevant and require cleaning because they do not refer to pages clicked by visitors[47]. A user's request to view a particular page often results in several log entries for a single web page access since graphics and scripts are downloaded. Removing these irrelevant items can reduce the data that will be analysed and increase the analysis's speed. It also can decrease the irrelevant items' negative influence to the mining process.

Feature Selection

Log files usually contain nonessential information from the analytical point of view. Thus the first data pre-processing step is the selection of features. Moreover, reducing the number of features at this stage decreases the memory usage and improves performance[32]. It is also beneficial from the computational point of view, since log files contain thousands of megabytes of data. The final output of the pre-processing must be divided into sessions. The key attributes to build sessions are page ID, computer's IP address (or host) and page request time. These are the main features to work with in web usage mining. Other features are less relevant unless participating in some specific tasks.

Selamat and Omatu [32] proposed a neural network based method to classify web pages and use principle component analysis to select the most relevant features for the classification In [19], Quick Reduct and Variable Precision Rough Set (VPRS) Algorithms are proposed for feature selection from the web log file. These feature selection algorithms are used for selecting significant attributes (features) for describing a session which is suitable for pattern discovery phase.

2.1.2 User's Identification

User's identification is, to identify who access web site and which pages are accessed. If users have logged in their information, it is easy to identify them. In fact, there is lot of users who do not register their information. What's more is there are several users who access web sites through agent, several users use the same computer, firewall's existence, one user use different browsers, and so forth. All of these problems make this task greatly complicated and it is very difficult, to identify every unique user accurately.

2.1.3 Session Identification

For logs that span long periods of time, it is very likely that users will visit the web site more than once. The goal of session identification is to divide the page accesses of each user at a time into individual sessions. A session is a series of web pages user browse in a single access. The simplest method of achieving session is through a timeout, where if the time between page requests exceeds a certain time limit, it is assumed that the user is starting a new session. Many commercial products use 30 minutes as a default timeout.

2.1.4 Data Formatting

Data formatting [6] consists of mapping the number of valid URLs on a website to distinct indices. A user's clickstream consists of accesses originating from the same IP address within a predefined time period. Each URL in the site is assigned a unique number. Thus the pages visited by the users are encoded as binary attribute vectors.

2.1.5 Path Completion

Another critical step in data preprocessing is path completion. There are some reasons which result in path's incompletion, for instance, local cache, agent cache, "post" technique and browser's "back" button can result in some important accesses not recorded in the access log file, and the number of URLs recorded in log may be less than the real one. This problem is referred to path completion, which will influence next steps' efficiency and accuracy if it is not solved properly.

2.2 *Pattern Discovery*

In the pattern discovery stage, machine learning and statistical methods are used to extract patterns of usage from the preprocessed web data[2]. A variety of machine learning methods have been used for pattern discovery in web usage mining. These methods represent the four approaches that most often appear in the data mining literature: clustering, classification, association discovery and sequential pattern discovery. Similar to most of the work in data mining, classification methods were the first to be applied to web usage mining tasks[9]. However, the difficulty of labeling large quantities of data for supervised learning has led to the adoption of unsupervised methods, especially clustering

2.3 *Pattern Analysis*

Pattern Analysis is the final stage of the whole web usage mining. The goal of this process is to eliminate the irrelative rules or patterns and to extract the interesting rules or patterns from the output of the pattern discovery process. The output of web mining algorithms is often not in the suitable form for direct human consumption, and thus need to be transformed to a format that can be assimilated easily. There are two most common approaches for the patter analysis. One is to use the knowledge query mechanism such as SQL, while another is to construct multi-dimensional data cube before perform Online Analytical Processing (OLAP) operations. All these methods assume that the output of the previous phase has been structured.

3   Clickstream Clustering–Literature Review

Clustering aims to discover sensible organization of objects in a given dataset by identifying similarities as well as dissimilarities between objects. It classifies a mass of data, without any prior knowledge, into clusters which are clear in space partition outside and highly similar inside. In web usage mining, clustering algorithms can be used in two ways: usage clickstream   clusters and page clickstream clusters[30].

3.1   Web User Clustering

An important research point in web usage mining is the clustering of web users based on their common properties. Clustering of users tends to establish groups of users exhibiting similar browsing patterns. Such knowledge is especially useful for inferring user demographics in order to perform market segmentation in E-commerce applications or provide personalized web content to the users. On the other hand, clustering of pages will discover groups of pages having related content. Several researchers have applied data mining techniques to web server logs, attempting to unlock the usage patterns of web users hidden in the log files [13].

User profiling on the web consists of studying important characteristics of the web visitors. Due to the ease of movement from one portal to another, web users can be highly mobile. If a particular web site doesn't satisfy the needs of a user in a relatively short period of time, the user will quickly move on to another web site[1]. Therefore, it is very important to understand the needs and characteristics of web users. The clustering process is an important step in establishing user profiles. In web usage mining, clustering algorithms can be used in two ways: usage clusters and page clusters. Fu et al., [15] have demonstrated that web users can be clustered into meaningful groups, which help webmasters to better understand the users and therefore to provide more suitable, customized services. Perkowitz and Etzioni [29] proposed adaptive web sites that improve themselves by learning from user access patterns. The main mining subject is the web server logs.

In [45], Yan et al., the authors use the First Leader clustering algorithm to create groups of sessions. In [13], Estivill-Castro and et al., proposed a derivative form of the K-Means algorithms which uses the median instead of the barycentre to compute the position of the centre of the groups after each affectation of an object to a cluster. Nevertheless, the main problem still remains: the analyst has to specify the number of expected clusters. To solve this problem, Heer and et al. proposed [18] to find automatically, the number of expected clusters by evaluating the stability of the partitions for different number of clusters. This method works well but is extremely time consuming. Thus, this method may not be used in the web mining context, where the data sets to explore contain a lot of sessions. Labroche and et al. described [16] a clustering algorithm called ANTCLUST that is inspired from the chemical recognition system of ants and that allows to find automatically the number of expected clusters. In[28], an enhanced version of ANTCLUST is used to handle the files of sessions extracted from the web server files. Lingras and West [22] provided a theoretical and experimental analysis of a modified K-Means clustering based on the properties of rough sets. In [16] , a common model-based clustering algorithm is used to result in clusters of web users' sessions.

The clustering methods presented in this chapter are restricted to generate user clusters. The terms user/session clusters is used interchangeably because the log file taken for the experiment consists of no date/time information and each user/session is identified only using user id and Page id.

3.2   Web Page Clustering

Clustering of pages will discover groups of pages having related content which could be useful for mass personalization and web site adaptation.

Flake et al., [14] use only link information to discover web communities (Groups of URLs). The web communities they discover however, merely reflect the viewpoint of web site developers. Mobasher et al., [26] proposed a technique of usage based clustering of URLs for the purpose of creating adaptive web site. They directly compute overlapping clusters of URL references based on their co-occurrence patterns across user transactions. But their URLs clustering is still based on frequent items and needs user session identification. Selamat and Omatu [32] proposed a neural network based method to classify web pages. Obviously, it is a content-based method, and it needs class-profile which contains the most regular words in each class. In [30], vector analysis based and fuzzy set theory based methods are used for the discovery of user clusters and page clusters.

4   Clickstream Analysis

On a web site, clickstream analysis is the process of collecting, analyzing, and reporting aggregate data about which pages visitors visit in what order - which are the result of the succession of mouse clicks of each visitor. Analyzing visitors' personal information gives us an idea of who might be a potential customer, but their current interests should be taken into consideration as well. Customers search for information about products they are interested in on the internet and the web server records every movement they make in a server log file. A visitor's changing interest is hidden in this file. Studies exploring customers' web page access patterns based on web log files are given in [35],[33],[40] and [46]. The volume of information stored in the web server log accumulates over time. The more data processed, the more time needed to calculate results. In order to capture customers' current interests and provide instant service, on-line analysis must be performed. Discovering patterns of loyal customers' click streams stored in the log file is the province of off-line analysis, which does not require instant results. Web server log files are simple text files that are automatically generated every time someone accesses the web site. Every "hit" to the web site is logged in the form of one line of text. Information in the raw web log file format includes who the visitor was, where the visitor came from, and what he was doing on the site.

4.1 Common Web Log Format

A web server log file contains requests made to the web server, recorded in chronological order. The most popular log file formats are the Common Log Format (CLF) and the extended CLF.

A common log format file is created by the web server to keep track of the requests that occur on a web site. The format of a common log file [43] is shown in Table 1.

Table 1 Common Log file Format

remotehost	Remote hostname
Rfc931	The remote log name of the user.
Authuser	The username as which the user has authenticated himself
date	Date and time of the request.
request	The request line exactly as it came from the client
status	The HTTP status code returned to the client
Bytes	The content-length of the document transferred

The experiments in this chapter are conducted on the web access logs of www.microsoft.com which is available in UCI repository [http://www.ics.uci.edu/]. This web log consists of only page id to identify pages and user id to identify users.

The purpose of analyzing web logs is to understand the user's browsing behavior. Based on the result of perceived user's behavior, user's page searching time may be reduced by recommending pages customers may be interested in. The most important information revealed by analyzing customer's clickstream is the user's current interest.

Clustering analysis is widely used to establish object pro-files on the basis of objects' variables. Objects can be customers, web documents, web users, or facilities[19]. In this chapter clustering algorithms are applied on web clickstrems to analyze user access trends. On one hand, the profiles can be used for predicting the navigation behaviour of current users, thus aiding in web personalization. On the other hand, webmasters can improve the design and organization of websites based on the acquired profiles.

A sample web log file in the common log file format is shown in Table 2.

Table 2 Sample web log file

124.49.105.224 - - [29/Nov/2000:18:02:26 +0200] "GET /index.html HTTP/1.0" 200 1159
124.49.105.224 - - [29/Nov/2000:18:02:27 +0200] "GET /PtitLirmm.gif HTTP/1. 0" 200 1137
124.49.105.224 - - [29/Nov/2000:18:02:28 +0200] "GET /acceuil_fr.html HTTP/1. 0" 200 1150
124.49.105.224 - - [29/Nov/2000:18:02:30 +0200] "GET /venir/venir.html HTTP/1.0" 200 1141
124.49.105.225 - - [29/Oct/2000:18:03:07 +0200] "GET /index.html HTTP/1.0" 200 1051
124.49.105.225 - - [16/Oct/2000:20:34:32 +0100] "GET /formation.html HTTP/1.0" 200 -
124.49.105.225 - - [31/Oct/2000:01:17:40 +0200] "GET /formation.html#d HTTP/1.0" 304 -
124.49.105.225 - - [31/Oct/2000:01:17:42 +0200] "GET /theses2000.html HTTP/1.0" 304 -
124.49.105.56 - - [22/Nov/2000:11:06:11 +0200] "GET /lirmm/bili/ HTTP/1.0" 200 4280
124.49.105.56 - - [22/Nov/2000:11:06:12 +0200] "GET /lirmm/rev_fr.html HTTP/1.0" 200 -
124.49.105.56 - - [07/Dec/2000:11:44:15 +0200] "GET /ress/ressources.html HTTP/1.0" 200

5 Related Work

One of the major issues in web log mining is to group all the users' page requests so to clearly identify the paths that users followed during navigation through the web site [1].

There are many methods applied in clustering analysis, such as hierarchical clustering, partition-based clustering, density-based clustering, and artificial intelligence based clustering. In this chapter, artificial intelligence based clustering algorithms and partition-based clustering algorithms are studied for clustering clickstream patterns. The traditional clustering algorithms such as K-Means require users to provide the correct (actual) number of clusters in a given pattern set at the beginning. However, in many applications, a priori knowledge of the actual number of clusters is unavailable, and the optimal number of clusters cannot be easily and intuitively estimated beforehand. If the number of clusters estimated is larger than the actual number of clusters, one good compact cluster in nature is divided into more compact clusters with inappropriate separations; in contrast, if the estimated number of clusters is smaller, two or more compact clusters in nature must be grouped into one loose cluster. Thus, how to determine the optimal number of clusters in a given pattern set is an important problem in cluster analysis. It is very difficult to identify the optimal number of clusters on the access sessions from the clustering results in an unsupervised way because this number determines how many representative navigation patterns will be extracted from user access sessions, and how many user profiles are supposed to be constructed next. The optimal number means that the partition of user access sessions can best reflect the distribution of sessions, and can also be validated by user's inspection [8]. All the existing algorithms discussed in this chapter except K-Means and Rough K-Means estimates optimal number of clusters from the given web log data set.

5.1 Artificial Intelligence Based Clustering Algorithms

Applying ant colony system in clustering analysis is still a very novel research area. Ant Colony Algorithm (ACA) is a meta-heuristic approach successfully applied to solve hard combinatorial optimization problems [48]. It is also feasible for clustering analysis in data mining. Many researchers use ant algorithms for clustering analysis and the result is better than other heuristic methods. Unlike traditional clustering methods, ACA is an intelligent approach, which is successfully used in discrete optimization. The ant-based clustering algorithm is inspired by the behavior of ant colonies in clustering their corpses and sorting their larvae and automatically finds the number of clusters .

Algorithm: ANTCLUST
---

    1)   Intitialisation of the ants.
    2) $Genetic_i \leftarrow X_{i,}$ i^{th} session of the data set.
    3) $Label_i \leftarrow 0$.
    4) $Template_i$ is initialized.
    5) $M_i \leftarrow 0$, $M_i^+ \leftarrow 0$, $A_i \leftarrow 0$.
    6)   Simulate N_{iter} iterations during which two ants that are randomly chosen meet.
    7)   Delete nests less than P * N (P<1) ants
    8)   Re-assign each ant having no more nest to the nest of the most similar ant found that have a nest.

Fig. 1 ANTCLUST

5.1.1 Antclust

This algorithm is based on real ants' collective behaviour, namely the construction of colonial odour and its use for determining the ants nest membership. Every day, the real ants have to solve a crucial recognition problem when they meet: they have to decide whether they belong to the same nest or not, in order to guaranty the survival of the nest. This phenomenon is called "colonial closure" [25]. It manly relies on continuous exchanges and updates of chemical cues on the ant's cuticle and in their post-pharyngeal gland, determining, as an identity card, their belonging to the nest. Thus, each ant has its own view of its colony odor at a given time, and updates it continuously. By this way, an ant preserves its nest from being attacked by predators or parasites and reinforces its integration in nest. The gathering of ants in a finite number of nests where nest-mates are more similar to each other than the ants of other colonies provides a cluster of the set of objects. In this chapter, ANTCLUST[26] and accelerated ant clustering algorithm [16] are used for comparative analysis.

5.1.2 ACCANTCLUST

In ANTCLUST algorithm, when meeting between two ants is simulated, if the meeting is between two ants with no nest, and if they accept each other, these two ants are placed in a new nest. If they do not accept each other, no nest is created. In ACCANTCLUST, if the ants do not accept each other, two new nests are created and the ants are placed in two different nests.

Algorithm: ACCANTCLUST

Initialization of the ants
1) $\forall$ants(sessions) i ε [1,n]
2) Genome i $\longleftarrow X_i$, i th session vector of the data set
3) Label$_i$ $\longleftarrow$ 0
4) Template is initialized
5) $M_i \longleftarrow 0 \ M_i^+ \longleftarrow 0 \ A_i \longleftarrow 0$
6) $N_{biter} \longleftarrow 50 * n$
7) Simulate Nbiter iterations during which two randomly chosen ants(sessions) meet.
    i)   If Label of ith and jth ant are zero and
             if the acceptance is true,
               Place them in the same nest
             Else
               Create two different nests and place them separately
    ii)   If Label of ith ant is zero and jth ant is not zero and
             If the acceptance is true
               Adding an ant with no label to an existing nest:
    iii)   Positive" meeting between two nestmates:
               Increase the values of Parameters M_i and M_i^+
    iv)   Negative" meeting between two nestmates:
               The worst integrated ant is removed from the nest and its label is set to zero.
    v)   Meeting between two ants of different nests:
               The ant x with the lowest M_x changes its nest and belongs now to the nest of the encountered ant.

Fig. 2 Accelerated Ant Colony Algorithm

5.2 Partitional Clustering Algorithms

In partitional clustering algorithms, each cluster can be represented by its center; thus, the solution of the partitional clustering algorithms can be represented by a set of clusters or a set of cluster centers [39]. Partitional algorithms construct a partition of a database D of n objects into a set of K clusters, where K is an input parameter for these algorithms. To set the value of K, some domain knowledge is required which unfortunately is not available in many applications such as clustering of web clickstream patterns.

5.2.1 K-Means Algorithm

The K-Means algorithm is the most well-known partitional clustering method due to its easy implementation and rapid convergence. This algorithm iteratively updates the solution in a deterministic manner such that their results are heavily influenced by the choice of initial solution. It is the simplest and most commonly used algorithm that employs a squared error criterion [25]. Provided with a set of n numeric objects and an integer number $K(K \leq n)$, it calculates a partition of patterns in K clusters. This process takes place in an iterative manner starting from a random initial partition and keeping on searching for a partition of n that minimizes the within groups sum of squared errors.

Algorithm: K-MEANS

---

1) Choose K initial cluster centres from the set of sessions $Z_1, Z_2 \ldots Z_k$.
2) At the k-th iterative step, distribute the user sessions $\{X_n\}$ among the K clusters using the relation

$$X_n \in C_j(k) \quad \text{if} \quad \| X_n - Z_j(k)\| \quad < \quad \| X_n - Z_i(k) |$$

for all i=1, 2, ... ,K; $i \neq j$; where $C_j(k)$ denotes the set of user sessions whose cluster centre is $Z_j(k)$.
3) Compute the new cluster centres $Z_j(k+1)$, j=1, 2...K such that the sum of the squared distances from all points in $C_j(k)$ to the new cluster centre is minimized. The measure which minimizes this is simply the sample mean of $C_j(k)$. Therefore, the new cluster centre is given by

$$Z(k+1) \quad = \quad \sum_{x \in Cj(k)} X_n \quad j \quad = \quad 1,2, \ldots K$$

where N_j is the number of samples in $C_j(k)$
4) If $Z_j(k+1) = Z_j(k)$ for j = 1,2,...K then the algorithm has converged and the procedure is terminated.
5) Otherwise go to step 2.

Fig. 3 K-Means algorithm

5.2.2 Rough Clustering

In rough clustering each cluster has two approximations, a lower and an upper approximation. The lower approximation is a subset of the upper approximation. The members of the lower approximation belong certainly to the cluster, therefore they cannot belong to any other cluster. The data objects in an upper approximation may belong to the cluster. Since their membership is uncertain they must be a member of an upper approximation of at least another cluster[17. There is a crucial difference to fuzzy set theory where we have overlapping clusters too: in fuzzy set theory an object can belong to many sets; in rough sets the memberships to two or more sets indicate that there is information missing to determine the actual membership to one and only one cluster.

Algorithm: Rough K-Means

---

1) Assign the session vectors to the approximations.
   (i) For a given session vector X_n ,determine its closest mean m_h

$$d(X_l,\ Z_h) \quad = \quad \min_{n,\ k}\ d(X_n, Z_k) \quad \Rightarrow \quad X_l \in \overline{C_k}\ \wedge\ X_l \in \underline{C_k}$$

Assign X_n to the upper approximation of the cluster h, $X_n \in \overline{C_h}$

(ii) Determine the means m_t that are also close to X_n they are not farther away from X_n than d (X_n, Z_h) + ε where ε is a given threshold:

$$d(X_l, Z_h) = \min_{n,k}\ d(X_n, Z_k) \Rightarrow X_l \in \overline{C_k} \wedge\ X_l \in \underline{C_k}$$

If $T \neq \emptyset$ (X_n is also close to at least one other mean Z_t besides Z_h)

Then $X_n \in \overline{C_l}\ \forall\ t \in T$

Else $X_n \in \underline{C_h}$

2) Compute new mean for each cluster C_i using the following equation

$$Z_k = W_l \sum_{X_n \in \underline{C_k}} \frac{\overline{X_n}}{|\underline{C_k}|} + W_b \sum_{X_n \in C_k^B} \frac{\overline{X_n}}{|C_k^B|} \quad \text{for}\ C_k^B \neq 0$$

$$W_l \sum_{X_n \in \underline{C_k}} \frac{\overline{X_n}}{|\underline{C_k}|} \qquad\qquad \text{otherwise}$$

with $W_l + W_u = 1$.

Fig. 4 Rough K-Means algorithm

As defined in [15], for stable results, the parameter values are taken as $W_1 = 0.7$, $W_u = 0.3$, and threshold is taken as 0.04 since the session matrix constructed from the web log file is a sparse matrix.

5.2.3 Multi Pass CSD Means Clustering Algorithm

This section proposes Multi Pass CSD Means algorithm for clustering web clickstreams. In this algorithm, web users are grouped into the corresponding available clusters when they are similar. On the other hand, the dissimilar or irrelevant sessions are placed into the new cluster created. This shows, the algorithm remains adaptive in response to significant events by existing clusters and yet remains stable to irrelevant events by creating new cluster[37]. It retains the centroid of the available clusters. This process is repeated till all the clickstream patterns in the web log data set residing are considered. The multi-pass CSD-means algorithm is described as follows:

Algorithm: Multi Pass CSD Means

---

1) Assign the first session vector as the initial centroid for the first cluster
2) Assign each session vector to the available clusters with nearest center or Create and assign to the new cluster

    i) For each session vector X_n
        Check class-set value for X_n in the available clusters
        Class-set = false;
    ii) For each cluster k do
        find the combined standard deviation of σ_{csd} of X_n and Z_k
        And the session with maximum standard deviation σ_{max} of X_n and Z_k
        if there exist σ_{max} which is less than σ_{csd} then it can be set to some classes break;
        if (class-set = true) then
          find the minimum distance and place it in the corresponding cluster such as K-Means algorithm
  3) if current centroid and old centroid are not very close repeat the process

4) find minimum distance to a center and place the session into the corresponding cluster

    Stop.

Fig. 5 Multi Pass CSD Means

6 Experimental Results

6.1 Data source

The data source of web usage mining is web log files, from which we can realize users' clickstream patterns by web usage mining. For a web site, user access information is generally gathered automatically by web server and recorded in the server logs. There are four different log files, the access log, agent log, error log, and referrer log. These log files are text files, and their sizes depend on the traffic at a particular site. Recorded in these files is the volume of activity at each page on a web site, the type of browser used to access each page, any errors that users may have experienced downloading pages from the web site, and where users were referred from when they accessed pages at the web site.

For the purpose of evaluating the performance and the effectiveness of the intelligent clustering algorithms, experiments were conducted with preprocessed web access logs of www.microsoft.com which is available in UCI repository [http://www.ics.uci.edu/]. This log file records the use of www.microsoft.com by 5000 anonymous, randomly-selected users who have visited the web site in a one week timeframe in February 1998 with an average of 5.7 page views per user. The file contains no personally identifiable information. This data set includes visits which are recorded are recorded in time order and no pre-processing is required since data set was given in sessions. The 294 web pages are identified by their title (e.g. "NetShow for PowerPoint") and URL (e.g. "/stream"). These algorithms are applied only for testing instances available in UCI repository by taking only 100 web pages and 5000 users. The five data sets with sizes of 1000, 2000,3000, 4000,5000 users were extracted from the log file of Microsoft web site.

6.2 Data Preparation

As the web log file of Microsoft web site available in UCI repository is a preprocessed one, the only preprocessing step needed is data formatting. The only fields available in this log file are user id and id of page visited . Since there is no date and time information in the given web log, all the pages visited by the user are considered as a single session. Using this information session (user access) matrix is constructed.

6.3 Web Traffic Patterns

Fig 6. shows the web traffic patterns of users who have visited the Microsoft web site in a one week timeframe in February 1998 and the average number of page views is per user is 5.7.

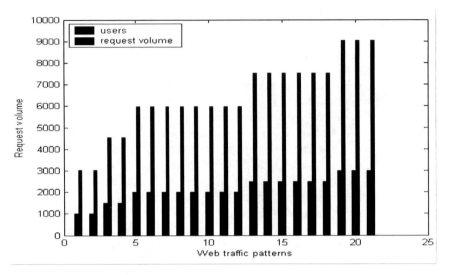

Fig. 6 Web Traffic Patterns of Microsoft web site

6.4 Results of the Proposed Algorithm

6.4.1 User Profiles

User profiling on the web consists of studying important characteristics of the web visitors. The clustering process is an important step in establishing user profiles.

Some of the discovered user profiles obtained using Multi Pass CSD Means algorithm are summarized in Table 3. These user profiles are obtained by clustering clickstreams only 100 users from the web log file.

The extracted user profiles reveal that most of the users who have visited the web site are interested in various Microsoft software's , some of the users are interested in internet tools and others interested in network related softwares. Single users in clusters show that those users are interested in country specific web pages. On one hand, the profiles can be used for predicting the navigation behaviour of current users, thus aiding in web personalization. On the other hand, webmasters can improve the design and organization of web sites based on the acquired profiles.

6.4.2 User and Page Distribution of Clusters

It is also very useful to have a view about the contents of each cluster, since a deeper knowledge for the inside of each cluster can draw useful and meaningful inferences for the users' navigation behavior.

We also extracted a brief summary for each cluster by computing a mean vector for each cluster and extracted the corresponding web pages accessed in each cluster. The results of applying Multi Pass K-Means algorithm for 5000 users

Table 3 Sample User Profiles obtained using Multi Pass CSD-Means algorithm

Cluster	Size	URLs/Profile	Profile Descriptions
1	19	/mexico /homeessentials /kids /msp /support /vstudio /publisher /activex /products /ntworkstation /proxy /kb /windowssupport /catlog /teammanager /officefreestuff /workshop /msdn /iis	Mexico, Microsoft Home Essentials, MSHome Kids Stuff, Microsoft Solution Providers, Support Desktop, Visual Studio, MS Publisher, ActiveX Technology Development, Products, Windows NT Workstation, MS Proxy Server, Knowledge Base, Windows95 Support, Product Catalog, MS TeamManager, Office Free Stuff, Developer Workshop, Developer Network, Internet Information Server
2	1	/Taiwan	Taiwan
3	8	/sbnmember /ie_intl /intdev /netmeeting /activex /java /workshop /msdn	SiteBuilder Network Membership, Internet Development, International IE content, NetMeeting, ActiveX Technology Development, Java Strategy and Info, Developer Workshop, Developer Network
4	8	/windowssupport /kb /supportnet /mspress /products /iesupport /security /support	Windows95 Support, Knowledge Base, Support Network Program Information , Products , Microsoft Press, IE Support, Internet Security Framework, Support Desktop
5	2	/intdev /workshop	Internet Development, Developer Workshop
6	2	/regwiz /support	Regwiz, Support Desktop
7	1	/msp	Microsoft Solution Providers
8	4	/education /support /truetype /catalog	MS in Education, Support Desktop, Typography Site, Product Catalog
9	1	/windows support	Windows95 Support
10	4	/promo /automap /organizations /frontpage	Promo, N. American Automap, Corporate Desktop Evaluation, FrontPage
11	4	/referral /frontpage /uk /msdn	SP Referral (ART), FrontPage, UK, Developer Network
12	1	/ie_intl	International IE content
13	1	/workssupport	Works Support
14	4	/office reference /organizations /regwiz /officefreestuff	Office Reference, Corporate Desktop Evaluation, regwiz, Office Free Stuff
15	2	/Canada /referral	Canada, SP Referral (ART)
16	3	/ntwkssupport /promo /ntwkssupport	NT Workstation Support, promo, NT Workstation Support

taken from web log file is summarized below. The optimum number of clusters generated by the proposed algorithm for 5000 users is 48. For 48 clusters, we computed the mean vector and extracted the corresponding web pages accessed in each cluster based on the values of its mean vector. Table 4 gives a specific description on each of the obtained, the number of users in each cluster and the percentage of users in the cluster. It can be seen from the table that the forty eighth cluster accounts for the largest proportion of users.

Table 5 gives a specific description on each of the obtained number of pages distributed in each user cluster and the percentage of pages visited by the users in the cluster. It can be observed from the table that the cluster four encounters larger number of web pages.

More specifically, Fig. 7 depicts the percentage frequency of requested users and web page categories observed in each cluster for 5000 users by applying proposed algorithm to help in understanding users' navigation behavior for the web clickstream patterns.

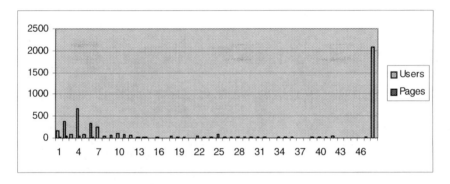

Fig. 7 User and Page request volume of clusters

6.5 Comparative Analysis

6.5.1 Cluster Validation

Cluster validation refers to procedures that evaluate the clustering results in a quantitative and objective function. The validation index is a single real value that can describe the quality of a complete cluster partition Some kinds of validity indices are usually adopted to measure the adequacy of a structure recovered through cluster analysis In fact, if cluster analysis is to make a significant contribution to engineering applications, much more attention must be paid to cluster validity issues that are concerned with determining the optimal number of clusters and checking the quality of clustering results. Many different cluster validity measures have been proposed such as the Dunn's separation measure [12], the Bezdek's partition coefficient [3], the Xie-Beni's separation measure [44], Davies-Bouldin's measure [8], *etc.* In this chapter , the popular validity measures

Table 4 User distribution of clusters

Cluster label	Number of Users	Percentage Of Users
1	167	0.3
2	376	0.39
3	79	0.07
4	675	0.43
5	81	0.04
6	343	0.24
7	244	0.07
8	36	0.02
9	72	0.09
10	97	0.08
11	88	0.08
12	61	0.08
13	23	0.19
14	17	0.15
15	9	0.02
16	14	0.01
17	9	0.02
18	36	0.01
19	25	0.02
20	23	0.06
21	8	0.02
22	33	0.02
23	27	0.01
24	13	0.02
25	74	0.05
26	14	0.02
27	11	0.03
28	21	0.01
29	31	0.06
30	13	0.02
31	16	0.1
32	17	0.01
33	6	0.01
34	11	0.02
35	13	0.07
36	12	0.02
37	9	0.03
38	4	0.01
39	13	0.01
40	12	0.01
41	13	0.02
42	33	0.01
43	4	0.01
44	1	0.01
45	2	0.01
46	5	0.01
47	25	0.01
48	2084	0.01

Table 5 Page distribution of clusters

Cluster label	Number of Pages	Percentage of Pages
1	30	0.3
2	39	0.39
3	7	0.07
4	43	0.43
5	4	0.04
6	24	0.24
7	7	0.07
8	2	0.02
9	9	0.09
10	8	0.08
11	8	0.08
12	8	0.08
13	19	0.19
14	15	0.15
15	2	0.02
16	1	0.01
17	2	0.02
18	1	0.01
19	2	0.02
20	6	0.06
21	2	0.02
22	2	0.02
23	1	0.01
24	2	0.02
25	5	0.05
26	2	0.02
27	3	0.03
28	1	0.01
29	6	0.06
30	2	0.02
31	10	0.1
32	1	0.01
33	1	0.01
34	2	0.02
35	7	0.07
36	2	0.02
37	3	0.03
38	1	0.01
39	1	0.01
40	1	0.01
41	2	0.02
42	1	0.01
43	1	0.01
44	1	0.01
45	1	0.01
46	1	0.01
47	1	0.01
48	1	0.01

such as the Davies-Bouldin's measure, Xie-Beni's separation measure and Dunn's index are used to evaluate the web clickstream clusters and the results are shown in Table 6. For the evaluation of Rough-K-means algorithm , rough version of Davies-Bouldin's measure and Dunn's index [36] is used.

Table 6 Cluster Validity Results

No.of Users	Page Request volume	Clustering Algorithm	No.of Clusters (Input)	No.of Clusters Generated	Davies Bouldin Index	Dunn's Index	Xie-Beni's Index
1000	3033	ANTCLUST	-	57	0.41	0.98	0.71
		ACCANTCLUST	-	41	0.39	1.13	0.68
		K-Means	5	-	0.307	1.289	0.549
		Rough K-Means	5	-	0.373	1.31	0.55
		Multi Pass CSD	-	37	0.32	1.33	0.67
2000	5992	ANTCLUST	-	62	0.403	1.296	0.59
		ACCANTCLUST	-	47	0.437	0.98	0.64
		K-Means	7	-	0.3	1.39	0.613
		Rough K-Means	7	-	0.362	1.01	0.62
		Multi Pass CSD	-	40	0.311	1.37	0.58
3000	9062	ANTCLUST	-	71	0.423	0.912	0.59
		ACCANTCLUST	-	52	0.414	1.10	0.54
		K-Means	9	-	0.519	1.117	0.642
		Rough K-Means	9	--	0.46	1.05	0.69
		Multi Pass CSD	-	41	0.324	1.279	0.578
4000	12066	ANTCLUST	-	74	0.517	0.928	0.73
		ACCANTCLUST	-	61	0.501	1.11	0.58
		K-Means	11	-	0.579	0.934	0.623
		Rough K-Means	11	-	0.410	1.19	0.69
		Multi Pass CSD	-	43	0.338	1.24	0.52
5000	14960	ANTCLUST	-	82	0.44	0.98	0.77
		ACCANTCLUST	-	63	0.41	0.95	0.78
		K-Means	13	-	0.567	0.99	0.623
		Rough K -Means	13	-	0.48	1.01	0.641
		Multi Pass CSD	-	45	0.38	1.227	0.593

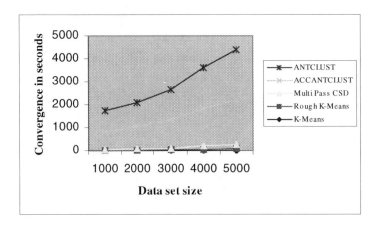

Fig. 8 CPU Time taken by Clustering algorithms

If a data set contains well-separated clusters, the distances among the clusters are usually large and the diameters of the clusters are expected to be small. Therefore larger value of Dunn's index (DI) means better cluster configuration. The Davies - Bouldin(DB) index is based on similarity measure of clusters whose bases are the dispersion measure of a cluster and the cluster dissimilarity measure. The

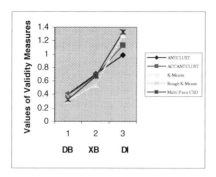

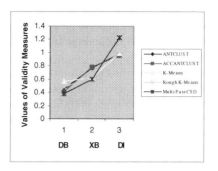

Fig. 9 Data set size = 1000

Fig. 10 Data set size = 2000

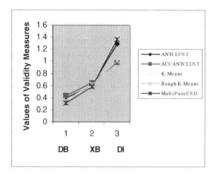

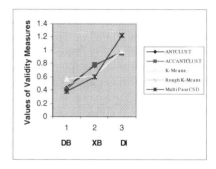

Fig. 11 Data set size = 3000

Fig. 12 Data set size = 4000

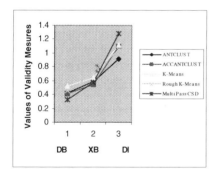

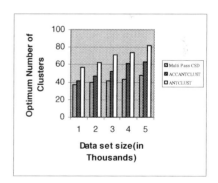

Fig. 13 Data set size = 5000

Fig. 14 Comparison of Algorithms

Davies–Boludin index measures the average of similarity between each cluster and its most similar one. As the clusters have to be compact and separated the lower Davies–Bouldin index means better cluster configuration. Xie-Beni (XB) validity index which measures the compactness and separation of clusters. Thus a smaller value of the index reflects that the clusters have greater separation from each other and are more compact.

Table 6 shows the effectiveness of the proposed algorithm. For each set of users taken, the value of Davies–Bouldin index for the proposed algorithm is lower than the value of Davies–Bouldin index for other algorithms. The value of Dunn's index for each set of users taken for the proposed algorithm is larger than the Dunn's index value of other algorithms. Figures 9 to 13 show the values of DB, DI and XB for the algorithms proposed for various sizes of the data sets. The optimum number of clusters generated by various algorithms is depicted in Fig 14.

In K-Means and Rough K-Means algorithm, the accuracy of the clusters obtained depends upon the number of clusters selected and the initial value of seed points. From the table 6, it can be observed that the proposed algorithm shows the accurate results since it estimates the number of clusters automatically from the given data set. Though ANTCLUST and ACCANTCLUST estimate the optimum number of clusters, these two algorithms take large amount of CPU time and the memory consumption is also very high. So it is not suitable for very large data sets. Fig 8 shows the CPU time taken for the covergence of clustering algorithms for each data set . We performed experiments on 2.4 GHz, 512MB RAM, Pentium-IV machine running on Microsoft Windows XP .

7 Conclusion

In this chapter, we investigated the proposed Multi-Pass CSD-Means algorithm for clustering web clickstream patterns from web access logs of www.microsoft.com which is available in UCI repository. A meaningful contribution for the clustering analysis of web logs can be made by using these algorithms. This algorithm estimates the optimum number of clusters automatically and is capable of manipulating efficiently very large data sets. A comparative analysis is made with K-Means algorithm, Rough K-Means and Model based algorithms ANTCLUST and ACCANTCLUST. The study of validity measure shows the effectiveness of the proposed algorithm for mining web clickstream patterns. Experimental results also show the stability and accuracy of the proposed algorithm. Though ANTCLUST and ACCANTCLUST estimates the number of clusters from the data set, experimental results clearly show that they are not stable and these algorithms take enormous amount of time for the convergence of clusters. Further research may be extended to computing optimum number of clusters for soft clusters.

References

1. Abraham, A.: Natural Computation for Business Intelligence from Web Usage Mining. In: Proceedings of the Seventh International Symposium on Symbolic and Numeric Algorithms for Scientific Computing (SYNASC 2005) (2005)
2. Baumgarten, M., Bchner, A.G., Anand, S.S., Mulvenna, M.D., Hughes, J.G.: Navigation Pattern Discovery from Internet Data. In: Masand, B., Spiliopoulou, M. (eds.) WebKDD 1999. LNCS, vol. 1836. Springer, Heidelberg (2000)

3. Bezdek, J.C.: Numerical Taxonomy with Fuzzy Sets. J. Math. Biol. 1, 57–71 (1974)

4. Bezdek, J.C.: Pattern Recognition with Fuzzy Objective Function Algorithms. Plenum, New York (1981)

5. Cooley, R.: Web Usage Mining: Discovery and Application of Interesting Patterns from web data, Ph.D. Thesis, University of Minnesota (2000)

6. Cooley, R., Mobasher, B., Srivastava, J.: Data Preparation for Mining World Wide Web Browsing Patterns. J. Knowledge and Information Systems 1(1), 5–32 (1999)

7. Cooley, R., Srivastava, J., Mobasher, B.: Web Mining: Information and Pattern Discovery on the World Wide Web. In: Proceedings of the 9th IEEE International Conference on Tools with Artificial Intelligence (ICTAI 1997), pp. 558–567 (1997b)

8. Davies, D.L., Bouldin, D.W.: A cluster separation measure. IEEE Trans. Pattern Analysis and Machine Intelligence 1(4), 224–227 (1979)

9. Pierrakos, D., Paliouras, G.O., Papatheodorou, C., Spyropoulos, C.D.: Web Usage Mining as a Tool for Personalization: A Survey. User Modeling and User-Adapted Interaction 13, 311–372 (2003)

10. Dubes, R., Jain, A.K.: Validity studies in clustering methodologies. Pattern Recognition 11(1), 235–253 (1979)

11. Duda, R., Hart, P.: Pattern Classification and Scene Analysis. Wiley Interscience, New York (1973)

12. Dunn, J.C.: A Fuzzy Relative of the ISODATA Process and its Use in Detecting Compact Well-Separated Clusters. Journal Cybern. 3(3), 32–57 (1973)

13. Estivill-Castro, V., Yang, J.: Fast and robust general purpose clustering algorithms. In: Pacific Rim International Conference on Artificial intelligence, pp. 208–218 (1979)

14. Flake, G.W., Lawrence, S., Lee Giles, C., Coetzee, F.M.: Self- organization and identification of Web communities. IEEE Computer 35(3), 66–71 (2002)

15. Fu, Y., Sandhu, K., Shi, M.: Clustering of web users based on access patterns. In: Masand, B., Spiliopoulou, M. (eds.) WebKDD 1999. LNCS (LNAI), vol. 1836, pp. 21–38. Springer, Heidelberg (2000)

16. Pallis, G., Angelis, L., Vakali, A.: Validation and interpretation of Web users' sessions clusters. In: Information Processing and Management (2006)

17. Peters, G.: Some refinements of rough k-means clustering. Pattern Recognition 39, 1481–1491 (2006)

18. Hannah Inbarani, H., Thangavel, K.: Clickstream Intelligent Clustering using Accelerated Ant Colony Algorithm. In: Advanced Computing and Communications, 2006. ADCOM 2006. International Conference I, pp. 129–134 (2006)

19. Hannah Inbarani, H., Thangavel, K., Pethalakshmi, A.: Rough Set Based Feature Selection for Web Usage Mining. In: Proceedings of the International Conference on Computational Intelligence and Multimedia Applications (ICCIMA 2007), pp. 33–38 (2007) ISBN:0-7695-3050-8

20. Heer, J., Chi, E.: Mining the structure of user activity using cluster stability. In: Proceedings of the Workshop on Web Analytics, SIAM Conference on Data Mining, Arlington, VA (April 2002)

21. Chang, H.-J., Hung, L.-P., Ho, C.-L.: An anticipation model of potential customers' purchasing behavior based on clustering analysis and association rules analysis. Expert Systems with Applications 32, 753–764 (2007)

22. Krishnapuram, R., Keller, J.: A possibilistic approach to clustering. IEEE Trans. Fuzzy Syst. 1(2), 98–110 (1993)

23. Kuo, R.J., Wang, H.S., Hu, T.-L., Chou, S.H.: Application of Ant K-Means on Clustering Analysis. Computers and Mathematics with Applications 50, 1709–1724 (2005)

24. Lingras, P., West, C.: Interval Set Clustering of Web Users with Rough K-means. Journal of Intelligent Information Systems (2002)

25. McQueen, J.: Some methods for classification and analysis of multivariate observations. In: Le Cam, L.M., Newman, J. (eds.) Proceedings of the Fifth Berkeley Symposium on Mathematical Statistics and Probability, vol. 1, pp. 281–297. University of California Press, Berkeley (1967)

26. Mobasher, B., Cooley, R., Srivastava, J.: Creating adaptive web sites through usage-based clustering of URLs. In: Proceedings of the 1999 IEEE Knowledge and Data Engineering Exchange Workshop (KDEX) (1999)

27. Labroche, N., Monmarche, N., Venturini, G.: A new clustering algorithm based on the chemical recognition system of ants. In: Proceedings of 15th European Conference on Artificial Intelligence (ECAI 2002), Lyon FRANCE, pp. 345–349 (2002)

28. Labroche, N., Monmarche, N., Venturini, G.: Web session clustering with artificial ants colonies'. In: Proc. of WWW 2003, May 20-24 (2003)

29. Perkowitz, M., Etzioni O.: Adaptive sites: automatically learning from user access patterns. In: Proceedings of WWW6 (1997),
    http://www.scope.gmd.de/info/www6/posters/722/index.html

30. Song, Q., Shepperd, M.: Mining web browsing patterns for E-commerce. Computers in Industry 57, 622–630 (2006)

31. Bucklin, R.E., Lattin, J.M., Ansari, A., Gupta, S., Bell, D., Coupey, E., Little, J.D.C., Mela, C., Montgomery, A., Steckel, J.: Choice And the Internet: From Clickstream to Research Stream. Marketing Letters 13(3), 245–258 (2002)

32. Selamat, A., Sigeru, O.: Web page feature selection and classification using neural networks. Information Sciences 158, 69–88 (2004)

33. Song, A.-B., Zhao, M.-X., Liang, Z.-P., Dong, Y.-S., Luo, J.-Z.: Discovering user profiles for Web personalization recommendation. Journal of Computer Science and Technology 19(3), 320–328 (2004)

34. Srivastava, J., Cooley, R., Deshpande, M., Tan, P.T.: Web Usage Mining: Discovery and Applications of Usage Patterns from Web Data. SIGKDD Explorations 1(2), 2–23 (2000)

35. Kumar De, S., Radha Krishna, P.: Clustering web transactions using rough approximation. Fuzzy Sets and Systems 148, 131–138 (2004)

36. Mitra, S.: Rough-Fuzzy Collaborative Clustering. IEEE Transactions on Systems, Man and Cybernetics 36(4) (2006)

37. Thangavel, K., Ashok Kumar, D.: Pattern Clustering using Neural Network, Vision 2020: The Strategic role of Operational Research. Allied Publishers PVT LTD, New Delhi, pp. 662–679 (2006)

38. Titterington, D., Smith, A., Makov, U.: Statistical analysis of finite mixture distributions. John Wiley and Sons, Chichester (1985)

39. Voges, K.E., Pope, N.K.L., Brown, M.R.: Cluster analysis of marketing data examining online shopping orientation: a comparison of k-means and rough clustering approaches. In: Abbass, H.A., Sarker, R.A., Newton, C.S. (eds.) Heuristics and Optimization for Knowledge Discovery, pp. 207–224. Idea Group Publishing, Hershey (2002)

40. Wang, X., Abraham, A., Smith, K.: Intelligent web traffic mining and analysis. Journal of Network and Computer Applications 28(2), 147–165 (2005)

41. WangBin, Liuzhijing: Web Mining Research. In: Proceedings of the Fifth nternational Conference on Computational Intelligence and Multimedia Applications (ICCIMA 2003) (2003)

42. Xing, W., Ghorbani, A.: Weighted PageRank Algorithm. In: Proceedings of the Second Annual Conference on Communication Networks and Services Research (CNSR 2004) (2004)

43. W.W.W. Consortium. The Common Log file Format (1995),
    http://www.w3.org/Daemon/User/Config/
    Logging.html#common-logfile-format

44. Xie, X.L., Beni, G.: A Validity Measure for fuzzy Clustering. IEEE Trans. on Pattern Analysis and MachineIntelligence 13(8), 841–847 (1991)

45. Yan, T.W., Jacobsen, M., Garcia-Molina, H., Dayal, U.: From user access patterns to dynamic hypertext linking. In: Proceedings of 5th WWW, pp. 1007–1014 (1996)

46. Zhang, X., Gong, W., Kawamura, Y.: Customer behavior pattern discovering with web mining. In: Proceedings of Asia Pacific web conference, Hangzhou, China, pp. 844–853 (2004)

47. Pabarskaite, Z., Raudys, A.: A process of knowledge discovery from web log data: Systematization and critical review. Journal of Intelligent Information Systems 28, 79–104 (2007)

An Overview on Mining Data Streams

João Gama and Pedro Pereira Rodrigues

Summary. The most challenging applications of knowledge discovery involve dynamic environments where data continuous flow at high-speed and exhibit non-stationary properties. In this chapter we discuss the main challenges and issues when learning from data streams. In this work, we discuss the most relevant issues in knowledge discovery from data streams: incremental learning, cost-performance management, change detection, and novelty detection. We present illustrative algorithms for these learning tasks, and a real-world application illustrating the advantages of stream processing. The chapter ends with some open issues that emerge from this new research area.

1 Introduction

Machine Learning algorithms are faced to new problems and challenges. Nowadays, we have sensors and computers sending information to other computers. In some applications, like those that emerge from sensor networks, the data is modeled best not as persistent tables but rather as transient data streams. Sometimes, it is not feasible to load the arriving data into a traditional DataBase Management Systems (DBMS), and traditional DBMS are not designed to directly support the continuous queries required in these application [Babcock et al., 2002]. These sources of data are called Data Streams.

João Gama
Faculty of Economics, LIAAD, INESC-Porto, University of Porto
Rua de Ceuta, 118, 6; 4050-190 Porto, Portugal
e-mail: jgama@fep.up.pt

Pedro Pereira Rodrigues
Faculty of Medicine, LIAAD, INESC-Porto, University of Porto
Rua de Ceuta, 118, 6; 4050-190 Porto, Portugal
e-mail: pprodrigues@fc.up.pt

A. Abraham et al. (Eds.): Foundations of Comput. Intel. Vol. 6, SCI 206, pp. 29–45.
springerlink.com   © Springer-Verlag Berlin Heidelberg 2009

Fig. 1 Illustrative example of the electrical grid
for the electrical load-
demand problem

In the last two decades, machine learning research and practice has focused
on batch learning usually with small datasets. In batch learning, the whole
training data is available to the algorithm that outputs a decision model
after processing the data eventually (or most of the times) multiple times.
The rationale behind this practice is that examples are generated at random
accordingly to some stationary probability distribution. Also, most learners
use a greedy, hill-climbing search in the space of models.

1.1 An Illustrative Problem

Electricity distribution companies usually set their management operators on
SCADA/DMS products (Supervisory Control and Data Acquisition / Distribution Management Systems). One of their important tasks is to forecast
the electrical load (electricity demand) for a given sub-network of consumers.
Load forecast is a relevant auxiliary tool for operational management of an
electricity distribution network, since it enables the identification of critical
points in load evolution, allowing necessary corrections within available time,
and planning strategies for different horizon.

In this application, data is collected from a set of sensors distributed all
around the network. Sensors can send information at different time scales,
speed, and granularity. Data continuously flow eventually at high-speed, in
a dynamic and time-changing environment. Data mining in this context requires continuously processing of the incoming data monitoring trends, and
detecting changes. Traditional one-shot systems, memory based, trained from
fixed training sets and generating static models are not prepared to process
the high detailed data available, they are not able to continuously maintain a
predictive model consistent with the actual state of the nature, nor are they
ready to quickly react to changes. Moreover, with the evolution of hardware
components, these sensors are acquiring computational power. The challenge
will be to run the predictive model in the sensors itself.

The challenge problem in learning from data streams is the ability to permanently maintain an accurate decision model. This issue requires learning algorithms that can modify the current model whenever new data is available at the rate of data arrival. In this context, the assumption that examples are generated at random according to a stationary probability distribution does not hold, at least in complex systems and for large time periods. In the presence of a non-stationary distribution, the learning system must incorporate some form of forgetting past and outdated information. Learning from data streams require incremental learning algorithms that take into account concept drift.

2   Key Issues in Learning from Data Streams

Learning from data flowing at high-speed in dynamic and non-stationary environments require new sampling and randomization techniques, and new approximate, incremental and decremental algorithms. In [Domingos & Hulten, 2001], the authors identify desirable properties of learning systems that are able to mine continuous, high-volume, open-ended data streams as they arrive: i) incrementality, ii) online learning, iii) constant time to process each example using fixed memory, iv) single scan over the training data, and v) tacking drift into account. Examples of learning algorithms designed to process open-ended streams include: predictive learning (decision trees [Domingos & Hulten, 2000, Gama et al., 2003, Jin & Agrawal, 2003], decision rules [Ferrer-Troyano et al., 2005]); descriptive learning (variants of k-means clustering [Zhang et al., 1996, Sheikholeslami et al., 1998], clustering [O'Callaghan et al., 2002, Aggarwal et al., 2003], hierarchical time-series clustering [Rodrigues et al., 2006]); association learning (frequent itemsets mining [Han et al., 2000], frequent pattern mining [Giannella et al., 2003]); novelty detection [Markou & Singh, 2003, Spinosa et al., 2008]; feature selection [Sousa et al., 2007], etc. All these algorithms share some common properties. They process examples at the rate they arrive using a single scan of data and fixed memory. They maintain a decision model at any time, and are able to adapt the model to the most recent data.

2.1   *Cost-Performance Management*

The ability to update the decision model whenever new information is available is an important property, but it is not enough. Another required operator is the ability to forget past information [Kifer et al., 2004]. Some data stream models allow delete and update operators. For example, sliding windows models require the forgetting of old information. In these situations the incremental property is not enough. Learning algorithms need forgetting operators that reverse learning: decremental unlearning [Cauwenberghs & Poggio, 2000].

The incremental and decremental issues requires a permanent maintenance and updating of the decision model as new data is available. Of course, there is a trade-off between the cost of update and the gain in performance we may obtain. Learning algorithms exhibit different profiles. Algorithms with strong variance management are quite efficient for small training sets. Very simple models, using few free-parameters, can be quite efficient in variance management, and effective in incremental and decremental operations (for example naive Bayes) being a natural choice in the sliding windows framework. The main problem with simple representation languages is the boundary in generalization performance they can achieve, since they are limited by high bias. Large volumes of data require efficient bias management. Complex tasks requiring more complex models increase the search space and the cost for structural updating. These models, require efficient control strategies for the trade-off between the gain in performance and the cost of updating.

2.2 *Monitoring Learning*

Whenever data flows over time, it is highly improvable the assumption that the examples are generated at random according to a stationary probability distribution [Basseville & Nikiforov, 1987]. At least in complex systems and for large time periods, we should expect changes (smooth or abrupt) in the distribution of the examples. A natural approach for these incremental tasks is adaptive learning algorithms, incremental learning algorithms that take into account concept drift.

Concept drift [Klinkenberg, 2004] means that the concept about which data is being collected may shift from time to time, each time after some minimum permanence. Changes occur over time. The evidence for changes in a concept is reflected in some way in the training examples. Old observations, that reflect the behavior of nature in the past, become irrelevant to the current state of the phenomena under observation and the learning agent must forget that information. The nature of change is diverse. Changes may occur in the context of learning, due to changes in hidden variables, or in the characteristic properties of the observed variables. Most learning algorithms use blind methods that adapt the decision model at regular intervals without considering whether changes have really occurred. Much more interesting are explicit change detection mechanisms. The advantage is that they can provide meaningful description (indicating change-points or small time-windows where the change occurs) and quantify the degree of change. They may follow two different approaches:

- Monitoring the evolution of performance indicators adapting techniques used in Statistical Process Control [Gama et al., 2004].

- Monitoring distributions on two different time windows [Kifer et al., 2004]. The method monitors the evolution of a distance function between two distributions: data in a reference window and in a current window of the most recent data points.

The main research issue is to develop methods with fast and accurate detection rates with few false alarms. A related problem is: how to incorporate change detection mechanisms inside learning algorithms. Also, the level of granularity of decision models is a relevant property [Gaber et al., 2004], because it can allow partial, fast and efficient updating in the decision model instead of rebuilding a complete new model whenever a change is detected. Finally, the ability to recognize seasonal and re-occurring patterns is an open issue.

2.3   Feature Selection and Pre-processing

Selection of relevant and informative features, discretization, noise and rare events detection are common tasks in Machine Learning and Data Mining. They are used in a one-shot process. In the streaming context the semantics of these tasks changes drastically. Consider the feature selection problem. In streaming data the concept of *irrelevant* or *redundant* features are now restricted to a certain period of time. Features previously considered *irrelevant* may become *relevant*, and vice-versa to reflect the dynamics of the process generating data. While in standard data mining, an irrelevant feature could be ignored forever, in the streaming setting we need still monitor the evolution of those features. Recent work based on the *fractal dimension* [Barbara & Chen, 2000] could point interesting directions for research.

3   Illustrative Learning Algorithms for Data Streams

To clarify the ideas on learning from data streams, we present some illustrative examples of learning algorithms that continuously maintain a model evolving over time. The illustrative examples covers histograms, the first one is a hierarchical clustering time series data streams, an algorithm designed to deal with thousands of time series that flow at high-speed, decision trees and the well-known multi-layer `perceptron` adapted to learn from high-speed data streams.

3.1   Histograms from Data Streams

Discretization of continuous attributes is an important task for certain types of machine learning algorithms. In Bayesian learning, discretization is the most common approach when data is described by continuous

Fig. 2 Illustrative example of the 2 layers in the
em Partition Incremental
Discretization algorithm

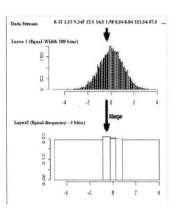

features [Domingos & Pazzani, 1997]. Although discretization is a well-
known topic in data analysis and machine learning, most of the works refer
to a batch discretization where all the examples are available for discretization. Few works refer to incremental discretization [Lin et al., 2003,
Gibbons & Matias, 1999, Guha & Harb, 2005].

The *Partition Incremental Discretization* algorithm [Gama & Pinto, 2006]
(*PiD* for short) is composed by two layers. The first layer simplifies and
summarizes the data; the second layer constructs the final histogram.

The first layer is initialized without seeing any data. The input for the
initialization phase is the number of intervals (that should be much larger
than the desired final number of intervals) and the range of the variable.
The range of the variable is only indicative. It is used to initialize the set
of breaks using a equal-width strategy. Each time we observe a value of
the random variable, we update $layer_1$. The update process determines the
interval corresponding to the observed value, and increments the counter of
this interval. Whenever the counter of an interval is above a user defined
threshold (a percentage of the total number of points seen so far), a split
operator triggers. The split operator generates new intervals in $layer_1$. If the
interval that triggers the split operator is the first or the last, a new interval
with the same step is inserted. In all the other cases, the interval is split into
two, generating a new interval.

The process of updating $layer_1$ works online, performing a single scan
over the data stream. It can process infinite sequences of data, processing
each example in constant time and space.

The second layer merges the set of intervals defined by the first layer.
It triggers whenever it is necessary (e.g. by user action). The input for the
second layer is the breaks and counters of $layer_1$, the type of histogram
(equal-width or equal-frequency) and the number of intervals. The algorithm
for the $layer_2$ is very simple. For equal-width histograms, it first computes
the breaks of the final histogram, from the actual range of the variable (estimated in $layer_1$). The algorithm traverses the vector of breaks once, adding
the counters corresponding to two consecutive breaks. For equal-frequency

histograms, we first compute the exact number F of points that should be in each final interval (from the total number of points and the number of desired intervals). The algorithm traverses the vector of counters of $layer_1$ adding the counts of consecutive intervals till F.

The two-layer architecture divides the histogram problem into two phases. In the first phase, the algorithm traverses the data stream and incrementally maintains an equal-width discretization. The second phase constructs the final histogram using only the discretization of the first phase. The computational costs of this phase can be ignored: it traverses once the discretization obtained in the first phase. We can construct several histograms using different number of intervals and different strategies: equal-width or equal-frequency. This is the main advantage of PiD in exploratory data stream analysis.

3.2 Clustering of Time Series Data Streams

Data streams usually consist of variables producing examples continuously over time at high-speed. The basic idea behind clustering time series is to find groups of variables that behave similarly through time. Most of the work in incremental clustering of data streams has been concentrated on example clustering rather than variable clustering. Clustering variables (e.g. time series) is a very useful tool for some applications, such as sensor networks, social networks, electrical power demand, stock market, etc. The standard approach to cluster variables in a batch scenario, uses the *transpose* of the working matrix. In a data stream scenario, this is not possible, because the transpose operator is a block operator [Barbará, 2002]. For the task of clustering variables in streams new algorithms are required.

In this section, we describe the ODAC clustering algorithm [Rodrigues et al., 2006] which includes an incremental dissimilarity measure based on the correlation between time series, calculated with sufficient statistics gathered continuously over time. There are two main operations in the hierarchical structure of clusters: *expansion* that splits one cluster into two new clusters; and *aggregation* that aggregates two clusters. Both operators are based on the diameters of the clusters, and supported by confidence levels given by the Hoeffding bounds. The system continuously monitors the evolution of the diameters. In ODAC, the dissimilarity between variables a and b is measured by: $rnomc(a, b) = \sqrt{(1 - corr(a, b))/2}$, where $corr(a, b)$ is the Pearson's correlation coefficient. For each cluster, the system chooses two variables that define the diameter of that cluster (those that are less correlated). If a given heuristic condition is met on this diameter, the system splits the cluster in two, assigning each of those variables to one of the two new clusters. Afterwards, the remaining variables are assigned to the cluster that has the closest pivot (first assigned variables). The newly

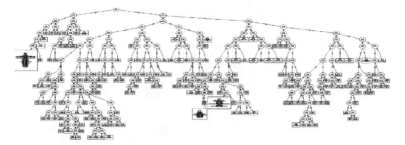

Fig. 3 Illustrative example of a cluster structure in the electrical load-demand problem

created leaves start new statistics, assuming that only the future information will be useful to decide if the cluster should be split.

A requirement to process data streams is change detection. In most applications, the correlation structure between variables evolves smoothly. The clustering structure must adapt to this type of changes. In a hierarchical structure of clusters, considering that the data streams are produced by a stable concept, the intra-cluster dissimilarity should decrease with each split. For each given cluster C_k, the system verifies if older split decision still represents the structure of data, testing the diameters of C_k, C_k's sibling and C_k's parent. If diameters are increasing above parent's diameter, changes have occurred, so the system aggregates the leaves, restarting the sufficient statistics for that group.

The presented clustering procedure is oriented towards processing high speed data streams. The main characteristics of the system are constant memory and constant time in respect to the number of examples. In ODAC, system space complexity is constant on the number of examples, even considering the infinite amount of examples usually present in data streams. An important feature of this algorithm is that every time a split is performed on a leaf with $a + b$ variables, where a and b are the number of variables that goes to each descendant node, the global number of dissimilarities needed to be computed at the next iteration diminishes by $a \times b$. The time complexity of each iteration of the system is constant given the number of examples, and decreases with every split occurrence.

3.3 Decision Trees for High-Speed Data Streams

Learning from large datasets may be more effective when using algorithms that place greater emphasis on bias management. One such algorithms is the VFDT [Domingos & Hulten, 2000] system. VFDT is a decision-tree learning algorithm that dynamically adjusts its bias whenever new examples are available. In decision tree induction, the main issue is the decision of when to expand the tree, installing a splitting-test and generating new leaves. The

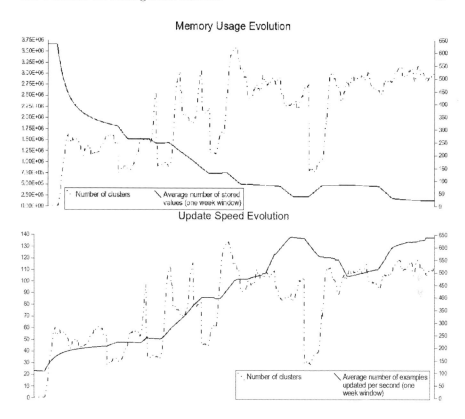

Fig. 4 ODAC memory usage and processing examples speed in the electrical load-demand problem

basic idea of VFDT consists of using a small set of examples to select the splitting-test to incorporate in a decision tree node. If after seeing a set of examples, the difference of the merit between the two best splitting-tests does not satisfy a statistical test (the Hoeffding bound), VFDT proceeds by examining more examples. VFDT only makes a decision (i.e., adds a splitting-test in that node), when there is enough statistical evidence in favor of a particular test. This strategy guarantees model stability (low variance), controls overfiting, while it may achieve an increased number of degrees of freedom (low bias) with increasing number of examples.

In VFDT a decision tree is learned by recursively replacing leaves with decision nodes. Each leaf stores the sufficient statistics about attribute-values. The sufficient statistics are those needed by a heuristic evaluation function that computes the merit of split-tests based on attribute-values. When an example is available, it traverses the tree from the root to a leaf, evaluating the appropriate attribute at each node, and following the branch corresponding to the attribute's value in the example. When the example reaches a leaf, the sufficient statistics are updated. Then, each possible condition based on

attribute-values is evaluated. If there is enough statistical support in favor
of one test over the others, the leaf is changed to a decision node. The new
decision node will have as many descendant leaves as the number of possible
values for the chosen attribute (therefore this tree is not necessarily binary).
The decision nodes only maintain the information about the split-test installed within them.

Algorithm 1. The Hoeffding tree algorithm.

---

input : S: A Sequence of examples
$\quad\quad\quad\quad$ X: A Set of nominal attributes
$\quad\quad\quad\quad$ Y: $Y = \{y_1, \ldots, y_k\}$ set of class values
$\quad\quad\quad\quad$ $H(.)$: Split evaluation function
$\quad\quad\quad\quad$ δ: 1 minus the desired probability
$\quad\quad\quad\quad$ of choosing the correct attribute at any node.
$\quad\quad\quad\quad$ τ: Constant used to break ties.
output: HT: A decision tree
begin
$\quad$ Let $HT \leftarrow$ Empty Leaf (Root)
$\quad$ **foreach** $y_k \in Y$, $x_i \in X$, value j of x_i **do**
$\quad\quad$ $x_{ijk} \leftarrow 0$
$\quad$ **foreach** example $(x, y_k) \in S$ **do**
$\quad\quad$ Traverse the tree HT from the root to a leaf l
$\quad\quad$ **foreach** value j of x_i **do**
$\quad\quad\quad$ Increment x_{ijk}
$\quad\quad$ **if** all examples in l are not of the same class **then**
$\quad\quad\quad$ Compute $G_l(X_i)$ for all the attributes
$\quad\quad\quad$ Let X_a be the attribute with highest H_l
$\quad\quad\quad$ Let X_b be the attribute with second highest H_l
$\quad\quad\quad$ Compute ϵ (Hoeffding bound)
$\quad\quad\quad$ **if** $(H(X_a) - H(X_b) > \epsilon)$ **then**
$\quad\quad\quad\quad$ Replace l with a splitting test based on attribute X_a
$\quad\quad\quad\quad$ Add a new empty leaf for each branch of the split
$\quad\quad\quad$ **else**
$\quad\quad\quad\quad$ **if** $\epsilon < \tau$ **then**
$\quad\quad\quad\quad\quad$ Replace l with a splitting test based on attribute X_a
$\quad\quad\quad\quad\quad$ Add a new empty leaf for each branch of the split

end

The main innovation of the VFDT system is the use of Hoeffding bounds to
decide how many examples must be observed before installing a split-test at a
leaf. Suppose we have made n independent observations of a random variable
r whose range is R. The Hoeffding bound states that the true average of r,
$\bar{r}$, is at least $\bar{r} - \epsilon$ where $\epsilon = \sqrt{R^2 \frac{\ln(\frac{1}{\delta})}{2n}}$, with probability $1 - \delta$.

Let $H(\cdot)$ be the evaluation function of an attribute. For the information
gain, the range R, of $H(\cdot)$ is $log_2(k)$ where k denotes the number of classes.

Fig. 5 Illustrative example comparing the performance evolution between C4.5 and VFDT

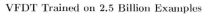
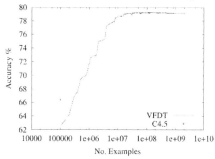

Let x_a be the attribute with the highest $H(\cdot)$, x_b the attribute with second-highest $H(\cdot)$ and $\overline{\Delta H} = \overline{H}(x_a) - \overline{H}(x_b)$, the difference between the two best attributes. Then if $\overline{\Delta H} > \epsilon$ with n examples observed in the leaf, the Hoeffding bound states that, with probability $1 - \delta$, x_a is really the attribute with highest value in the evaluation function. In this case the leaf must be transformed into a decision node that splits on x_a.

The evaluation of the merit function for each example could be very expensive. It turns out that it is not efficient to compute $H(\cdot)$ every time that an example arrives. VFDT only computes the attribute evaluation function $H(\cdot)$ when a minimum number of examples has been observed since the last evaluation. This minimum number of examples is a user-defined parameter. When two or more attributes continuously have very similar values of $H(\cdot)$, even with a large number of examples, the Hoeffding bound will not decide between them. To solve this problem the VFDT uses a constant τ introduced by the user for run-off, e.g., if $\overline{\Delta H} < \epsilon < \tau$ then the leaf is transformed into a decision node. The split test is based on the best attribute.

Later, the same authors presented the CVFDT algorithm [Hulten et al., 2001], an extension to VFDT designed for time-changing data streams. CVFDT generates alternative decision trees at nodes where there is evidence that the splitting test is no longer appropriate. The system replaces the old tree with the new one when the latter becomes more accurate. A similar algorithm but for regression problems was presented in [Ikonomovska & Gama, 2008].

3.4 Neural Networks and Data Streams

In previous sections we described a hierarchical clustering algorithm and a decision tree designed for processing high-speed data streams. While in these cases the characteristic properties of streams implies new algorithms, this is not always the case. In this section we will illustrate how a well known algorithm, a multi-layer percepton, can be applied in high-speed streams.

Fig. 6 Illustrative example comparing the
mape evolution between
a static neural network
and a neural network
trained using the stochastic incremental train

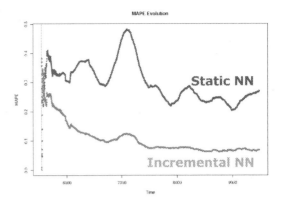

ANNs are powerful models that can approximate any continuous function [Craven & Shavlik, 1997] with arbitrary small error with a three layer
network. The *mauvaise reputation* of ANNs comes from slower learning times.
Two other known problems of the generalization capacity of neural networks
are overfitting and large variance. At a first glance, these aspects would for-
bidden the application of neural nets in the data stream scenario.

The standard process to train a neural net consists of processing training
examples in epochs. Why would we need to process the same set of examples several times? The only reason is the lack of examples. One of the key
characteristics of data streams is that data is abundant. A neural network
can be trained processing each incoming example once. Whenever a training
example is available, it is propagated and the error backpropagated through
the network only once, as data is abundant and flow continuously. This is the
stochastic sequential training of neural networks.

This is a main advantage, allowing the neural network to process an infinite
number of examples at high speed. Another advantage is the smooth adaptation in dynamic data streams where the target function evolves over time.
Craven and Shavlik [Craven & Shavlik, 1997] point out that the inductive
bias of neural networks is the most appropriate for sequential and temporal
prediction tasks.

3.5 Illustrative Application

The electrical network problem described in section 1.1 contains around 4000
sensors spreaded out over the network. Sensors continuous measure quantities of interest. Here, we only consider *Current Intensity* as the measures
of interest. One predictive task of great economical interest consist of predict the electrical demand for short horizon: next hour, next 24 hours, next
week. For this problem, we have compared the *stochastic sequential training* of a 3 layer feed-forward neural network (to which we refer as RETI-
NAE as acknowledgment to the research/industry consortium project during

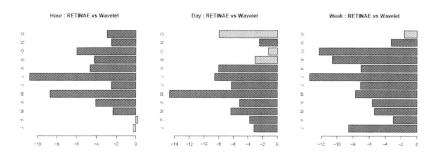

Fig. 7 Difference between the median of errors of predictions using RETINAE against predictions using Wavelets. Darker bars represent error distributions which are significantly different, according to the *Wilcoxon Sign Ranks Tests* [Bauer, 1972] with a confidence level of 95%

which it was mainly developed) against Wavelets [Rauschenbach, 2002]. Wavelets are the standard method used in the company we are working with, but no exhaustive and complete analysis could be done due to slow batch procedure used in Wavelet analysis. Full details of this application appear in [Rodrigues & Gama, 2009 to appear].

All evaluation measures are computed as follow. At time i the system makes a prediction for the specific measure and sensor for time $i + k$. k hours later, we observed the real value provided by the sensor. The quality measure usually considered in electricity load forecast is the MAPE *(Mean Absolute Percentage Error)* defined as $MAPE = \sum_{i=1}^{n} \frac{|(\hat{y}_i - y_i)/y_i|}{n}$, where y_i is the real value of variable y at time i and $\hat{y}_i$ is the corresponding predicted value.

For each sensor, we consider around three years of data. The system RETINAE makes forecasts for next hour, one day ahead, and one week ahead. At each time point t, the user can consult the forecast for next hour, next 24 hours and all hours for the next week, for each sensor. We have conducted experiments where, for the given year, the quality of the system in each month is compared with Wavelets on two precise variables each month, chosen, by an expert, as relevant predictable streams but exhibiting either low or high error. Results are shown on Figure 7 for the three different horizons. Practically all sensors reported lower median errors using RETINAE than using Wavelets. The *Wilcoxon Signed Ranks Test* [Bauer, 1972] was applied to compare the error distributions, and darker bars represent significantly different distributions (we consider a significance level of 5%). The relevance of the on-line system using neural networks is exposed, with lower error values on the majority of the studied variables. Moreover, in precise analysis of results, it was noticed an improvement on the performance of the system, compared to the predictions made using Wavelets, after failures or abnormal behavior in the streams.

4 Future Trends in Learning from Data Streams

Streaming data offers a symbiosis between Streaming Data Management Systems and Machine Learning. On one hand, the techniques developed in data streams management systems, like synopsis and sketches require counts over very high dimensions both in the number of examples and in the domain of the variables, can provide tools for designing Machine Learning algorithms in these domains. On the other hand, Machine Learning provides compact descriptions of the data that can be useful for answering queries in DSMS.

What are the current trends and directions for future research in learning from data streams? Issues on sampling, incremental learning and forgetting, bias and cost-performance management are the basic challenges in stream mining. In most applications, we are interested in maintaining a decision model consistent with the current status of the nature. This has led us to the sliding window models where data is continuously inserted and deleted from a window. So, learning algorithms must have operators for incremental learning and forgetting. Closed related are change detection issues. Concept drift in the predictive setting is a well studied topic [Klinkenberg, 2004]. In other scenarios, like clustering, very few works address the problem. The main research issue is how to incorporate change detection mechanisms in the learning algorithm for different paradigms. Another relevant aspect of any learning algorithm is the hypothesis evaluation criteria and metrics. Most of evaluation methods and metrics were designed for the static case and provide a single measurement about the quality of the hypothesis. In the streaming context, we are much more interested in how the evaluation metric evolves over time. Results from the sequential statistics [Wald, 1947] may be much more appropriate.

5 Conclusion Remarks

Learning from data streams is an increasing research area with challenging applications and contributions from fields like data bases, learning theory, machine learning, and data mining. Sensor networks, scientific data, monitoring processes, web analysis, traffic logs, are examples of real-world applications were stream algorithms have been successful applied. Continuously learning, forgetting, self-adaptation, and self-reaction are main characteristics of any intelligent system. They are characteristic properties of stream learning algorithms.

References

Aggarwal et al., 2003. Aggarwal, C., Han, J., Wang, J., Yu, P.: A framework for clustering evolving data streams. In: Proceedings of Twenty-Ninth International Conference on Very Large Data Bases, pp. 81–92. Morgan Kaufmann, San Francisco (2003)

Babcock et al., 2002. Babcock, B., Babu, S., Datar, M., Motwani, R., Widom, J.: Models and issues in data stream systems. In: Proceedings of the 21st Symposium on Principles of Database Systems, pp. 1–16. ACM Press, New York (2002)

Barbará, 2002. Barbará, D.: Requirements for clustering data streams. SIGKDD Explorations 3, 23–27 (2002)

Barbara & Chen, 2000. Barbara, D., Chen, P.: Using the fractal dimension to cluster datasets. In: Proc. of the 6th International Conference on Knowledge Discovery and Data Mining, pp. 260–264. ACM Press, New York (2000)

Basseville & Nikiforov, 1987. Basseville, M., Nikiforov, I.: Detection of abrupt changes: Theory and applications. Prentice-Hall Inc., Englewood Cliffs (1987)

Bauer, 1972. Bauer, D.F.: Constructing confidence sets using rank statistics. Journal of American Statistical Association, 687–690 (1972)

Cauwenberghs & Poggio, 2000. Cauwenberghs, G., Poggio, T.: Incremental and decremental support vector machine learning. In: Proceedings of the 13th Neural Information Processing Systems (2000)

Craven & Shavlik, 1997. Craven, M., Shavlik, J.: Using neural networks for data mining. Future Generation Computer Systems 13, 211–229 (1997)

Domingos & Hulten, 2000. Domingos, P., Hulten, G.: Mining High-Speed Data Streams. In: Proceedings of the ACM Sixth International Conference on Knowledge Discovery and Data Mining, pp. 71–80. ACM Press, New York (2000)

Domingos & Hulten, 2001. Domingos, P., Hulten, G.: A general method for scaling up machine learning algorithms and its application to clustering. In: Proceedings of the Eighteenth International Conference on Machine Learning, pp. 106–113. Morgan Kaufmann, San Francisco (2001)

Domingos & Pazzani, 1997. Domingos, P., Pazzani, M.: On the optimality of the simple Bayesian classifier under zero-one loss. Machine Learning 29, 103–129 (1997)

Ferrer-Troyano et al., 2005. Ferrer-Troyano, F., Aguilar-Ruiz, J., Riquelme, J.: Incremental rule learning and border examples selection from numerical data streams. Journal of Universal Computer Science 11, 1426–1439 (2005)

Gaber et al., 2004. Gaber, M.M., Krishnaswamy, S., Zaslavsky, A.: Cost-efficient mining techniques for data streams. In: Proceedings of the second workshop on Australasian information security, pp. 109–114. Australian Computer Society, Inc. (2004)

Gama et al., 2004. Gama, J., Medas, P., Castillo, G., Rodrigues, P.: Learning with drift detection. In: Bazzan, A.L.C., Labidi, S. (eds.) SBIA 2004. LNCS, vol. 3171, pp. 286–295. Springer, Heidelberg (2004)

Gama & Pinto, 2006. Gama, J., Pinto, C.: Discretization from data streams: applications to histograms and data mining. In: SAC, pp. 662–667. ACM Press, New York (2006)

Gama et al., 2003. Gama, J., Rocha, R., Medas, P.: Accurate decision trees for mining high-speed data streams. In: Proceedings of the ninth ACM SIGKDD international conference on Knowledge discovery and data mining, pp. 523–528. ACM Press, Washington (2003)

Giannella et al., 2003. Giannella, C., Han, J., Pei, J., Yan, X., Yu, P.: Mining frequent patterns in data streams at multiple time granularities. In: Next Generation Data Mining. AAAI/MIT (2003)

Gibbons & Matias, 1999. Gibbons, P.B., Matias, Y.: Synopsis data structures for massive data sets. In: ACM-SIAM Symposium on Discrete Algorithms (SODA), pp. 909–910. Society for Industrial and Applied Mathematics (1999)

Guha & Harb, 2005. Guha, S., Harb, B.: Wavelet synopsis for data streams: minimizing non-euclidean error. In: Proceeding of the eleventh ACM SIGKDD international conference on Knowledge discovery in data mining, pp. 88–97. ACM Press, New York (2005)

Han et al., 2000. Han, J., Pei, J., Yin, Y.: Mining frequent patterns without candidate generation. In: SIGMOD 2000: Proceedings of the 2000 ACM SIGMOD international conference on Management of data, pp. 1–12. ACM Press, New York (2000)

Hulten et al., 2001. Hulten, G., Spencer, L., Domingos, P.: Mining time-changing data streams. In: Proceedings of the 7th ACM SIGKDD International conference on Knowledge discovery and data mining, pp. 97–106. ACM Press, San Francisco (2001)

Ikonomovska & Gama, 2008. Ikonomovska, E., Gama, J.: Learning model trees from data streams. In: Discovery Science, (no prelo). Springer, Heidelberg (2008)

Jin & Agrawal, 2003. Jin, R., Agrawal, G.: Efficient decision tree construction on streaming data. In: Proceedings of the Ninth International Conference on Knowledge Discovery and Data Mining. ACM Press, New York (2003)

Kifer et al., 2004. Kifer, D., Ben-David, S., Gehrke, J.: Detecting change in data streams. In: VLDB 2004: Proceedings of the 30th International Conference on Very Large Data Bases, pp. 180–191. Morgan Kaufmann Publishers Inc., San Francisco (2004)

Klinkenberg, 2004. Klinkenberg, R.: Learning drifting concepts: Example selection vs. example weighting. Intelligent Data Analysis 8, 281–300 (2004)

Lin et al., 2003. Lin, J., Keogh, E., Lonardi, S., Chiu, B.: A symbolic representation of time series, with implications for streaming algorithms. In: Proceedings of the 8th ACM SIGMOD Workshop on Research Issues in Data Mining and Knowledge Discovery (DMKD 2003), pp. 2–11 (2003)

Markou & Singh, 2003. Markou, M., Singh, S.: Novelty detection: a review-part 1: neural network based approaches (2003)

O'Callaghan et al., 2002. O'Callaghan, L., Mishra, N., Meyerson, A., Guha, S., Motwani, R.: Streaming-data algorithms for high-quality clustering. In: Proceedings of IEEE International Conference on Data Engineering. IEEE Press, Los Alamitos (2002)

Rauschenbach, 2002. Rauschenbach, T.: Short-term load forecast using wavelet transformation. Proceeding (362) Artificial Intelligence and Applications (2002)

Rodrigues & Gama, 2009 to appear. Rodrigues, P., Gama, J.: A system for analysis and prediction of electricity-load streams. Intelligent Data Analysis 13 (to appear, 2009)

Rodrigues et al., 2006. Rodrigues, P., Gama, J., Pedroso, J.: Odac: Hierarchical clustering of time series data streams. In: Proceedings of the Sixth SIAM International Conference on Data Mining, pp. 499–503. Society for Industrial and Applied Mathematics, Bethesda (2006)

Sheikholeslami et al., 1998. Sheikholeslami, G., Chatterjee, S., Zhang, A.: WaveCluster: A multi-resolution clustering approach for very large spatial databases. In: Proceedings of the Twenty-Fourth International Conference on Very Large Data Bases, pp. 428–439. ACM Press, New York (1998)

Sousa et al., 2007. Sousa, E., Traina, A., Traina, J.C., Faloutsos, C.: Evaluating the intrinsic dimension of evolving data streams. New Generation Computing 25 (2007)

Spinosa et al., 2008. Spinosa, E., Gama, J., Carvalho, A.: Cluster-based novel concept detection in data streams applied to intrusion detection in computer networks. In: Proceedings of the 2008 ACM Symposium on Applied computing, pp. 976–980. ACM Press, New York (2008)

Wald, 1947. Wald, A.: Sequential analysis. John Wiley and Sons, Chichester (1947)

Zhang et al., 1996. Zhang, T., Ramakrishnan, R., Livny, M.: Birch: an eficient data clustering method for very large databases. In: Proceedings of the 1996 ACM SIGMOD International Conference on Management of Data, pp. 103–114. ACM Press, New York (1996)

Data Stream Mining Using Granularity-Based Approach

Mohamed Medhat Gaber

Summary. Significant applications require data stream mining algorithms to run in resource-constrained environments. Thus, adaptation is a key process to ensure the consistency and continuity of the running algorithms. This chapter provides a theoretical framework for applying the granularity-based approach in mining data streams. Our *Algorithm Output Granularity (AOG)* is explained in details providing practitioners the ability to use it for enabling resource-awareness and adaptability for their algorithms. Theoretically, *AOG* has been formalized using the *Probably Approximately Correct (PAC)* learning model allowing researchers to formalize the adaptability of their techniques. Finally, the integration of *AOG* with other adaptation strategies is provided.

1 Introduction

Several techniques have been proposed to address the challenges posed by continuously generated data streams [10]. There are a large number of sources from which data streams are produced including sensor networks, web logs and click streams, network traffic, scientific simulations and ATM transactions. Sampling, projection, approximation, group testing and embeddings have been used in different ways as possible solutions to facilitate processing of high speed data streams. These techniques have successfully reduced the space and time complexity of the mining algorithms. The rate of producing the output of any of these algorithms is affected by the rate of the incoming data stream. The amount of memory to store the resultant knowledge affects

Mohamed Medhat Gaber
Caulfield School of Information Technology
Monash University
900 Dandenong Rd, Caulfield East, VIC 3145, Australia
e-mail: Mohamed.Gaber@infotech.monash.edu.au

A. Abraham et al. (Eds.): Foundations of Comput. Intel. Vol. 6, SCI 206, pp. 47–66.
springerlink.com   © Springer-Verlag Berlin Heidelberg 2009

the performance of the mining algorithm in both space and time. It affects the space by generating knowledge structures that consume memory. On the other hand, it affects the time by increasing the search space of the mining results resident in memory. Although the proposed techniques have reduced the complexity of algorithms, the resource constraints in a wide range of data stream applications may lead to the failure of the running algorithms. Thus, the previously proposed techniques have not addressed the issue of output rate with regard to the input speed given the resource availability in essential data stream applications. To clarify this, we can use clustering as an example. If the algorithm generates a larger number of clusters, this will result in higher memory consumption and longer running time for the assignment of any data record to an existing cluster. The same analogy applies to other mining techniques. Thus, the key research problem to be addressed can be stated as follows: *Limited availability of resources of the computational devices that process data streams in a wide range of applications such as wireless sensor networks and small mobile devices may result in possibility of failure of current data stream mining algorithms due to lack of adaptation of the running algorithm to available resources.* For example, if the battery charge of a mobile device is getting low while running some data analysis algorithm, the user would prefer to get results with lower precision rather than having no results because the device has run out of its battery charge. The same analogy could be applied to memory and processing power. Furthermore, if resource levels are not considered, there is always the possibility that the application may cause an extremely undesirable event such as a device shutdown. Thus, there is a need for an approach that can adapt to available resources and data stream input rate for a wide range of data stream applications. This need is based on the following arguments:

- it has been proven experimentally [2] that running mining techniques onboard resource-constrained devices consumes less energy than transmitting data streams to a high performance central processing power,
- extremely large amounts of flowing data streams from 1 Mb/Second for oil drills [12] to 1.5 TB/day for astronomical applications [4], and
- the resource constraints of data stream sources and processing units in a wide range of stream applications including wireless sensor networks sensor networks and mobile devices,

We have proposed to adapt the mining algorithm output according to resource availability and data stream rate. We have termed this approach as *Algorithm Output Granularity* (AOG). This chapter discusses the theoretical framework of AOG in terms of mathematical formalization, and procedural steps. The generalization of the approach using the *Probably Approximately Correct* (PAC) machine learning model [20] is also discussed as a basis for theoretically applying AOG to other stream mining techniques. The choice of PAC learning is based on its acceptance as the fundamental machine learning model [13].

AOG [7, 9] can be used along with other input and processing adaptation approaches in order to increase its adaptation performance. Techniques that can use such an integrated holistic approach can benefit from its ability to adapt from different end-points (input, processing, and output) as required by the current status of resource availability (memory, processing power, and battery charge). We also propose a framework for the integration of the three granularitybased approaches as adaptation techniques that can cope with the high speed data streams.

AOG forms the basis for a set of data stream mining algorithms that we have proposed and developed [7]. This class/suite of lightweight algorithms underpins the fundamental principles of adaptation to data rate and available computational resources.

2   AOG: An Overview

AOG operates using three factors to enable the adaptation of the mining algorithm to the available memory. The rate of the incoming data stream is the first factor. The rate of the algorithm output represents the second one. From these two, an estimated time to fill the available memory according to the logged history of data rate and algorithm output rate is calculated. This represents the last factor. These three factors are used to adjust what we call the algorithm threshold. This threshold can control the output rate of the algorithm according to the mining technique. This point further explained in this chapter. Figure 1 shows how the algorithm threshold can control the output rate of a mining algorithm according to the three factors that AOG operates on. The data arrives sequentially and its rate is calculated. The algorithm runs with an initial threshold value, and the rate of the output is calculated. The threshold is adjusted periodically to conserve the available memory according to the relationship among the three factors. AOG is a three-stage, resource-aware threshold-based data stream mining approach. The process of mining data streams using AOG starts with a mining phase.

In this step, the algorithm threshold that can control the algorithm output rate is determined as an initial value set by the user of the system (or preset to a default initial value). In this stage, there are three variations in using this threshold according to the mining technique:

- Clustering: the threshold is used to specify the minimum distance between the cluster center and the data stream record.
- Classification: in addition to using the threshold in specifying the distance, the class label is checked. If the class label of the stored records and the new item/record that are close (within the accepted distance) is the same, the weight of the stored item is increased and stored along with the weighted average of the other attributes, otherwise the weight is decreased and the new record is ignored.

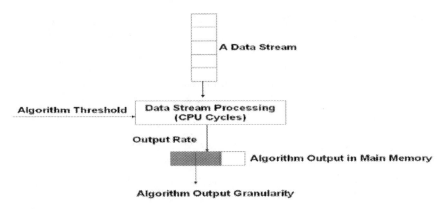

Fig. 1 AOG Approach

- Frequent patterns: the threshold is used to determine the number of counters for the frequent items.

The second stage in the AOG approach is the adaptation phase. In this phase, the threshold value is adjusted to cope with the data rate of the incoming stream, the available memory, and time constraints to fill the available memory with resultant knowledge structures.

The third and final stage in AOG approach is the knowledge integration phase. This stage represents the merging of produced results when the computational device is running out of memory. In clustering, we use the merging of clusters that are within short proximity. The merging of class representatives is used in classification. Releasing the least frequent items from memory is our strategy for frequent pattern mining. This integration allows the continuity of the data mining process. Otherwise the computational device would run out of memory even with adapting the algorithm threshold to its highest possible value that results in the lowest possible generation of knowledge structures. Figure 2 shows a flowchart of AOG-mining process. It shows the sequence of the three stages of AOG.

The algorithm output granularity approach is based on the following axioms:

- The algorithm output rate (AR) is a function of the data rate (DR), i.e., $AR = f(DR)$. Thus, the higher the input rate, the higher the output rate. For example the number of clusters created over a period of time should depend on the number of data records received over this period of time.
- The time needed to fill the available memory by the algorithm results (knowledge structures namely: clusters, classification models and frequent items) (TM) is a function of (AR), i.e., $TM = f(AR)$. Thus, a higher output rate would result in shorter time to fill the available memory assigned for the application.
- The algorithm accuracy (AC) is a function of (TM), i.e., $AC = f(TM)$. If the time of filling the available memory is considerably short, that would

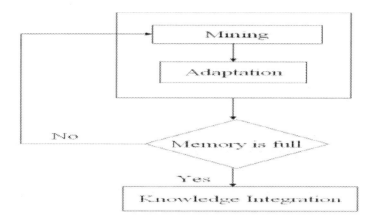

Fig. 2 AOG Approach

result in a higher frequency of knowledge integration such as cluster mergings. The higher the frequency of knowledge integration, the less accurate the results are.

Having discussed a general overview of AOG and the main motivation behind the idea, the following section provides details about the main concepts and mathematical formalization of our proposed approach.

3 Formalization of AOG

This section provides the concepts and definitions of our AOG approach. This is followed by the mathematical formalization of AOG. AOG primitives are given as a guide for the use of AOG in developing adaptive data stream mining techniques.

3.1 Definitions and Procedure

AOG is based on a number of new concepts which are introduced and discussed below:

Definition 1
Algorithm threshold is a major parameter incorporated in the algorithm logic that controls the process of producing resultant knowledge structures, namely clusters, classification models, counters of frequent items, according to three factors that vary over time:
a) Available memory.
b) Remaining time to fill the available memory.
c) Data stream rate.

The algorithm threshold is the maximum acceptable distance between the group mean value (e.g., center of a cluster) and a new incoming data record of a data stream. The higher the threshold, the lower is the size of the produced output. The algorithm threshold can use Euclidean or Manhattan distance functions and a normalization process could be done online in the case of a multidimensional data stream. Also higher norms can be used according to the application requirements. Applying the algorithm threshold may vary according to the data mining technique. Definition 1 determines the algorithm threshold that could be adopted in any data analysis technique. However we use the similarity measures in developing our lightweight mining techniques based on the AOG concept.

Definition 2
Threshold Lower Bound is the minimum acceptable distance (similarity measure) that could be used according to the application requirements.

The less the threshold is the higher the algorithm accuracy will be. If the distance measure is very small, it has two major drawbacks. It is meaningless in some applications to set the distance measure to a very small value. For example, in astronomical applications distance measure should make sense relative to physical distances between astronomical objects and could be measured in millions of kilometers. Secondly, the smaller the threshold the higher the running time for the model use. That is because with a smaller threshold, the number of knowledge structures produce would increase. This in turn would result in a longer running time for a data record to be assigned to an existing cluster or inducing the class label of a created classification model that would have larger number of representatives of each class. The class representative is the mean value of a group that falls within the same class label.

Definition 3
Threshold Upper Bound: is the maximum acceptable similarity measure that could produce meaningful results.

If the distance measure is high, the model building would be faster. However, a very high distance measure can also produce meaningless results in some applications. In other words, this may result in the grouping of records into the same cluster or class even though they are far apart in similarity. It is worth mentioning that setting the upper and lower bounds of the algorithm is application dependent. In each application, there are bounds of the distance measure that are pertinent to that application.

Definition 4
Output Granularity is the size of produced results that are acceptable according to a pre-specified accuracy measure. Memory of that size should be available before doing any incremental integration.

One of the main concepts used in AOG is the output granularity, i.e., the memory of the specified size should be reserved before merging of the results. The concept of output granularity works in conjunction with the concept of time threshold that is given in Definition 5 as follows.

Definition 5
Time Threshold is the required time period to produce the results before any incremental integration.

This time threshold could be specified by the user or calculated adaptively based on the history of running the algorithm. It is an application-dependent parameter. For some applications like business ones, this time period should be long enough to capture the patterns and extracted knowledge from the data. In applications such as abnormality detection in security applications, it should be short. It also varies from one data stream mining technique to the other. For example, in clustering, it is easier to merge close clusters. However, in classification; the results would be less accurate when merging class representatives.

Definition 6
Time Frame: is the time period between two consecutive data rate measurements. This period varies from one application to another and from one mining technique to another.

AOG adapts to available resources and the adaptation period remains unchanged within the same application. However this time period changes across different applications. The greater the size of the time frame, the smaller the AOG overhead would be. However this tradeoff may result in an application failure because the computational device would run out of memory during a longer time frame. This failure is the result of not running the adaptation at appropriate intervals. The smaller the size of the time frame, the lower the likelihood of the failure though this come at the cost of a higher overhead. This time frame should be chosen carefully to achieve a balance between the risk of algorithm failure and the overheads of the adaptation process.

Having introduced the main definitions used in AOG, we now describe a procedure of how to use AOG to enable adaptability in a mining algorithm. The main procedure for mining data streams using our proposed approach could be summarized in the following steps:

1. Determine the frequency of adaptation and mining.
2. According to the data rate, calculate the algorithm output rate and the algorithm threshold.
3. Mine the incoming stream using the calculated algorithm threshold.
4. Adjust the threshold after a time frame to adapt with the change in the data rate using linear regression or direct interaction between input and output rates over the most recent time frame.
5. Repeat the last three steps till the algorithm reaches time threshold.
6. Perform knowledge integration of the results.
7. Repeat the last five steps according to application requirements.

The above procedure could be adjusted according to the algorithm. The main objective of AOG is to use resource-awareness to guarantee the continuity of the mining process using adaptation. We have used the above procedure in our lightweight mining techniques. However, the notion of resource-awareness and adaptability could be applied using other techniques. Having discussed the main concepts and procedural steps in AOG-based data stream mining approach, the mathematical formalization of the above concepts is explained in depth in the following subsection.

3.2 Formalization

The concepts discussed in the previous section form the basis of the AOG approach for mining data streams. The development of data stream mining algorithms using AOG requires a rigorous mathematical formalization. This formalization of AOG is discussed in this section using the notation in Table 1.

The main idea behind our approach is to continuously change/adapt the algorithm threshold value introduced in Definition 1. This process of changing the algorithm threshold in turn changes the algorithm rate according to the following three factors that vary over time:

1. History of data rate/algorithm rate ratio.
2. Remaining time to fill the available memory.
3. Remaining size of available memory.

The objective is to keep the balance between the algorithm output rate/ data rate ratio from one end, and the remaining time for filling the memory up to the size of remaining memory from the other end as shown in equation (1).

The adaptation process of AOG calculates the algorithm output rate based on the current observed data stream rate, size of the remaining memory,

Table 1 AOG Symbols

Symbol	Meaning
AAO	Atomic Algorithm Output size. The memory size for storing the smallest element (knowledge structure) produced from the data mining algorithm. In clustering the AAO represents the memory size of storing the cluster center and the weight of the cluster. In classification, it represents the group mean values and class label of this group. In frequency counting, it represents the value of the frequent item along with the current frequency counter value.
D	Duration of the time frame.
M_i	The size of remaining memory by the end of time frame i. ($M_i = M_{i-1}(AAO \times O(TF_i))$)
TF_i	Time frame i by which the threshold is adjusted to cope with the data rate.
$N(TF_i)$	Number of data elements/records arrived during the time frame i.
$O(TF_i)$	Number of knowledge structures produced during the time frame i.
AR_i	The average algorithm rate during TFi. ($O(TF_i)/D$)
DR_i	The average data rate during TFi. ($N(TF_i)/D$)
T_i	Remaining time from the time threshold needed by the algorithm to fill the memory allocated for the task. ($T_i = T_{i-1}D$)
Th_i	Threshold value during the time frame i.
N	Number of time frames lasted

remaining time for filling the available memory. Equation (2) shows the calculation of the algorithm output rate.

$$\frac{\frac{AR_{i+1}}{DR_{i+1}}}{\frac{AR_i}{DR_i}} = \frac{\frac{M_i}{AR_i}}{T_i} \tag{1}$$

$$AR_{i+1} = \frac{M_i}{T_i} \times \frac{DRi+1}{DR_i} \tag{2}$$

The threshold adjustment for the new time frame is done according to the current threshold value, and the ratios of algorithm output rates to data stream rates. We use the AR_{i+1} calculated in equation (2) in equation (3) to determine the new threshold value:

$$Th_{i+1} = \frac{\frac{AR_{i+1}}{DR_{i+1}} \times Th_i}{\frac{AR_i}{DR_i}} \tag{3}$$

After the algorithm runs for a sufficient number of time frames enough to stabilize the relationship between the algorithm output rate and the algorithm threshold, we use linear regression as simplified in [15] to estimate the algorithm threshold value using the logged history of values obtained from AR_i and Th_i. Equations (4) through to (6) show the calculation of the simplified linear regression coefficients (a and b).

$$Th_i = a \times AR_i + b \qquad (4)$$

$$b = \frac{\sum_{i=1}^{n} Th_i \times AR_i}{\sum_{i=1}^{n} AR_i^2} \qquad (5)$$

$$a = \frac{\sum_{i=1}^{n} Th_i}{\sum_{i=1}^{n} n} - \frac{b \sum_{i=1}^{n} Th_i}{n} \qquad (6)$$

Linear regression is used because of the possible fluctuating distribution of the incoming evolving data stream records. Data stream distribution is an effective factor in determining the algorithm output rate. For example, uniform distribution is likely to result in a higher number of clusters than normal distribution. Linear regression is used to capture the different factors that affect the algorithm output rate in the linear regression coefficients (a and b) as approximation parameters of the behavior of the effect of the threshold value on the output rate. It is worth mentioning that the use of linear regression should be in its simplest form in order to decrease the overhead of the adaptation process. The other alternative is the use of direct interaction between the threshold value and the algorithm output rate as shown in Equations (1) through to (3).

The use of AOG is guided by a number of parameters, settings and operations that affect the whole process. The user settings of an AOG-based data mining algorithm are essential to the success of the data mining session according to the application requirements. The following section discusses AOG parameters, settings and operations. AOG parameters are the variables computed by the system which the user can not control. The settings are the user specified variables. The AOG operations are the three stages of AOG: mining, adapt, and integrate. We term all these as AOG primitives.

4   AOG Primitives

The algorithm output granularity in mining data streams involves parameters and operations on these parameters. AOG primitives are concerned with defining these parameters and operations. The development of AOG-based mining techniques should be guided by these primitives depending on application-based empirical studies. This implies that determining the temporal settings of these parameters will result in getting the output with acceptable accuracy. These parameters depend on the application and the used data mining algorithm used. For example, we can set different settings for the same clustering technique when we use it in astronomical applications that require higher accuracy when compared with business applications that require less accuracy. Table 2 describes different AOG parameters, operations and settings. AOG parameters are variables computed by the system in the course of running the data mining algorithm. Operations refer to the three operational stages in AOG: mining, adaptation, and integration. Settings refer to the user specified variables to set the required accuracy of AOG-based mining results.

Table 2 AOG Primitives

Primitive	Meaning	Type
TF_i	the time frame i	Parameter
D_i	Input data stream during the time frame i	Parameter
$I(D_i)$	Average data rate of the input stream D_i	Parameter
$O(D_i)$	Average output rate resulting from mining the stream D_i	Parameter
$\alpha(D_i)$	Mining process of the D_i stream	Operation
$\beta([I(D_1), O(D_1)], ..., [I(D_i), O(D_i)])$	Adaptation process of the algorithm threshold at the end of the time frame i	Operation
$\Omega(O_i, ..., O_x)$	Knowledge integration process done on the output i to the output x	Operation
$D(TF)$	Time duration of each time frame	Setting
$D(\Omega)$	Time duration between two consecutive knowledge integration processes	Setting

Figure 3 and Figure 4 show the conceptual framework of AOG in summarized and detailed views respectively. Figure 3 shows the sequence of mining and adaptation (α and β operations) followed by the knowledge integration (Ω operation). This shows that the adaptation process is more frequent than the knowledge integration process. This frequency preserves the accuracy of the running data stream mining algorithm.

Figure 4 shows a detailed sequence of AOG operations in each time frame. The input data stream D_1 has passed through the data mining operation $\alpha(D_1)$ and then the input and output rates ($I(D_2) and O(D_1)$) are used in the adaptation process $\beta([I(D_1), O(D_1)])$ to change the algorithm threshold value. The same process is repeated with D_2. The process continues till the knowledge integration process takes place using $\Omega(O_1, ..., O_x)$ operation. This sequence is controlled using the AOG settings $D(TF)$ and $D(\Omega)$ shown in

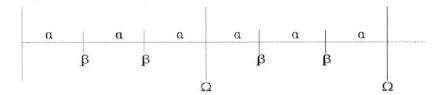

Fig. 3 AOG-based Mining

Fig. 4 AOG-based Min-
ing - Detailed

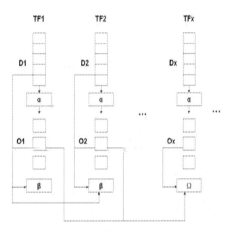

Table 2. In other words, these settings control the frequency of adaptation and knowledge integration of the AOG-based data stream mining process.

Having discussed the theoretical framework of our AOG adaptation approach, the generality of AOG to data stream mining techniques is discussed in the following section. The section demonstrates this generality using the PAC learning model [20, 13] supported by examples from existing mining data stream techniques. This section provides theoretical evidence of the generality feature of AOG. Practical evidence using one of the examples listed in the following section is used to establish the applicability of the adoption of AOG to other existing data stream mining algorithms.

5   AOG Generalization

Enabling resource-awareness and adaptation to data mining algorithms using a generic approach has advantages due to a number of factors that we have presented and discussed earlier. These factors are summarized as follows:

- The increasing number of applications use data streams as the source of incoming data and as result of data processing;

- The need for data stream mining is demonstrated in more and more applications;
- The resource constraints of data stream sources and processing units can not be totally eliminated;
- A large number of data mining techniques that are not adaptive and resource-aware.

In the following subsections, we are going to discuss the formalization of AOG using the PAC learning model. The choice of the PAC learning model is based on the following two arguments:

- The goal of both the PAC learning model and AOG is to reduce the error rate using algorithm parameters.
- The PAC learning model is considered a fundamental theoretical model in machine learning [20, 13].

We now briefly discuss the PAC learning model followed by a formalization of using the PAC model within our AOG approach. Two examples that use the PAC learning model follow the discussion to demonstrate the applicability of our AOG model in these techniques.

5.1 PAC Learnable

The idea of the PAC learning model is to measure the learnability of the algorithm. It is analogous to determining whether an algorithm satisfies Turing machine computability [16] in theoretical computer science. The same idea is applied to the machine learning using the PAC learning model.

Given the learning problem, it is required to find a solution that gives an exact answer. Since most learning problems are computationally hard, approximation techniques are used to find the solution. The PAC model of Valiant [20, 13] defines a learning algorithm as follows:

Definition 7

The learning algorithm L outputs a Model M on any input distribution D such that the probability of error is less than or equal a pre-specified setting (ϵ) with probability $(1 - \delta)$, where δ is a pre-specified probability.

This definition provides the basis for any learning problem. It is a combination of the complexity concept in theoretical computer science and the learning problem. The theory is used to formally find a solution to a learning problem that guarantees a minimum error rate with some pre-specified probability. If such a formalization is found for a learning algorithm, we call this problem learnable and the algorithm is called an approximate learning technique [13]. The use of the model within AOG as a generic resource-aware adaptive approach is discussed in the following subsection.

5.2 PAC-Based AOG

AOG has two main categories of parameters that are used in different ways according to the data mining technique. The first one is the resource monitoring and the other is the accuracy. It tries to tradeoff between the accuracy of the output and the availability of resources. For the sake of generalization, we can denote the available resources by R and the accuracy by the error rate symbol introduced in the PAC model ϵ. The following definition will help specifying an approximate learning algorithm:

Definition 8

The learning algorithm L is an adaptive and resource-aware technique iff the output model M at any time t, is within an error upper bound ϵ_u and lower bound ϵ_l given the available resources R at time t with a probability of $(1 - \delta)$, where δ is a pre-specified parameter, that also could be bounded. The resources R are freed using β (i.e., adaptation) and/or Ω (i.e., integration) of AOG.

This definition generalizes the AOG idea for any resource-aware data mining algorithm. If the learning algorithm could be formalized within this theoretical approach, it is considered an adaptive and resource-aware algorithm. The idea is that the algorithm can work within error bounds given the available resources with some given probability of error that could be bounded or given as a single value. The procedure of using this generalization could be summarized in the following steps:

1. Monitor resources R.
2. Use user-specified ϵ_u and ϵ_l and δ.
3. Compute ϵ according to R (β process).
4. Mine data streams if $\epsilon_l < \epsilon < \epsilon_u$ else free R (Ω process).
5. Repeat the above steps according to the application needs.

This generalization of the AOG using the PAC learning model provides a theoretical reference model for formalizing the current streaming algorithms to enable resource-awareness using AOG. This process is essential given the resource constraints of a wide range of data stream processing environments.

VFML (Very Fast Machine Learning) represents the first generation of algorithms in the area of mining data streams. The adoption of the resource-awareness to such a general framework would strengthen the approach in terms of applicability to run in resource-constrained environments that serve essential applications.

The following subsections discuss two examples from VFML, namely VFKM and VFDT, and discuss how the above formalization in developing AOG-based techniques can be applied to these algorithms. VFKM and VFDT are two data stream mining techniques that use the concepts introduced in PAC learning model.

VFKM

VFKM [5] stands for Very Fast K-Means. It uses a statistical bounding called Hoeffding bound [11] to find out the number of data points/records to be examined in each step in K-means clustering algorithm in order to satisfy some pre-specified value of the above mentioned statistical measure. The notion of finding the number of data records that satisfy a statistical bound could be combined with AOG as follows. The process starts with monitoring the available resources and then estimating the Hoeffding bound that could be achieved according to the measured availability of resources. We run VFKM to find out the number of records that satisfy the Hoeffding bound. This process is repeated at each iteration of running VFKM (each run of K-means algorithm using the current calculated number of records). We summarize the AOG-based VFKM in the following steps:

1. Monitor the available resources.
2. Estimate the Hoeffding bound parameters (ϵ and δ) (β process).
3. Run VFKM for a time frame to satisfy the above bound.
4. Repeat the above steps according to the application needs.

We have proposed, implemented, and evaluated the combination of VFKM with our AOG approach. We have termed our AOG-based VFKM as RA-VFKM [19] to refer to the added resource-awareness feature of the new technique.

VFDT

VFDT [6] stands for Very Fast Decision Trees. It splits the decision tree according to the current best attribute (the attribute that has the highest information gain) with the number of records that satisfy the Hoeffding bound. Similar to VFKM, we summarize the steps of using AOG in developing resource-aware VFDT in the following:

1. Monitor the available resources.
2. Estimate the Hoeffding bound (ϵ and δ) (β process).
3. Calculate the information gain of each attribute.
4. Choose the root attribute according to the number of records that satisfy the Hoeffding measure.
5. Choose the next best attribute in the tree.
6. Repeat 1-5 (except 4) according to the application needs.

In the previous sections, AOG has been discussed in details. As pointed out in the introduction of this chapter, AOG can be integrated with other proposed adaptation approaches in data stream mining such as sampling and randomization to increase its efficiency of adaptation. A classification of these approaches can fall in the two broad categories: input and processing granularities. A short introduction to these two categories is given in the following sections.

6   Algorithm Input Granularity (AIG)

AIG represents the process of changing the data rates that feed the data stream mining algorithm according to resource availability. Sampling, load shedding and computing data synopsis represent the techniques under this category. The following are the definitions of these techniques under this class:

Sampling is the process of statistically choosing a number of data records to be processed. Sampling basically depends on the probability. Thus, every data record in the data stream has the chance to be processed depending on the probability distribution of the used sampling technique. Different applications may require different sampling distributions [14].

Load shedding is the process of dropping a chunk of data records from being processed. This could be an appropriate technique to stop the processing enabling some optimization process to be done during this time. It is considered to be a direct solution if there is a burst in data streams. We shed the load and continue the processing after the burst [1, 3, 17, 18].

Computing data synopsis is the process of summarizing and/or compressing the incoming data records on the fly before processing them. Wavelets and simple statistical summarization techniques represent the typical strategies in this category. It should be noted that the process of creating this synopsis of data should be a lightweight one compared to the data mining technique that will use them [1].

7   Algorithm Processing Granularity (APG)

APG is the process of changing the algorithm parameters in order to reduce the CPU load. Randomization and approximation techniques represent the solution strategies in this category.

The bulk of the data stream algorithms mainly depend on result approximation using different strategies. The main idea behind this approach is to randomize through the solution space in order to get the results in shorter running time. The accuracy of the results is measured with this randomization factor. The higher the randomization is, the lower the accuracy will be and vice versa. It should be noted that these three granularity based approaches complement each other. AIG mainly affects the data rate and it is associated with bandwidth consumption and battery. On the other hand, AOG is associated with memory and APG is associated with CPU load. The integration of these three approaches is proposed and discussed in the following section.

8 Integration of Granularity-Based Approaches

Integration among the three granularity approaches has its advantages for using one or more of them according to the current consumption pattern of the affected resources [8]. For example, if the memory consumption pattern is stable, there is no need for AOG-based adaptation. However, if the battery consumption pattern is not stable, there might be a need for applying AIG. Figure 5 shows the input, processing, and output end-points with their corresponding granularity settings. Examples of algorithm input granularity are sampling, load shedding and computing data synopsis. Algorithm processing granularity could be achieved through approximation and randomization techniques.

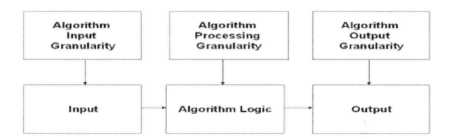

Fig. 5 AOG-based Mining - Detailed

The generalized system architecture of our framework for integration of AOG with the input and processing granularity approaches is shown in Figure 6. The system is composed from the following components: resource monitoring, algorithm granularity settings (input, processing, and output), and the data mining technique. The following is a brief description of each component.

Resource monitoring is the component responsible for computing all the required statistical measurements about resource consumption over the most recent time frame. This process is done periodically over fixed time frames as used in AOG discussed earlier in this chapter. The resources that are required to be monitored are: memory, CPU and battery. Algorithm granularity settings component is responsible for changing AIG, AOG, or APG parameters according to the output of the resource monitoring component. This change is done according to the pattern of resource consumption and the status of the

Fig. 6 AOG-based Mining - Detailed

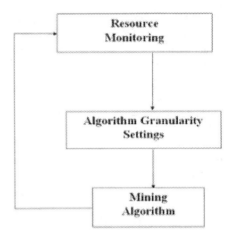

resource in terms of degree of availability. Thus, there are two main factors that affect the decision regarding the granularity settings:

- The pattern of consumption (whether increasing or decreasing).
- The measure of how critical the resource availability is.

The last component is the data mining algorithm. It is the technique that adapts to the algorithm granularity settings. We have used our novel resource-aware clustering RA-Cluster [8]. However, any data mining algorithm could be combined with resource-awareness using our framework. The general adaptation framework adopted in our RA-Cluster technique [8] has been presented in this section.

9 Summary

In this chapter, we have presented our novel approach to solving the problem of mining data streams in resource constrained environments that are the typical representatives for data stream sources and processing units in many applications. The proposed AOG approach adapts the output rate of a data mining algorithm according to available resources and data rate.

AOG has been formalized and the main concepts and definitions have been introduced followed by a rigorous discussion. AOG approach has its own operations and settings that could be chosen according to the algorithm that uses AOG or the application area. We call this set of operations and settings AOG primitives. Generalization of AOG using the PAC learning model has been discussed. The main idea is for a data mining algorithm to acquire its resource-awareness and adaptive capability once it follows the procedures proposed in this chapter. This generalization provides the basis for applying the idea to different data stream mining techniques. Augmenting VFKM and VFDT data stream mining algorithms with our PAC-based AOG has been proposed and discussed.

AOG can be integrated with other granularity-based approaches from the input and processing end-points to form a holistic approach that can adapt to different availability of resources. This integration has been also discussed in this chapter.

References

1. Babcock, B., Datar, M., Motwani, R.: Load Shedding Techniques for Data Stream Systems (short paper). In: Proc. of the 2003 Workshop on Management and Processing of Data Streams (MPDS 2003) (June 2003)
2. Bhargava, R., Kargupta, H., Powers, M.: Energy Consumption in Data Analysis for On-board and Distributed Applications. In: Proceedings of the ICML 2003 workshop on Machine Learning Technologies for Autonomous Space Applications (2003)
3. Chi, Y., Yu, P.S., Wang, H., Muntz, R.R.: Loadstar: A Load Shedding Scheme for Classifying Data Streams. In: The 2005 SIAM International Conference on Data Mining (SIAM SDM 2005) (2005)
4. Coughlan, J.: Accelerating Scientific Discovery at NASA. In: Jonker, W., Petković, M. (eds.) SDM 2004. LNCS, vol. 3178. Springer, Heidelberg (2004)
5. Domingos, P., Hulten, G.: A General Method for Scaling Up Machine Learning Algorithms and its Application to Clustering. In: Proceedings of the Eighteenth International Conference on Machine Learning, 2001, pp. 106–113. Morgan Kaufmann, Williamstown (2001)
6. Domingos, P., Hulten, G.: Mining High-Speed Data Streams. In: Proceedings of the Association for Computing Machinery Sixth International Conference on Knowledge Discovery and Data Mining, pp. 71–80 (2000)
7. Gaber, M.M., Krishnaswamy, S., Zaslavsky, A.: On-board Mining of Data Streams in Sensor Networks. In: Badhyopadhyay, S., Maulik, U., Holder, L., Cook, D. (eds.) Advanced Methods of Knowledge Discovery from Complex Data, pp. 307–336. Springer, Heidelberg (2005) (forthcoming)
8. Gaber, M.M., Yu, P.S.: A Holistic Approach for Resource-aware Adaptive Data Stream Mining. Journal of New Generation Computing, Special Issue on Knowledge Discovery from Data Streams (2006)
9. Gaber, M.M., Krishnaswamy, M., Zaslavsky, S.: Resource- Aware Mining of Data Streams. In: Aguilar-Ruiz, J.S., Gama, J. (eds.) Journal of Universal Computer Science, Special Issue on Knowledge Discovery in Data Streams, pp. 1440–1453 (August 2005)
10. Gaber, M.M., Zaslavsky, A., Krishnaswamy, S.: Mining Data Streams: A Review. ACM SIGMOD Record 34(1) (June 2005) ISSN: 0163-5808
11. Hoeffding, W.: Probability inequalities for sums of bounded random variables. Journal of the American Statistical Association (58), 13–30 (1963)
12. Muthukrishnan, S.: Data streams: algorithms and applications. In: Proceedings of the fourteenth annual ACM-SIAM symposium on discrete algorithms (2003)
13. Natarajan, B.K.: Machine learning: a theoretical approach. M. Kaufmann, San Mateo (1991)
14. Park, B.-H., Ostrouchov, G., Samatova, N.F., Geist, A.: Reservoir-Based Random Sampling with Replacement from Data Stream. In: Proceedings of SIAM International Conference on Data Mining 2004 (2004)

15. Roiger, R., Geatz, M.: Data mining: a tutorial-based primer. Addison Wesley, Boston (2003)
16. Sipser, M.: Introduction to the Theory of Computation. In: Part Two: Computability Theory, chs. 3-6, pp. 123–222. PWS Publishing (1997) ISBN 0-534-94728-X
17. Tatbul, N., Cetintemel, U., Zdonik, S., Cherniack, M., Stonebraker, M.: Load Shedding in a Data Stream Manager. In: Proceedings of the 29th International Conference on Very Large Data Bases, VLDB (September 2003)
18. Tatbul, N., Cetintemel, U., Zdonik, S., Cherniack, M., Stonebraker, M.: Load Shedding on Data Streams. In: Proceedings of the Workshop on Management and Processing of Data Streams (MPDS 2003), San Diego, CA, USA, June 8 (2003)
19. Shah, R., Krishnaswamy, S., Gaber, M.M.: Resource-Aware Very Fast K-Means for Ubiquitous Data Stream Mining. In: Proceedings of Second International Workshop on Knowledge Discovery in Data Streams, to be held in conjunction with the 16th European Conference on Machine Learning (ECML 2005) and the 9th European Conference on the Principals and Practice of Knowledge Discovery in Databases (PKDD 2005), Porto, Portugal, October 3-7 (2005)
20. Valiant, L.G.: A theory of the learnable. Communications of the ACM 27(11), 1134–1142 (1984)

Time Granularity in Temporal Data Mining

Paul Cotofrei and Kilian Stoffel

Summary. In this paper, a formalism for a specific temporal data mining task (the discovery of rules, inferred from databases of events having a temporal dimension), is defined. The proposed theoretical framework, based on first-order temporal logic, allows the definition of the main notions (*event, temporal rule, confidence*) in a formal way. This formalism is then extended to include the notion of temporal granularity and a detailed study is made to investigate the formal relationships between the support measures of the same event in linear time structures with different granularities. Finally, based on the concept of *consistency*, a strong result concerning the independence of the confidence measure for a temporal rule, over the worlds with different granularities, is proved.

1 Introduction

The domain of temporal data mining focuses on the discovery of causal relationships among events that are ordered in time and may be causally related. The contributions in this domain encompass the discovery of temporal rule, of sequences and of patterns. However, in many respects this is just a terminological heterogeneity among researchers that are, nevertheless, addressing the same problem, albeit from different starting points and domains.

Although there is a rich bibliography concerning formalism for temporal databases, there are very few articles on this topic for temporal data mining.

Paul Cotofrei

Information Management Institute, University of Neuchâtel, Pierre-à-Mazel, 7, 2000, Neuchâtel, Switzerland

e-mail: `paul.cotofrei@unine.ch`

Kilian Stoffel

Information Management Institute, University of Neuchâtel, Pierre-à-Mazel, 7, 2000, Neuchâtel, Switzerland

e-mail: `kilian.stoffel@unine.ch`

A. Abraham et al. (Eds.): Foundations of Comput. Intel. Vol. 6, SCI 206, pp. 67–96.
springerlink.com © Springer-Verlag Berlin Heidelberg 2009

In [1, 5, 21] general frameworks for temporal mining are proposed, but usually the researches on causal and temporal rules are more concentrated on the methodological/algorithmic aspect, and less on the theoretical aspect. Based on a methodology for temporal rule extraction, described in [9], we proposed in [10, 11] an innovative formalism based on first-order temporal logic, which permits an abstract view on temporal rules. An important concept defined in this formalism is the property of *consistency*, which guarantees the preserving over time of the confidence/support of a temporal rule. The formalism is developed around a time model for which the events are those that describe system evolution (event-based temporal logic). Each formula expresses what the system does at each event, events are referring to other events, and so on: this results in specifying relationships of precedence and cause-effect among events. But the real systems are systems whose components (events) have dynamic behavior regulated by very different – even by orders of magnitude – time granularities. Analyzing such systems (hereinafter granular systems) means to approach theories, methodologies, techniques and tools that make use of granules (or groups, classes, clusters of a universe) in the process of problem solving. Granular computing (the label which covers this approach) is a way of thinking that relies on our ability to perceive the real world under various grain sizes, to abstract and to consider only those things that serve our present interest, and to switch among different granularities. By focusing on different levels of granularities, one can obtain various levels of knowledge, as well as inherent knowledge structure. Granular computing is essential to human problem solving, and hence has a very significant impact on the design and implementation of intelligent systems [28, 27, 29, 20].

2 State of Art

The notions of granularity and abstraction are used in many subfields of artificial intelligence. The granulation of time and space leads naturally to temporal and spatial granularities. They play an important role in temporal and spatial reasoning [13, 18, 26]. Based on granularity and abstraction, many authors studied some fundamental topics of artificial intelligence, such as knowledge representation [30], theorem proving [15], search [31], planning [19], natural language understanding [22], intelligent tutoring systems [23], machine learning [25], and data mining [16].

Despite the widespread recognition of its relevance in the fields of formal specifications, knowledge representation and temporal databases, there is a lack of a systematic framework for time granularity. Hobbs [17] proposed a formal characterization of the general notion of granularity, but gives no special attention to time granularity. Clifford et al. [8] provide a set-theoretic formalization of time granularity, but they do not attempt to relate the truth value of assertions to time granularity. Extensions to existing languages for formal specifications, knowledge representation and temporal databases that do support a limited concept of time granularity are proposed in [24, 14, 7]. Finally, Bettini et al. [2, 4] provide a formal framework for expressing data mining tasks involving time granularities, investigate the formal relationships among event structures that have temporal constraints, define the pattern-discovery problem with these structures and study effective algorithms to solve it.

The purpose of this paper is to extend our formalism to include the concept of time granularity. We define the process by which a given structure of time granules μ (called temporal type) induces a first-order linear time structure M_μ (called granular world) on the basic (or absolute) linear time structure M. The major change for the temporal logic based on M_μ is at the semantic level: for a formula p, the interpretation does not assign a meaning of truth (one of the values $\{true, false\}$), but a degree of truth (a real value from $[0, 1]$). Consequently, we can give an answer to the following question: if the temporal type μ is *finer than* temporal type ν, what is the relationship between the support of the same temporal rule T_p in the linear time structures M_μ and M_ν. We also study the variation process for the set of satisfiable events (degree of truth equal one) during the transition between two time structures with different granularity. By an extension at the syntactic and semantic level we are able to define an aggregation mechanism for events, reflecting the following intuitive phenomenon: in a coarser world, not all events inherited from a finer world are satisfied, but in exchange there are new events which become satisfiable. Finally, using an extension of the concept of consistency for a granular time structure M_μ, we prove a strong result concerning the invariance of the confidence measure for a temporal rule during the process of information transfer between worlds with different granularities.

The rest of the paper is structured as follows. In the next section, the first-order temporal logic formalism is extensively described (the main terms and concepts). The definitions and theorems concerning the extension of the formalism towards a temporal granular logic are presented in Sect. 4. Finally, the last section summarizes our work, followed by an appendix containing the proofs of the theorems in the paper.

3 Formalism of Temporal Rules

Time is ubiquitous in information systems, but the mode of representation/perception varies in function of the purpose of the analysis [6, 12]. Firstly, there is a choice of a *temporal ontology*, which can be based either on *time points* (instants) or on *intervals* (periods). Secondly, time may have a *discrete* or a *continuous* structure. Finally, there is a choice of *linear* vs. *nonlinear* time (e.g. acyclic graph). Our selection, imposed by the discrete representation of all databases, is a temporal domain represented by linearly ordered discrete instants.

Databases being first-order structures, the first-order logic represents a natural formalism for their description. Consequently, the first-order temporal logic expresses the formalism of temporal databases. For the purpose of our approach we consider a restricted first-order temporal language L which contains only constant symbols $\{c, d, ..\}$, n-ary $(n \geq 1)$ function symbols $\{f, g, ..\}$, variable symbols $\{y_1, y_2, ...\}$, n-ary predicate symbols $(n \geq 1$, so no proposition symbols), the set of relational symbols $\{=, <, \leq, >, \geq\}$, the logical connective $\wedge$ and a temporal connective of the form ∇_k, $k \in \mathbb{Z}$, where k strictly positive means *after k time instants*, k strictly negative means *before k time instant* and $k = 0$ means *now*.

The syntax of L defines terms, atomic formulae and compound formulae. The *terms* of L are defined inductively by the following rules:

T1. Each constant is a term.
T2. Each variable is a term.
T3. If $t_1, t_2, \ldots, t_n$ are terms and f is an n-ary function symbol then $f(t_1, \ldots, t_n)$ is a term.

The atomic formulae (or atoms) of L are defined by the following rules:

A1. If $t_1, \ldots, t_n$ are terms and P is an n-ary predicate symbol then $P(t_1, \ldots, t_n)$ is an atom.
A2. If t_1, t_2 are terms and ρ is a relational symbol then $t_1 \rho t_2$ is an atom (also called relational atom).

Finally, the (compound) formulae of L are defined inductively as follow:

F1. Each atomic formula is a formula.
F2. If p, q are formulae then $(p \wedge q)$, $\nabla_k p$ are formulae.

A Horn clause is a formula of the form $B_1 \wedge \cdots \wedge B_m \rightarrow B_{m+1}$ where each B_i is a positive (non-negated) atom. The atoms B_i, $i = 1, \ldots, m$ are called implication clauses, whereas B_{m+1} is known as the implicated clause. Syntactically, we cannot express Horn clauses in our language L because the logical connective $\rightarrow$ is not included. However, to allow the description of rules, which formally look like a Horn clause, we introduce a new logical connective, $\mapsto$, representing practically a rewrite of the connective $\wedge$. Therefore, a formula in L of the form $p \mapsto q$ is syntactically equivalent to the formula $p \wedge q$. When and under what conditions we may use the new connective, is explained in the next definitions.

Definition 1. *An event (or temporal atom) is an atom formed by the predicate symbol E followed by a bracketed n-tuple of terms $(n \geq 1)$ $E(t_1, t_2, \ldots, t_n)$. The first term of the tuple, t_1, is a constant symbol representing the name of the event and all others terms are expressed according to the rule $t_i = f(t_{i1}, \ldots, t_{ik_i})$. A short temporal atom (or the event's head) is the atom $E(t_1)$.*

Definition 2. *A constraint formula for the event $E(t_1, t_2, \ldots t_n)$ is a conjunctive compound formula, $E(t_1, t_2, \ldots t_n) \wedge C_1 \wedge C_2 \wedge \cdots \wedge C_k$. Each C_j is a* *relational atom* $t\rho c$, *where the first term* t *is one of the terms* t_i, $i = 1 \ldots n$ *and the second term is a constant symbol.*

For a short temporal atom $E(t_1)$, the only constraint formula that is permitted is $E(t_1) \wedge (t_1 = c)$. We denote such constraint formula as *short constraint formula.*

Definition 3. *A temporal rule is a formula of the form* $H_1 \wedge \cdots \wedge H_m \mapsto H_{m+1}$, *where* H_{m+1} *is a short constraint formula and* $H_i, i = 1..m$ *are constraint formulae, prefixed by the temporal connectives* ∇_{-k}, $k \geq 0$. *The maximum value of the index* k *is called the time window of the temporal rule.*

Remark. The reason for which we did not permit the expression of the implication connective in our language is related to the truth table for a formula $p \rightarrow q$: even if p is false, the formula is still true, which is unacceptable for a temporal rationing of the form *cause→ effect*.

If we change in Definition 1 the conditions imposed on the terms $t_i, i = 1...n$, into "each term t_i is a variable symbol", we obtain the definition of a temporal atom template. We denote such a template as $E(y_1, \ldots, y_n)$. Following the same rationing, a constraint formula template for $E(y_1, \ldots, y_n)$ is defined as a conjunctive compound formula, $C_1 \wedge C_2 \wedge \cdots \wedge C_k$, where the first term of each relational atom C_j is one of the variables y_i, $i = 1 \ldots n$. Consequently, a short constraint formula template is the relational atom $y_1 = c$. Finally, by replacing in Definition 3 the notion "constraint formula" with "constraint formula template" we obtain the definition of a temporal rule template. Practically, the only formulae constructed in L are temporal atoms, constraint formulae, temporal rules and the corresponding templates.

The semantics of L is provided by an interpretation I over a domain D (in our formalism, D is always a linearly ordered domain). The interpretation assigns an appropriate meaning over D to the (non-logical) symbols of L. Usually, the domain D is imposed during the discretisation phase, which is a pre-processing phase used in almost all knowledge extraction methodologies. Based on Definition 1, an event can be seen as a labelled (constant symbol t_1) sequence of points extracted from raw data and characterized by a finite set of features (terms $t_2, \cdots, t_n$). Consequently, the domain D is the union $D_e \cup D_f$, where the set D_e contains all the strings used as event names and the set D_f represents the union of all domains corresponding to chosen features.

Example 1. Consider a database containing daily price variations of a given stock. Suppose that a particular methodology for event detection was applied, which revealed three types of events (shape patterns in this case), potentially useful for a final user. Each event is labelled with one of the strings from the set $\{peak, flat, valley\}$ and is characterized by two features, f_1 and f_2, representing the output of the statistical functions mean and standard error. These statistics are calculated using daily prices, supposed to be subsequences of length $w = 12$. In the frame of our formalism the language L will include a 3-ary predicate symbol E, three variable symbols $y_i, i = 1..3$, two 12-ary function symbols f and g, two sets of constant symbols – $\{d_1, \ldots, d_3\}$ and $\{c_1, \ldots, c_n\}$ – and the usual set of relational symbols and logical(temporal) connectives. Consequently, a temporal atom in L is defined as $E(d_i, f(c_{j_1}, \ldots, c_{j_{12}}), g(c_{k_1}, \ldots, c_{k_{12}}))$, whereas an event template is defined as $E(y_1, y_2, y_3)$. Finally, the domain D is the union of the set $D_e = \{peak, flat, valley\}$ and of the set $D_f = \Re^+$ (as the stock prices are positives real numbers and the features are statistical functions).

To define a first-order linear temporal logic based on L, we need a structure having a temporal dimension and capable to capture the relationship between a time moment and the interpretation I at this moment.

Definition 4. *Given L and a domain D, a (first order) linear time structure is a triple $M = (S, x, \boldsymbol{I})$, where S is a set of states, $x : \mathbb{N} \to S$ is an infinite sequence of states $(s_1, s_2, \ldots, s_n, \ldots)$ and $\boldsymbol{I}$ is a function that associates with each state s an interpretation $\boldsymbol{I}_s$ of all symbols from L.*

In the framework of linear temporal logic, the set of symbols is divided into two classes, the class of global symbols and the class of local symbols. Intuitively, a global symbol w has the same interpretation in each state, i.e. $\boldsymbol{I}_s(w) = \boldsymbol{I}_{s'}(w) = \boldsymbol{I}(w)$, for all $s, s' \in S$; the interpretation of a local symbol may vary, depending on the state at which is evaluated. The formalism of temporal rules assumes that all function symbols (including constants) and all relational symbols are global, whereas the predicate symbols and variable symbols are local. Consequently, as the temporal atoms, constraint formulae, temporal rules and the corresponding templates are expressed using the predicate symbol E or the variable symbols y_i, the meaning of truth for these formulae depend on the state at which they are evaluated. Given a first order time structure M and a formula p, we denote the instant i (or equivalently, the state s_i) for which $\boldsymbol{I}_{s_i}(p) = true$ by $i \models p$, i.e. at time instant i the formula p is true. Therefore, $i \models E(t_1, \ldots, t_n)$ means that at time i an event with the name $\boldsymbol{I}(t_1)$ and characterized by the global features $\boldsymbol{I}(t_2), \ldots, \boldsymbol{I}(t_n)$ occurs. Concerning the event template $E(y_1, \ldots, y_n)$, the interpretation of the variable symbols y_j at the state s_i, $\boldsymbol{I}_{s_i}(y_j)$, is chosen such that $i \models E(y_1, \ldots, y_n)$ for each time moment i. Because

- $i \models p \wedge q$ if and only if $i \models q$ and $i \models q$, and
- $i \models \nabla_k p$ if and only if $i + k \models p$,

a constraint formula (template) is true at time i if and only if all relational atoms are true at time i and $i \models E(t_1, \ldots, t_n)$, whereas a temporal rule (template) is true at time i if and only if $i \models H_{m+1}$ and $i \models (H_1 \wedge \cdots \wedge H_m)$.

Remark. The fact that the symbols of language L are divided into two sets (local and global), according to the persistence of their interpretation along the infinite sequence of states $s_1, s_2, \ldots$, is the main reason for the introduction of the notion of *template*. Consider, as example, the temporal atom

$E(t_1, t_2, t_3)$ and its corresponding template $E(y_1, y_2, y_3)$. In our vision, the event template is a kind of event pattern and because there is a real event which matches the pattern (an event with name $I(t_1)$ and features $I(t_2)$ and $I(t_3)$), the interpretation of the template must be true at each moment. For this reason we imposed the condition that the interpretation of variable symbols must be chosen such that $i \models E(y_1, y_2, y_3)$ for each time moment i. On the other hand, we expect that in the real word, an event occurs only at certain moments, i.e. the interpretation of the event is evaluated *true* only at these moments. Because the terms t_i, $i = 1..3$ are global symbols (as constant symbols and function symbols) and $I(E(t_1, t_2, t_3)) = I(E)(I(t_1), I(t_2), I(t_3))$, the only way to achieve the variability in time for the event interpretation is to include the predicate symbol E in the set of local symbols.

Example 2. The database of events, obtained after applying the specific methodology for event detection (see Example 1), contains tuples with three values, (v_1, v_2, v_3). For a tuple with a recording index i, the first value expresses the name of the event – *peak, flat, valley* – which occurs at time moment i and the two other values express the values of the two characterizing features. Therefore, to define a linear time structure $M = (S, x, \mathbf{I})$, we may consider a state s as a triple (v_1, v_2, v_3), the set S as the set of all tuples from the database and the sequence x as the ordered sequence of tuples in database (see Table 1).

Table 1 The first eighteen states of the linear time structure M (example)

Index	State	Index	State	Index	State
1	$(peak, 11.2, 3.91)$	7	$(peak, 9.15, 4.03)$	13	$(flat, 7.14, 0.89)$
2	$(peak, 10.5, 4.87)$	8	$(peak, 11.52, 3.91)$	14	$(peak, 10.31, 4.42)$
3	$(peak, 14.03, 4.23)$	9	$(flat, 1.5, 1.81)$	15	$(peak, 12.8, 5.26)$
4	$(flat, 4.75, 1.42)$	10	$(valley, 3.08, 1.84)$	16	$(flat, 3.13, 1.44)$
5	$(peak, 9.49, 3.18)$	11	$(valley, 2.72, 1.58)$	17	$(flat, 5.08, 1.12)$
6	$(valley, 2.21, 1.12)$	12	$(valley, 4.42, 2.91)$	18	$(valley, 3.31, 3.20)$

At this stage the interpretation of all symbols (global and local symbols) can be defined. For the global symbols (function symbols and relational symbols), the interpretation is quite intuitive. Therefore, the meaning $\mathbf{I}(d_j)$ is an element of D_e, the meaning $\mathbf{I}(c_j)$, $j = 1..n$, is a positive real number, whereas the meaning $\mathbf{I}(f)$, respectively $\mathbf{I}(g)$, is the function $f : D_f^{12} \to D_f$, $f(\mathbf{x}) = \overline{\mathbf{x}}$, respectively the function $g : D_f^{12} \to D_f$, $g(\mathbf{x}) = se(\mathbf{x})$ – we used the standard notations in statistics for the mean and standard error estimators.

The interpretation of a local symbol (the variable symbols y_i and the predicate symbol E) depends on the state at which is evaluated. For this example suppose that the function $\mathbf{I}_{s_i}(E)$ defined on D^3 with values in $B = \{true, false\}$ is provided by a finite algorithm. This algorithm will receive at

Table 2 The temporal atoms evaluated *true* at the states s_2, s_4 and s_{10} of M (example)

State	Temporal atom
2	$E(peak, f(3, 5, 7, 9, 15, 18, 19, 14, 12, 9, 8, 7), g(3, 5, 7, 9, 15, 18, 19, 14, 12, 9, 8, 7))$
4	$E(flat, f(3, 3, 4, 4, 5, 6, 6, 7, 7, 5, 4, 3), g(3, 3, 4, 4, 5, 6, 6, 7, 7, 5, 4, 3))$
10	$E(valley, f(5, 4, 2, 1, 1, 2, 2, 1, 3, 4, 5, 7), g(5, 4, 2, 1, 1, 2, 2, 1, 3, 4, 5, 7))$

input at least the state s_i and will provide at output one of the values from B. Therefore, the interpretation of $E(t_1, t_2, t_3)$ evaluated at s_i is defined as:

ALGORITHM 1 *Temporal atom evaluation*

> *Consider the state* $s_i = (v_1, v_2, v_3)$
> *If* $(\mathbf{I}_{s_i}(t_1) = v_1)$ *and* $(\mathbf{I}_{s_i}(t_2) = v_2)$ *and* $(\mathbf{I}_{s_i}(t_3) = v_3)$
>   *Then* $\mathbf{I}_{s_i}(E(t_1, t_2, t_3)) = true$
>   *Else* $\mathbf{I}_{s_i}(E(t_1, t_2, t_3)) = false$

Finally, the interpretation of the variable symbol y_j at the state s_i is given by $\mathbf{I}_{s_i}(y_j) = v_j, j = 1..3$, which satisfies the condition imposed to the interpretation of temporal atom template, which is $\mathbf{I}_{s_i} E(y_1, y_2, y_3) = true$ for each state s_i. Having well-defined the language L, the syntax and the semantics of L, as well as the linear time structure M, we can construct the temporal atoms evaluated as *true* at time moment i (see Table 2)

3.1   Consistency

The connection between the restricted first-order temporal logic we defined and the temporal data mining task this logic tries to formalize (temporal rules extraction) is given by the following assumptions:

A. For each formula p in L, there is an algorithm that calculates the value of the interpretation $\mathbf{I}_s(p)$, for each state s, in a finite number of steps.
B. There are states (called incomplete states) that do not contain enough information to calculate the interpretation for all formulae defined at these states.
C. It is possible to establish a measure, (called *general interpretation*) about the degree of truth of a compound formula along the entire sequence of states $(s_0, s_1, \ldots, s_n, \ldots)$.

The first assumption express the calculability of the interpretation **I** (we already considered this assumption in Example 2). The second assumption express the situation when only the body of a temporal rule can be evaluated at a time moment i, but not the head of the rule. Therefore, for the state s_i, we cannot calculate the interpretation of the temporal rule and the only solution is to estimate it using a general interpretation. This solution is expressed by the third assumption. (*Remark:* The second assumption violates the condition about the existence of an interpretation in each state s_i, as defined in Definition 4. But it is well known that in data mining sometimes data is incomplete or is missing. Therefore, we must modify this condition as "*I is a function that associates with almost each state s an interpretation I_s of all symbols from L*").

However, to ensure that this general interpretation is well defined, the linear time structure must present some property of consistency. Practically, this means that if we take any sufficiently large subset of time instants, the conclusions we may infer from this subset are sufficiently close from those inferred from the entire set of time instants. Therefore,

Definition 5. *Given L and a linear time structure M, we say that M is a consistent time structure for L if, for every formula p, the limit $supp(p) = \lim_{n \to \infty} n^{-1} \# A$ exists, where $\#$ means "cardinality" and $A = \{i = 1..n \,|\, i \models p\}$. The notation $supp(p)$ denotes the support (of truth) of p.*

Now we define the general interpretation for an n-ary predicate symbol P as:

Definition 6. *Given L and a consistent linear time structure M for L, the general interpretation I_G for an n-ary predicate P is a function $D^n \to [0,1]$, such that, for each n-tuple of terms $\{t_1, \ldots, t_n\}$, $I_G(P(t_1, \ldots, t_n)) = supp(P(t_1, \ldots, t_n))$.*

The general interpretation is naturally extended to constraint formulae, temporal rules and the corresponding templates. There is another useful measure, called *confidence*, but available only for temporal rules (templates). This measure is calculated as a limit ratio between the number of certain applications (time instants where both the body and the head of the rule are true) and the number of potential applications (time instants where only the body of the rule is true).

Definition 7. *The confidence of a temporal rule (template) $H_1 \wedge \cdots \wedge H_m \mapsto H_{m+1}$ is the limit $\lim_{n \to \infty} (\#B)^{-1} \# A$, where $A = \{i = 1 \ldots n \,|\, i \models H_1 \wedge \cdots \wedge H_m \wedge H_{m+1}\}$ and $B = \{i = 1 \ldots n | i \models H_1 \wedge \cdots \wedge H_m\}$.*

The relation between the property of consistency and the existence of the confidence for a temporal rule is expressed in the following lemma.

Lemma 1. *If M is a consistent linear time structure for L then every temporal rule (template) $H_1 \wedge \cdots \wedge H_m \mapsto H_{m+1}$ for which $supp(H_1 \wedge \cdots \wedge H_m) \neq 0$ has a well-defined confidence.*

For different reasons, (the user has not access to the entire sequence of states, or the states he has access to are incomplete), the general interpretation cannot be calculated. A solution is to estimate I_G using a finite linear time structure, i.e. a model.

Definition 8. *Given L and a consistent time structure $M = (S, x, \boldsymbol{I})$, a model for M is a structure $\tilde{M} = (\tilde{T}, \tilde{x})$ where $\tilde{T}$ is a finite temporal domain $\{i_1, \ldots, i_n\}$, $\tilde{x}$ is the subsequence of states $\{x_{i_1}, \ldots, x_{i_n}\}$ (the restriction of x to the temporal domain $\tilde{T}$) and for each $i_j, j = 1, \ldots, n$, the state x_{i_j} is a complete state.*

Now we may define the estimator for a general interpretation:

Definition 9. *Given L and a model $\tilde{M}$ for M, an estimator of the general interpretation for an n-ary predicate P, $I_{G(\tilde{M})}(P)$, is a function $D^n \to [0, 1]$, assigning to each atomic formula $p = P(t_1, \ldots, t_n)$ the value defined as the ratio $\dfrac{\#A}{\#\tilde{T}}$, where $A = \{i \in \tilde{T} \mid i \models p\}$. The notation $supp(p, \tilde{M})$ will denote the estimated support of p, given $\tilde{M}$.*

The extension of this definition to the other types of formulae in L demands a deeper analysis. Consider, as example, the model $\tilde{M}$ induced by the sequence of $n > 1$ states $\tilde{x} = x_1, \ldots, x_n$. The interpretation of a formula $\nabla_1 p$ at the state x_n can not be calculated, because $n \models \nabla_1 p$ if $(n+1) \models p$, but $x_n + 1 \notin \tilde{x}$. Therefore, the cardinality of the set $A = \{i \leq n \mid i \models X_1 p\}$ is strictly smaller than n, which means that, for p a global formula having the meaning of truth *true*, the estimated support is

$$supp(\nabla_1 p, \tilde{M}) = (n - 1)/n \neq 1 = supp(\nabla_1 p).$$

The fact that the support estimator is biased seems at first glance without importance, especially when, as in this case, the bias (n^{-1}) tends to zero for $n \uparrow \infty$. But considering a formula of type $\nabla_n p$, it is evidently that the interpretation can not be calculated at none of the states from $\tilde{x}$, and so the support estimator is not even defined. Before indicating how the expression $\dfrac{\#A}{\#\tilde{T}}$ must be adjusted to avoid this kind of problem, we start by defining the standard form of a formula in L.

Definition 10. *A formula $\nabla_{k_1} p_1 \wedge \nabla_{k_2} p_2 \wedge \ldots \wedge \nabla_{k_n} p_n$, where $n \geq 1$ and p_i are atoms of L, is in standard form if exists $i_0 \in \{1, \ldots, n\}$ such that $k_{i_0} = 0$ and for all $i = 1..n$, $k_i \leq 0$.*

For an atomic formula p, it is clearly that its standard form is $X_0 p$. Another example of formula in standard form is a temporal rule (template), where the head of the rule is prefixed by ∇_0 and all other constraint formulae are prefixed by $\nabla_{-k}, k \geq 0$. It is obviously that, for M a consistent time structure, the support of a formula does not change if it is prefixed with a temporal connective ∇_k, $k \in \mathbb{Z}$. Therefore, to each formula p in L corresponds an equivalent formula (under the measure supp) having a standard form (denoted $\mathcal{F}(p)$). Based on this concept, we can now give a non equivocal definition for *time windows*:

Definition 11. *Let be p a formula in L having the standard form $\nabla_{k_1} p_1 \wedge \nabla_{k_2} p_2 \wedge \ldots \wedge \nabla_{k_n} p_n$. The time window of p – denoted $w(p)$ – is defined as $\max\{\,|k_i| : i = 1..n\}$*

In the following, a formula having a time window equal zero will be called *temporal free formula*, whereas a formula with a strictly positive time window will be called a *temporal formula*. The concept of time window allows us to define a non biased estimator for the support measure.

Definition 12. *Given L and a model $\tilde{M}$ for M, the estimator of the support for a formula p in L and having $w(p) < \#\tilde{T} = m$, denoted $supp(p, \tilde{M})$, is the ratio*

$$\frac{\#A}{m - w(p)}, \quad \text{where } A = \{i \in \tilde{T} \mid i \models \mathcal{F}(p)\}.$$

According to this definition, if $w(p) \geq m$ the estimator $supp(p, \tilde{M})$ is not defined. The use of the standard form of the formula, in the construction of the set A, eliminates the interpretation problem for a formula of type $\nabla_k p$, $k \geq m$. Moreover, it is easy to see that $supp(\nabla_k p, \tilde{M}) = supp(p, \tilde{M})$, for all $k \in \mathbb{Z}$.

Definition 13. *Given L and a model $\tilde{M}$ for M, an estimate of the general interpretation for a formula p is given by*

$$I_{G(\tilde{M})}(p) = \begin{cases} supp(p, \tilde{M}), & \text{if } w(p) < \#\tilde{T}, \\ 0 & \text{if } w(p) \geq \#\tilde{T} \end{cases} \tag{1}$$

Once again, the estimation of the confidence for a temporal rule (template) is defined as:

Definition 14. *Given a model $\tilde{M} = (\tilde{T}, \tilde{x})$ for M, the estimation of the confidence for the temporal rule (template) $H_1 \wedge \cdots \wedge H_m \mapsto H_{m+1}$ is the ratio $\frac{\#A}{\#B}$, where $A = \{i \in \tilde{T} \mid i \models H_1 \wedge \cdots \wedge H_m \wedge H_{m+1}\}$ and $B = \{i \in \tilde{T} \mid i \models H_1 \wedge \cdots \wedge H_m\}$. The notation $conf(H, \tilde{M})$ will denote the estimated confidence of the temporal rule (template) H given $\tilde{M}$.*

According to the same arguments used in the definition of a correct support estimator, the existence of a confidence estimator for a temporal rule H is guaranteed only for models having a number of states greater than the time window of the rule. Moreover, if $\tilde{\tilde{T}}$ is the set obtained from $\tilde{T}$ by deleting the first $w(H_1 \wedge \ldots \wedge H_{m+1}) - w(H_1 \wedge \ldots \wedge H_m)$ states, then we can obtain a non biased confidence estimator if in the expression of the set $B = \{i \in \tilde{T} \mid i \models H_1 \wedge \cdots \wedge H_m\}$ the set $\tilde{T}$ is replaced with $\tilde{\tilde{T}}$.

Example 3. Consider the following temporal rule template T (for the moment we are not concerned on how it was discovered):

$\nabla_{-2}(y_1 = peak) \wedge \nabla_{-2}(y_2 < 11) \wedge \nabla_{-1}(y_1 = peak) \wedge \nabla_{-1}(y_3 > 3) \mapsto \nabla_0(y_1 = flat)$

which may be "translated" in a natural language as:

IF at time $t - 2$ an event "peak" occurred with a mean less than 11 AND at time $t - 1$ another event "peak" occurred with a standard error greater than 3 THEN at time t an event type "flat" occurs.

If M is a consistent linear time structure for the temporal language L defined in Example 1, then the model $\tilde{M}$ given by the sequence of states $\{s_1, \ldots, s_{18}\}$ (see Table 1) can be used to estimate the confidence of the temporal rule T. If the local support for the rule is 0.125 (*true* at states 2 and 14, among 17 states) and the local support for the body of the rule is 0.176 (*true* at states 2, 7 and 14, among 16 states) then the estimated confidence for the rule, based on model $\tilde{M}$, is $0.125/0.176 = 0.71$. And, due to the consistency property, this estimation is a reliable information about the success rate for this rule when applied on future data.

4   The Granularity Model

We start with the concept of a temporal type to formalize the notion of time granularities, as described in [3].

Definition 15. *Let $(T, <)$ (index) be a linearly ordered temporal domain isomorphic to a subset of the integers with the usual order relation, and let $(A, <)$ (absolute time) be a linearly ordered set. Then, a temporal type on (T, A) is a mapping μ from T to 2^A such that*

1. *$\mu(i) \neq \emptyset$ and $\mu(j) \neq \emptyset$, where $i < j$, imply that each element in $\mu(i)$ is less than all the elements in $\mu(j)$, and*
2. *for all $i < j$, if $\mu(i) \neq \emptyset$ and $\mu(j) \neq \emptyset$, then $\forall k , i < k < j$ implies $\mu(k) \neq \emptyset$.*

Each set $\mu(i)$, if non-empty, is called a granule of μ. Property (1) says that granules do not overlap and that the order on indexes follows the order on the corresponding granules. Property (2) disallows an empty set to be the value of a mapping for a certain index value if a lower index and a higher index are mapped to non-empty sets.

When considering a particular application or formal context, we can specialize this very general model along the following dimensions:

- choice of the index set T,
- choice of the absolute time set A,
- restrictions on the structure of granules,
- restrictions on the temporal types by using relationships.

We call the resulting formalization a temporal type system. Consider some possibilities for each of the above four dimensions. Convenient choices for the index set are natural numbers, integers, and any finite subset of them. The choice for absolute time is typically between dense and discrete. In general, if the application imposes a fixed basic granularity, then a discrete absolute time in terms of the basic granularity is probably the appropriate choice. However, if one is interested in being able to represent arbitrary finer temporal types, a dense absolute time is required. In both cases, specific applications could impose left/right boundedness on the absolute time set. The structure of ticks could be restricted in several ways:

(1) disallow types with *gaps* in a granule,
(2) disallow types with *non-contiguous* granules,
(3) disallow types whose granules do *not cover all* the absolute time, or
(4) disallow types with *nonuniform* granules (only types with granules having the same size are allowed).

4.1   Relationships and Formal Properties

Following [3], we define a number of interesting relationships among temporal types.

Definition 16. *Let be μ and ν be temporal types on $(\mathcal{T}, \mathcal{A})$.*

- Finer-than*: μ is said to be finer than ν, denoted $\mu \preccurlyeq \nu$, if for each $i \in \mathcal{T}$, there exists $j \in \mathcal{T}$ such that $\mu(i) \subseteq \nu(j)$.*
- Groups-into*: μ is said to group into ν, denoted $\mu \trianglelefteq \nu$, if for each non-empty granule $\nu(j)$, there is a subset S of $\mathcal{T}$ such that $\nu(j) = \bigcup_{i \in S} \mu(i)$.*
- Subtype*: μ is said to be a subtype of ν, denoted $\mu \sqsubseteq \nu$, if for each $i \in \mathcal{T}$, there exists $j \in \mathcal{T}$ such that $\mu(i) = \nu(j)$.*
- Shifting*: μ and ν are said to be shifting equivalent, denoted $\mu_1 \rightleftharpoons \mu_2$, if $\mu \sqsubseteq \nu$ and $\nu \sqsubseteq \mu$.*

When a temporal type μ is finer than a temporal type ν, we also say that ν is *coarser* than μ. The *finer-than* relationship formalizes the notion of finer partitions of the absolute time. By definition, this relation is reflexive, i.e. $\mu \preccurlyeq \mu$ for each temporal type μ. Furthermore, the finer-than relation is obviously transitive. However, if no restrictions are given, it is not antisymmetric, and hence it is not a partial order. Indeed, $\mu \preccurlyeq \nu$ and $\nu \preccurlyeq \mu$ do not imply $\mu = \nu$, but only $\mu \rightleftharpoons \nu$. Considering the *groups-into* relation, $\mu \trianglelefteq \nu$ ensures that for each granule of μ there exists a set of granules of ν covering exactly the same span of time. The relation is useful, for example, in applications where attribute values are associated with time granules; sometimes it is possible to obtain the value associated with a granule of ν from the values associated with the granules of μ whose union covers the same time. The groups-into relation has the same two properties as the finer-than relation, but generally $\mu \preccurlyeq \nu$ does not imply $\mu \trianglelefteq \nu$ or viceversa. The *subtype* relation intuitively identifies a type corresponding to subsets of granules of another type. Similar to the two previous relations, subtype is reflexive and transitive, and satisfies

$\mu \sqsubseteq \nu \Rightarrow \mu \preccurlyeq \nu$. Finally, *shifting* is clearly an equivalence relation. Concerning this last relation, an equivalent, more useful and practical definition, is:

Definition 17. *Two temporal types μ_1 and μ_2 are said to be* shifting *equivalent (denoted $\mu_1 \rightleftharpoons \mu_2$) if there is a bijective function $h : \mathcal{T} \to \mathcal{T}$ such that $\mu_1(i) = \mu_2(h(i))$, for all $i \in \mathcal{T}$.*

In the following we consider only temporal type systems which satisfy the restriction that no pair of different types can be shifting equivalent, i.e.

$$\mu_1 \rightleftharpoons \mu_2 \Rightarrow \mu_1 = \mu_2. \tag{2}$$

For this class of systems, the three relationships $\preccurlyeq, \trianglelefteq$ and $\sqsubseteq$ are reflexive, transitive and antisymmetric and, hence, each relationship is a partial order. Therefore, for the relation we are particulary interested in, *finer-than*, there exists a unique least upper bound of the set of all temporal types, denoted by $\mu_\top$, and a unique greatest lower bound, denoted by $\mu_\perp$. These top and bottom elements are defined as follows: $\mu_\top(i) = \mathcal{A}$ for some $i \in \mathcal{T}$ and $\mu_\top(j) = \emptyset$ for each $j \neq i$, and $\mu_\perp(i) = \emptyset$ for each $i \in \mathcal{T}$. Moreover, for each pair of temporal types μ_1, μ_2, there exist a unique least upper bound $\overline{(\mu_1, \mu_2)}$ and a unique greatest lower bound $\underline{(\mu_1, \mu_2)}$ of the two types, with respect to $\preccurlyeq$. We formalize this result in the following theorem, proved by [3]:

Theorem 1. *Any temporal type system having an infinite index, and satisfying (2), is a lattice with respect to the finer-than relationship.*

Let denote $\mathcal{G}_0$ the set of temporal types for which the index set and the absolute time set are isomorphic with the set of positive natural numbers, i.e. $\mathcal{A} = \mathcal{T} = \mathbb{N}$. Consider now the following particular subsets of $\mathcal{G}_0$ (temporal types with non-empty granules, with granules covering all the absolute time and with constant size granules):

$$\mathcal{G}_1 = \{\mu \in \mathcal{G}_0 \,|\, \forall i \in \mathbb{N}, \, 0 < \#\mu(i)\} \tag{3}$$

$$\mathcal{G}_2 = \{\mu \in \mathcal{G}_1 \,|\, \forall i \in \mathbb{N}, \, \mu(i)^{-1} \neq 0\} \tag{4}$$

$$\mathcal{G}_3 = \{\mu \in \mathcal{G}_2 \,|\, \forall i \in \mathbb{N}, \, \mu(i) = c_\mu\} \tag{5}$$

The membership of a temporal type defined by one of these subsets implies very useful properties, a first result being expressed in the following lemma.

Lemma 2. *If μ_1, μ_2 are temporal types from $\mathcal{G}_1$, then $\mu_1 \rightleftharpoons \mu_2 \Rightarrow \mu_1 = \mu_2$.*

Therefore, the set $\mathcal{G}_1$ of temporal types is a lattice with respect to the *finer-than* relationship. The temporal type system $\mathcal{G}_1$ is not closed, because $\overline{(\mu_1, \mu_2)}$ is not always in $\mathcal{G}_1$. In exchange it can be shown that the temporal type system $\mathcal{G}_2$ (see (4), where $\mu^{-1}(i) = \{j \in \mathbb{N} : i \in \mu(j)\}$) is a closed system having a unique greatest lower bound, $\mu_\perp(i) = i, \, \forall i \in \mathbb{N}$, but no least upper bound $\mu_\top$. Furthermore, the membership of $\mathcal{G}_2$ is a sufficient condition for the equivalence of the relationships *finer-than* and *groups-into*, according to the following lemma:

Lemma 3. *If μ and ν are temporal types from $\mathcal{G}_2$, then $\mu \preceq \nu \Leftrightarrow \mu \trianglelefteq \nu$.*

4.2 Linear Granular Time Structure

If $M = (S, x, \mathbf{I})$ is a first-order linear time structure, then let the absolute time $\mathcal{A}$ be given by the sequence x, by identifying the time moment i with the state $s_{(i)}$ (on the ith position in the sequence). If μ is a temporal type from $\mathcal{G}_2$, then the temporal granule $\mu(i)$ may be identified with the set $\{s_j \in S \mid j \in \mu(i)\}$. Therefore, the temporal type μ induces a new sequence, x_μ, defined as $x_\mu : \mathbb{N} \to 2^S$, $x_\mu(i) = \mu(i)$. (*Remark:* In the following the set $\mu(i)$ will be considered, depending of the context, either as a set of states or as a set of natural numbers, the indexes of these states).

Consider now the linear time structure derived from M, $M_\mu = (2^S, x_\mu, \mathbf{I}^\mu)$. To be well defined, we must give the interpretation $\mathbf{I}^\mu_{\mu(i)}$, for each $i \in \mathbb{N}$. Because for a fixed i the set $\mu(i)$ is a finite sequence of states, it defines (if all the states are complete states) a model $\tilde{M}_{\mu(i)}$ for M. Therefore the estimated general interpretation $I_{G(\tilde{M}_{\mu(i)})}$ is well defined and we consider, by definition, that for a temporal free formula (e.g. a temporal atom) p in L,

$$\mathbf{I}^\mu_{\mu(i)}(p) = I_{G(\tilde{M}_{\mu(i)})}(p) = \mathrm{supp}(p, \tilde{M}_{\mu(i)}) \qquad (6)$$

This interpretation is extended to any temporal formula in L according to the rule:

$$\mathbf{I}^\mu_{\mu(i)}(\nabla_{k_1} p_1 \wedge \ldots \wedge \nabla_{k_n} p_n) = \frac{1}{n} \sum_{j=1}^{n} \mathbf{I}^\mu_{\mu(i+k_j)}(p_j) \qquad (7)$$

where p_i are temporal free formulae and $k_i \in \mathbb{Z}, i = 1 \ldots n$.

Definition 18. *If $M = (S, x, \mathbf{I})$ is a first-order linear time structure and μ is a temporal type from $\mathcal{G}_2$, then the linear granular time structure induced by μ on M is the triple $M_\mu = (2^S, x_\mu, \mathbf{I}^\mu)$, where $x_\mu : \mathbb{N} \to 2^S$, $x_\mu(i) = \mu(i)$ and $\mathbf{I}^\mu$ is a function that associates with almost each set of states $\mu(i)$ an interpretation $\mathbf{I}^\mu_{\mu(i)}$ according to the rules (6) and (7).*

Of a particular interest is the linear granular time structure induced by the greatest lower bound temporal type of the lattice $\mathcal{G}_2$, $\mu_\perp(i) = i$. In this case, $M_{\mu_\perp} = (S, x, \mathbf{I}^{\mu_\perp})$, where the only difference from the initial time structure M is at the interpretation level: for $p = P(t_1, \ldots, t_n)$ a formula in L, if the interpretation $\mathbf{I}_s(p)$ is a function defined on D^n with values in $\{true, false\}$ – giving so the meaning of truth – the interpretation $\mathbf{I}^{\mu_\perp}_s(p)$ is a function defined on D^n with values in $[0,1]$ – giving so the degree of truth. The relation linking the two interpretations is given by $\mathbf{I}_s(p) = true$ if and only if $\mathbf{I}^{\mu_\perp}_s(p) = 1$. Indeed, supposing the state $s_{(i)}$ is a complete state, it defines the model $\tilde{M}_i = (i, s_{(i)})$ and we have, for p a temporal free formula,

$$\mathbf{I}_{\mu_\perp(i)}^{\mu_\perp}(p) = \operatorname{supp}(p, \tilde{M}_i) = \begin{cases} 1, & \text{if } \mathbf{I}_{s(i)}(p) = true, \\ 0, & \text{if } \mathbf{I}_{s(i)}(p) = false \end{cases} \tag{8}$$

For a formula $\pi = \nabla_{k_1} p_1 \wedge \ldots \wedge \nabla_{k_n} p_n$, we have $\mathbf{I}_{s_i}(\pi) = true$ iff $\forall j \in \{1 \ldots n\}, i + k_j \models p_j$, which is equivalent with

$$\operatorname{supp}(p_1, \tilde{M}_{i+k_1}) = \cdots = \operatorname{supp}(p_n, \tilde{M}_{i+k_n}) = 1$$

$$\Leftrightarrow \frac{1}{n} \sum_{j=1}^{n} \mathbf{I}_{\mu_\perp(i+k_j)}^{\mu_\perp}(p_j) = \mathbf{I}_{\mu_\perp(i)}^{\mu_\perp}(\pi) = 1. \tag{9}$$

Consequently, the linear granular time structure $M_{\mu_\perp}$ can be seen as an extension, at the interpretation level, of the classical linear time structure M.

Example 4. Consider the first three granules of the linear granular time structure M_μ, which cover the first nine states of the linear time structure M (represented graphically in Fig. 1).

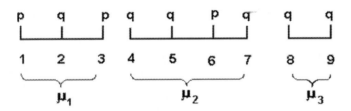

Fig. 1 Graphical representation of the first nine states from the time structure M and of the first three granules of temporal type μ

If p and q are two temporal free formulae in L, interpreted as *true* in the states $\{1, 3, 6\}$ (respectively $\{2, 4, 5, 7, 8, 9\}$) from M, then the following relations occur:

$$\mathbf{I}_{\mu(1)}(p) = \operatorname{supp}(p, \tilde{M}_{\mu_1}) = \frac{2}{3}, \quad \mathbf{I}_{\mu(2)}(p) = \operatorname{supp}(p, \tilde{M}_{\mu_2}) = \frac{1}{4}$$

$$\mathbf{I}_{\mu(3)}(p) = \operatorname{supp}(p, \tilde{M}_{\mu_3}) = \frac{0}{2}, \quad \mathbf{I}_{\mu(1)}(q) = \operatorname{supp}(q, \tilde{M}_{\mu_1}) = \frac{1}{3}$$

$$\mathbf{I}_{\mu(2)}(q) = \operatorname{supp}(p, \tilde{M}_{\mu_2}) = \frac{3}{4}, \quad \mathbf{I}_{\mu(3)}(q) = \operatorname{supp}(p, \tilde{M}_{\mu_3}) = \frac{2}{2}$$

$$\mathbf{I}_{\mu(1)}(p \wedge \nabla q) = \frac{1}{2} \left(\mathbf{I}_{\mu(1)}(p) + \mathbf{I}_{\mu(2)}(q) \right) = \frac{17}{24}$$

$$\mathbf{I}_{\mu(1)}(q \wedge \nabla p \wedge \nabla_2 q) = \frac{1}{3} \left(\mathbf{I}_{\mu(1)}(q) + \mathbf{I}_{\mu(2)}(p) + \mathbf{I}_{\mu(3)}(q) \right) = \frac{11}{36}$$

4.3 Linking Two Granular Time Structures

All the granular time structures induced by a temporal type have in common interpretations which take values in $[0, 1]$ if applied on predicate symbols in L. This observation allows us to establish the relation linking the interpretations $\mathbf{I}^\mu$ and $\mathbf{I}^\nu$, from two linear granular time structures induced by μ and ν, when there exists a relationship *finer-than* between these two temporal types. According to the lemma 3, for each $i \in \mathbb{N}$ there is a subset $N_i \subset \mathbb{N}$ such that $\nu(i) = \bigcup_{j \in N_i} \mu(j)$. If p is a temporal free formula in L, then the interpretation $\mathbf{I}^\nu$ for p at $\nu(i)$ is the weighted sum of the interpretations $\mathbf{I}^\mu_{\mu(j)}(p)$, where $j \in N_i$. We formalize this result in the following theorem:

Theorem 2. *If μ, ν are temporal types from $\mathcal{G}_2$, such that $\mu \preccurlyeq \nu$, and $\mathbf{I}^\mu, \mathbf{I}^\nu$ are the interpretations from the induced linear time structures M_μ and M_ν on M, then for each $i \in \mathbb{N}$,*

$$\mathbf{I}^\nu_{\nu(i)}(p) = \frac{1}{\#\nu(i)} \sum_{j \in N_i} \#\mu(j) \mathbf{I}^\mu_{\mu(j)}(p), \tag{10}$$

where N_i is the subset of $\mathbb{N}$ which satisfies $\nu(i) = \bigcup_{j \in N_i} \mu(j)$ and p is a temporal free formula in L.

If we consider $\mu = \mu_\perp \preccurlyeq \nu$ then $\#\mu(j) = 1$, for all $j \in \mathbb{N}$ and $\mathbf{I}^\mu_{\mu(j)}(p) = \mathrm{supp}(p, \tilde{M}_j)$. Therefore,

$$\mathbf{I}^\nu_{\nu(i)}(p) = \frac{1}{\#\nu(i)} \sum_{j \in \nu(i)} \mathrm{supp}(p, \tilde{M}_j)$$

$$= \frac{1}{\#\nu(i)} \#\{j \in \nu(i) \mid j \models p\} = \mathrm{supp}(p, \tilde{M}_{\nu(i)}) = \mathbf{I}_{G(\tilde{M}_{\nu(i)})}(p) \tag{11}$$

result which is consistent with Definition 18. But the significance of the theorem 2 is revealed in a particular context. If $\mu, \nu \in \mathcal{G}_3$ and $\mu \preccurlyeq \nu$, it can be shown that $\#N_i = \frac{c_\nu}{c_\mu}, \forall i \in \mathbb{N}$ and so the relation (10) becomes

$$\mathbf{I}^\nu_{\nu(i)}(p) = \frac{1}{\#N_i} \sum_{j \in N_i} \mathbf{I}^\mu_{\mu(j)}(p). \tag{12}$$

Generally speaking, consider three worlds, W_1, W_2 and W_3 – defined as sets of granules of information – where W_1 is finer than W_2 which is finer than W_3. Suppose also that the conversion between granules from two different worlds is given by a constant factor. If the **independent** part of information in each granule is transferred from W_1 to W_2 and then the world W_1 is "lost", the theorem 2 under the form (12) affirms that it is possible to transfer the independent information from W_2 to W_3 and to obtain the same result as for the transfer from W_1 to W_3.

Example 5. Consider a linear time structure M (here, the world W_1) and a temporal free formula p such that, for the first six time moments, we have $i \models p$ for $i \in \{1, 3, 5, 6\}$. The concept of independence, in this example, means that the interpretation of p in the state s_i does not depend on the interpretation of p in the state s_j. Let be $\mu, \nu \in \mathcal{G}_3$, $\mu \preccurlyeq \nu$, with $\mu(i) = \{2i - 1, 2i\}$ and $\nu(i) = \{6i - 5, \ldots, 6i\}$. Therefore, $\nu(1) = \mu(1) \cup \mu(2) \cup \mu(3)$. According to the definition 18, $\mathbf{I}^{\mu}_{\mu(1)}(p) = \mathrm{supp}(p, \{1, 2\}) = 0.5$, $\mathbf{I}^{\mu}_{\mu(2)}(p) = \mathrm{supp}(p, \{3, 4\}) = 0.5$, $\mathbf{I}^{\mu}_{\mu(3)}(p) = \mathrm{supp}(p, \{5, 6\}) = 1$, whereas $\mathbf{I}^{\nu}_{\nu(1)}(p) = \mathrm{supp}(p, \{1, .., 6\}) = 0.66$. If the linear time structure M is "lost", the temporal types μ and ν are "lost" too (we don't know the absolute time $\mathcal{A}$ given by M). But if we know the induced time structure M_μ (world W_2) and the relation between μ and ν

$$\nu(k) = \mu(3k - 2) \cup \mu(3k - 1) \cup \mu(3k), \ \forall k \in \mathbb{N}$$

then we can completely deduce the time structure M_ν (world W_3). As example, according to (12), $\mathbf{I}^{\nu}_{\nu(1)}(p) = \frac{1}{3} \sum_{i=1}^{3} \mathbf{I}^{\mu}_{\mu(i)}(p) = 0.66$. The condition about a constant conversion factor between granules is necessary because the information about the size of granules, as it appears in expression 10, is "lost" when the time structure M is "lost"

The theorem 2 is not effective for temporal formulae (which can be seen as the dependent part of the information of a temporal granule). In this case we can prove that the interpretation, in the coarser world, of a temporal formula with a given time window is linked with the interpretation, in the finer world, of a similar formula but having a larger time window.

Theorem 3. *If μ and ν are temporal types from $\mathcal{G}_3$ such that $\mu \preccurlyeq \nu$ and $\mathbf{I}^{\mu}, \mathbf{I}^{\nu}$ are the interpretations from the induced linear time structures M_μ and M_ν on M, then for each $i \in \mathbb{N}$,*

$$\mathbf{I}^{\nu}_{\nu(i)}(p \wedge \nabla q) = \frac{1}{k} \sum_{j \in N_i} \mathbf{I}^{\mu}_{\mu(j)}(p \wedge \nabla_k q) \tag{13}$$

where $k = c_\nu / c_\mu$, $\nu(i) = \bigcup_{j \in N_i} \mu(j)$ and p, q are temporal free formulae in L.

If we define the operator $zoom_k$ over the set of formulae in L as

$$zoom_k(\nabla_{k_1} p_1 \wedge \ldots \wedge \nabla_{k_n} p_n) = \nabla_{k \cdot k_1} p_1 \wedge \ldots \wedge \nabla_{k \cdot k_n} p_n$$

then an obvious corollary of this theorem is

Corollary 1. *If μ and ν are temporal types from $\mathcal{G}_3$ such that $\mu \preccurlyeq \nu$ and $\mathbf{I}^{\mu}, \mathbf{I}^{\nu}$ are the interpretations from the induced linear time structures M_μ and M_ν on M, then for each $i \in \mathbb{N}$,*

$$\mathbf{I}^{\nu}_{\nu(i)}(\nabla_{k_1} p_1 \wedge \ldots \wedge \nabla_{k_n} p_n) = \frac{1}{k} \sum_{j \in N_i} \mathbf{I}^{\mu}_{\mu(j)}(zoom_k(\nabla_{k_1} p_1 \wedge \ldots \wedge \nabla_{k_n} p_n)) \tag{14}$$

where $k = c_\nu/c_\mu$, $\nu(i) = \bigcup_{j \in N_i} \mu(j)$, $k_i \in \mathbb{N}$ and p_i, $i = 1..n$ are temporal free formulae in L.

According to this corollary, if we know the degree of truth of a temporal rule (template) in the world W_1, we can say nothing about the degree of truth of the same rule in the world W_2, coarser than W_1. The information is only transferred from the temporal rule $zoom_k(H)$ in W_1 (which has a time window greater than $k - 1$) to the temporal rule H in W_2, where k is the coefficient of conversion between the two worlds. Consequently, all the information related to temporal formulae having a time window less than k is lost during the transition to the coarser world W_2.

4.4   The Consistency Problem

The importance of the concepts of consistency, support and confidence, (see Subsect. 3.1), for the process of information transfer between worlds with different granularity may be highlighted by analyzing the analogous expressions for a linear granular time structure M_μ.

Definition 19. *Given L and a linear granular time structure M_μ on M, we say that M_μ is a consistent granular time structure for L if, for every formula p, the limit*

$$supp(p, M_\mu) = \lim_{n \to \infty} \frac{\sum_{i=1}^{n} I_{\mu(i)}^{\mu}(p)}{n} \qquad (15)$$

exists. The notation $supp(p, M_\mu)$ denotes the support (degree of truth) of p under M_μ.

A natural question concern the inheritance of the consistency property from the basic linear time structure M by the induced time structure M_μ. The answer is formalized in the following theorem.

Theorem 4. *If M is a consistent linear time structure and $\mu \in \mathcal{G}_3$ then the granular time structure M_μ is also consistent.*

The proof of the theorem (see Appendix 5) is based on the relation between the support of a formula p in M, respectively in M_μ, which is:

$$supp(p, M_\mu) = supp(p, M) \qquad (16)$$

$$supp(\nabla_{k_1} p_1 \wedge \ldots \wedge \nabla_{k_m} p_m, M_\mu) = \frac{1}{m} \sum_{j=1}^{m} supp(p_j, M) \qquad (17)$$

The implications of Theorem 4 are extremely important. It is easy to show, starting from Definition 14, that the confidence of a temporal rule (template) may be expressed using only the support measure if the linear time structure M is consistent. Therefore, considering that by definition the confidence of a temporal rule (template) $\mathcal{H}$, $H_1 \wedge \ldots \wedge H_m \mapsto H_{m+1}$, giving a consistent granular time structure M_μ, is

$$conf(\mathcal{H}, M_\mu) = \begin{cases} \frac{\text{supp}(H_1 \wedge \ldots \wedge H_m \wedge H_{m+1}, M_\mu)}{\text{supp}(H_1 \wedge \ldots \wedge H_m, M_\mu)} & \text{if } \text{supp}(H_1 \wedge \ldots \wedge H_m, M_\mu) > 0, \\ 0 & \text{if not} \end{cases}$$

(18)

we can deduce, by applying Theorem 4 and the relations (16) and (17), that the confidence of $\mathcal{H}$, for a granular time structure M_μ induced on a consistent time structure M by a temporal type $\mu \in \mathcal{G}_3$, is independent of μ. In other words,

   "*The property of consistency is a sufficient condition for the independence of the measure of support/confidence, during the process of information transfer between worlds with different granularities, all derived from an absolute world using constant conversion factors. In practice, this means that even we are not able to find, for a given world M_μ, the granules where a temporal rule $\mathcal{H}$ apply (according to Theorem 3), we are sure that the confidence of $\mathcal{H}$ is the same in each world $M_\mu, \forall \mu \in \mathcal{G}_3$.*"

4.5   Event Aggregation

All the deduction processes made until now were conducted to obtain an answer to the following question: how is changing the degree of truth of a formula p if we pass from a linear time structure with a given granularity to a coarser one. And we proved that we can give a proper expression if we impose some restrictions on the temporal types which induce these time structures. But there is another phenomenon which follows the process of transition between two real worlds with different time granularities: new kinds of events appear, some kinds of events disappear.

Definition 20. *An event type (denoted $E[t]$) is the set of all temporal atoms from L having the same name (or head).*

All the temporal atoms of a given type $E[t]$ are constructed using the same symbol predicate and we denote by $N[t]$ the arity of this symbol. Consider $E(t, t_2, \ldots, t_n) \in E[t]$ (where $n = N[t]$). According to Definition 1, a term $t_i, i \in \{2, .., n\}$ has the form $t_i = f(t_{i1}, \ldots, t_{ik_i})$. Suppose now that for each index i the function symbol f from the expression of t_i belongs to a family of function symbols with different arities, denoted $\mathcal{F}_i[t]$ (so different sets for different event types $E[t]$ and different index i). This family has the property that the interpretation for each of its member is given by a real functions which

- is applied on a variable number of arguments, and
- is invariant in the order of the arguments.

A good example of a such real function is a statistical function, e.g. mean $(x_1, .., x_n)$. Based on the set $\mathcal{F}_i[t]$ we consider the set of terms expressed as

$f_k(c_1,\ldots,c_k)$, where f_k is a $k-$ary function symbol from $\mathcal{F}_i[t]$ and $c_i, i = 1..k$ are constant symbols. We denote such a set as $T_i[t]$. Consider now the operator $\oplus$ defined on $T_i[t] \times T_i[t] \to T_i[t]$ such that

$$f_n(c_1,..,c_n) \oplus f_m(d_1,..,d_m) = f_{n+m}(c_1,..,c_n,d_1,..,d_m)$$

Of course, because the interpretation of any function symbol from $\mathcal{F}_i[t]$ is invariant in the order of arguments, we have

$$f_n(c_1,\ldots,c_n) \oplus f_m(d_1,\ldots,d_m) = f_n(c_{\sigma(1)},\ldots,c_{\sigma(n)}) \oplus f_m(d_{\varphi(1)},\ldots,c_{\varphi(n)})$$

where σ (respectively φ) is a permutation of the set $\{1,\ldots,n\}$ (respectively $\{1,\ldots,m\}$). Furthermore, it is evident that the operator $\oplus$ is commutative and associative.

We introduce now a new operator (denoted $\boxplus$) defined on $E[t] \times E[t] \to E[t]$, such that, for $E(t, t_2, .., t_i, .., t_n) \in E[t]$, $E(t, t_2', .., t_i', .., t_n') \in E[t]$ we have:

$$E(t, t_2, \ldots, t_n) \boxplus E(t, t_2', \ldots, t_n') = E(t, t_2 \oplus t_2', \ldots, t_n \oplus t_n') \qquad (19)$$

Once again, it is obviously that the operator $\boxplus$ is commutative and associative. Therefore, we can apply this operator on a subset $\mathcal{E}$ of temporal atoms from $E[t]$ and we denote the result as $\boxplus_{e_i \in \mathcal{E}} e_i$.

By definition, a formula p is satisfied by a linear time structure $M = (S, x, \mathbf{I})$ (respectively by a model $\tilde{M} = (\tilde{T}, \tilde{x})$ of M) if there is at least a state $s_i \in x$ (respectively $s_i \in \tilde{x}$) such that $\mathbf{I}_{s_i}(p) = true$. Therefore, the set of events of type t satisfied by M (respectively $\tilde{M}$) is given by:

$$E[t]_M = \{e \in E[t] \,|\, \exists s_i \in x \text{ such that } \mathbf{I}_{s_i}(e) = true\} \qquad (20)$$

respectively by:

$$E[t]_{\tilde{M}} = \{e \in E[t] \,|\, \exists s_i \in \tilde{x} \text{ such that } \mathbf{I}_{s_i}(e) = true\} \qquad (21)$$

If we consider now M_μ, the linear time structure induced by the temporal type μ on M, the definition of $E[t]_{M_\mu}$ is derived from (20) by changing the condition $\mathbf{I}_{s_i}(e) = true$ with $\mathbf{I}^\mu_{\mu(i)}(e) = 1$. Of course, only for $\mu = \mu_\perp$ we have $E[t]_M = E[t]_{M_\mu}$ (we proved that $\mathbf{I}_{s_i}(p) = true \Leftrightarrow \mathbf{I}^{\mu_\perp}_{\mu_\perp(i)}(p) = 1$). Generally $E[t]_M \supset E[t]_{M_\mu} \supset E[t]_{M_\nu}$, for $\mu \preccurlyeq \nu$, which is a consequence of the fact that a coarser world satisfies less temporal events than a finer one.

Example 6. If M is a linear time structure such that for the event $e \in E[t]$ we have $i \models e$ if and only if i is odd, and μ is a temporal type given by $\mu(i) = \{2i - 1, 2i\}$, then it is obviously that $e \in E[t]_M$ but $e \notin E[t]_{M_\mu}$ (for all $i \in \mathbb{N}, \mathbf{I}^\mu_{\mu(i)}(e) = \text{supp}(e, \{2i - 1, 2i\}) = 0.5$).

In the same time a coarser world may satisfies new events, representing a kind of aggregation of local, "finer" events.

Definition 21. *If μ is a temporal type from $\mathcal{G}_2$, we call the aggregate event of type t induced by the granule $\mu(i)$ (denoted $e[t]_{\mu(i)}$) the event obtained by applying the operator $\boxplus$ on the set of events of type t which are satisfied by the model $\tilde{M}_{\mu(i)}$, i.e.*

$$e[t]_{\mu(i)} = \underset{e_i \in E[t]_{\tilde{M}_{\mu(i)}}}{\boxplus} e_i \qquad (22)$$

According to (6), the interpretation of an event e in any world M_μ depends on the interpretation of the same event in M. Therefore, if e is not satisfied by M it is obvious that $\mathbf{I}^\mu_{\mu(i)}(e) = 0$, for all μ and all $i \in \mathbb{N}$. Because an aggregate event (conceived of a new, "federative" event), usually is not satisfied by M, the relation (6) is not appropriate to give the degree of truth for $e[t]_{\mu(i)}$. But before to give the rule expressing the interpretation for an aggregate temporal atom, we must impose on M the following restriction: two different events of type t can not be evaluated as *true* at the same state $s \in S$, or:

$$\exists h : E[t]_M \to S, \ h \text{ injective, such that } h(e) = s \text{ where } \mathbf{I}_s(e) = true \qquad (23)$$

Definition 22. *If M_ν is a linear granular time structure and $e[t]_{\mu(i_0)}$ is an aggregate event induced by the granule μ_{i_0} ($\mu, \nu \in \mathcal{G}_2$), then the interpretation of $e[t]_{\mu(i_0)}$ in the state $\nu(i)$ is defined as:*

$$\mathbf{I}^\nu_{\nu(i)}(e[t]_{\mu(i_0)}) = \frac{\#(\mathcal{E}_i \cap \mathcal{E})}{\#\mathcal{E}} \sum_{e_j \in \mathcal{E}} \mathbf{I}^\nu_{\nu(i)}(e_j) \qquad (24)$$

where $\mathcal{E} = E[t]_{\tilde{M}_{\mu(i_0)}}$, $\mathcal{E}_i = E[t]_{\tilde{M}_{\nu(i)}}$.

The restriction (23) is given to assure that $\sum_{e_j \in \mathcal{E}} \mathbf{I}^\nu_{\nu(i)}(e_j) \le 1$, for all $i, i_0 \in \mathbb{N}$. Indeed, let be $e_1, \ldots, e_n$ the events from $\mathcal{E}$. If $h(e_j) = s_j, j = 1..n$, then consider the sets $A_j = \{k \in \nu(i) \,|\, k \models e_j\} = \{s \in \nu(i) \,|\, s = s_j\}$. The function h being injective, the sets A_j are disjoint and therefore $\sum_{j=1}^n \#A_j \le \#\nu(i)$. Consequently, we have

$$\sum_{j=1}^n \mathbf{I}^\nu_{\nu(i)}(e_j) = \sum_{j=1}^n \text{supp}(e_j, M_{\nu(i)}) = \frac{1}{\#\nu(i)} \sum_{j=1}^n \#A_j \le \frac{\#\nu(i)}{\#\nu(i)} = 1. \qquad (25)$$

The relation (25) and the fact that the coefficient $\frac{\#(\mathcal{E}_i \cap \mathcal{E})}{\#\mathcal{E}}$ is less or equal one guarantee that the interpretation of an aggregate event is well-defined, i.e. $\mathbf{I}^\nu_{\nu(i)}(e[t]_{\mu(i_0)}) \le 1$. Furthermore, the interpretation is equal one if and only if:

$$(i) \quad \frac{\#(\mathcal{E}_i \cap \mathcal{E})}{\#\mathcal{E}} = 1 \Leftrightarrow \mathcal{E} = \mathcal{E}_i \qquad (26)$$

meaning that all the events of type t satisfied by $\tilde{M}_{\mu_{i_0}}$ are also satisfied by $\tilde{M}_{\nu_i}$, and

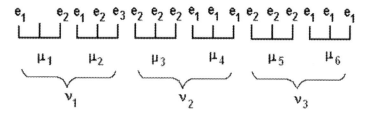

Fig. 2 Graphical representation of the first eighteen states of the time structure M and of the first six (respectively three) granules of temporal types μ and ν

$$(ii)\quad \sum_{j=1}^{n}\mathbf{I}^{\nu}_{\nu(i)}(e_j) = 1 \Leftrightarrow \frac{1}{\#\nu(i)}\sum_{j=1}^{n}\#A_j = 1 \Leftrightarrow \sum_{j=1}^{n}\#A_j = \#\nu(i) \quad (27)$$

meaning that the sets A_j form a partition of $\nu(i)$ (or equivalently $h^{-1}(\nu(i)) = \mathcal{E}$).

Example 7. Let be M a linear time structure, $e_1, e_2, e_3 \in E[t]$ such that (see Fig. 2)

$$1 \models e_1,\ 4 \models e_1 \text{ and } i \models e_1 \text{ for } i \in \{6k-2, 6k-1, 6k \,|\, k \geq 2\}$$
$$3 \models e_2,\ 5 \models e_2 \text{ and } i \models e_2 \text{ for } i \in \{6k+1, 6k+2, 6k+3 \,|\, k \geq 1\}$$
$$6 \models e_3$$

Consider two temporal types $\mu, \nu \in \mathcal{G}_3$ such that $\mu(i) = \{3i-2, 3i-1, 3i \,|\, i \geq 1\}$ and $\nu(i) = \{6i-5, \ldots, 6i \,|\, i \geq 1\}$. The different aggregate events induced by granules of temporal type μ and ν are:

$$e[t]_{\mu(1)} = e_1 \boxplus e_2,\ e[t]_{\mu(2)} = e_1 \boxplus e_2 \boxplus e_3,\ e[t]_{\mu(3)} = e_2,\ e[t]_{\mu(4)} = e_1$$
$$e[t]_{\nu(1)} = e_1 \boxplus e_2 \boxplus e_3,\ e[t]_{\nu(i)} = e_1 \boxplus e_2 \text{ (for } i > 1)$$

Lets denote e_{12} the aggregate event induced by $\mu(1)$. Evidently $\mathcal{E} = E[t]_{\tilde{M}_{\mu(1)}} = \{e_1, e_2\}$ so $\#\mathcal{E} = 2$. The interpretation of this event in different granules of types μ and ν can be calculated according to relation (24):

$$\mathbf{I}^{\mu}_{\mu(1)}(e_{12}) = \frac{\#(\mathcal{E} \cap \mathcal{E})}{\#\mathcal{E}}\sum_{j=1}^{2}\mathbf{I}^{\mu}_{\mu(1)}(e_j) = \sum_{j=1}^{2}\text{supp}(e_j, \tilde{M}_{\mu(1)}) = \frac{1}{3} + \frac{1}{3} = \frac{2}{3}$$

$$\mathbf{I}^{\mu}_{\mu(2)}(e_{12}) = \frac{\#(\mathcal{E}_2 \cap \mathcal{E})}{\#\mathcal{E}}\sum_{j=1}^{2}\mathbf{I}^{\mu}_{\mu(2)}(e_j) = \frac{\#(\{e_1, e_2, e_3\} \cap \{e_1, e_2\})}{2}\sum_{j=1}^{2}\text{supp}(e_j, \tilde{M}_{\mu(2)})$$

$$= \frac{1}{3} + \frac{1}{3} = \frac{2}{3}$$

$$\mathbf{I}^{\mu}_{\mu(3)}(e_{12}) = \frac{\#(\mathcal{E}_3 \cap \mathcal{E})}{\#\mathcal{E}} \sum_{j=1}^{2} \mathbf{I}^{\mu}_{\mu(2)}(e_j) = \frac{\#(\{e_2\} \cap \{e_1, e_2\})}{2} \sum_{j=1}^{2} \text{supp}(e_j, \tilde{M}_{\mu(3)})$$

$$= \frac{1}{2}\left(\frac{0}{3} + \frac{3}{3}\right) = \frac{1}{2}$$

$$\mathbf{I}^{\mu}_{\mu(4)}(e_{12}) = \frac{\#(\mathcal{E}_4 \cap \mathcal{E})}{\#\mathcal{E}} \sum_{j=1}^{2} \mathbf{I}^{\mu}_{\mu(2)}(e_j) = \frac{\#(\{e_1\} \cap \{e_1, e_2\})}{2} \sum_{j=1}^{2} \text{supp}(e_j, \tilde{M}_{\mu(4)})$$

$$= \frac{1}{2}\left(\frac{3}{3} + \frac{0}{3}\right) = \frac{1}{2}$$

$$\mathbf{I}^{\mu}_{\mu(2k+1)}(e_{12}) = \frac{1}{2}, \ \mathbf{I}^{\mu}_{\mu(2k+2)}(e_{12}) = \frac{1}{2}, \ \text{for all } k \geq 1.$$

According to these results, the event e_{12} obviously is not satisfied by M_{μ}.

$$\mathbf{I}^{\nu}_{\nu(1)}(e_{12}) = \frac{\#(\mathcal{E}_1 \cap \mathcal{E})}{\#\mathcal{E}} \sum_{j=1}^{2} \mathbf{I}^{\nu}_{\nu(1)}(e_j) = \frac{\#(\{e_1, e_2, e_3\} \cap \{e_1, e_2\})}{2} \sum_{j=1}^{2} \text{supp}(e_j, \tilde{M}_{\nu(1)})$$

$$= \frac{2}{6} + \frac{2}{6} = \frac{2}{3}$$

$$\mathbf{I}^{\nu}_{\nu(2)}(e_{12}) = \frac{\#(\mathcal{E}_2 \cap \mathcal{E})}{\#\mathcal{E}} \sum_{j=1}^{2} \mathbf{I}^{\nu}_{\nu(2)}(e_j) = \frac{\#(\{e_1, e_2\} \cap \{e_1, e_2\})}{2} \sum_{j=1}^{2} \text{supp}(e_j, \tilde{M}_{\nu(2)})$$

$$= \frac{3}{6} + \frac{3}{6} = 1$$

$$\mathbf{I}^{\nu}_{\nu(k)}(e_{12}) = 1 \text{ for all } k \geq 1,$$

which means that e_{12} is satisfied by M_{ν}, a coarser world than M_{μ}. As a general rule, the degree of truth for an aggregate event e is equal one in a given granule $\mu(i)$ if all individual events composing e (and only these events) are satisfied by $\mu(i)$.

5   Conclusions

In this article we developed a formalism for a specific temporal data mining task: the discovery of knowledge, represented in the form of general Horn clauses, inferred from databases with a temporal dimension. The theoretical framework we proposed, based on first-order temporal logic, permits to define the main notions (*event, temporal rule, constraint*) in a formal way. The concept of a consistent linear time structure allows us to introduce the notions of *general interpretation*, of *support* and of *confidence*, the lasts two measure being the expression of the two similar concepts used in data mining.

Starting from the inherent behavior of temporal systems – the perception of events and of their interactions is determined, in a large measure, by the temporal scale – we extended the capability of our formalism to "capture" the concept of time granularity. To keep an unitary viewpoint on the meaning of the same formula at different scales of time, we changed the usual definition of the interpretation $\mathbf{I}^\mu$ for a formula in the frame of a first-order temporal granular logic: it return the degree of truth (a real value between zero and one) and not only the meaning of truth (*true* or *false*).

The consequence of the definition for $\mathbf{I}^\mu$ is formalized in Theorem 2 : only the independent information (here, the degree of truth for a temporal free formula) may be transferred without loss between worlds with different granularities. Concerning the temporal rules (scale dependent information), we proved that the interpretation of a rule in a coarser world is linked with the interpretation of a similar rule in a finer world, rule obtained by applying the operator $zoom_k$ on the initial temporal rule.

By defining a similar concept of consistency for a granular time structure M_μ, we could proved that this property is inherited from the basic time structure M if the temporal type μ is of type $\mathcal{G}_2$ (granules with constant size). The major consequence of Theorem 4 is that the confidence of a temporal rule (template) is preserved in all granular time structures derived from the same consistent time structure.

We defined also a mechanism to aggregate events of the same type, that reflects the following intuitive phenomenon: in a coarser world, not all events inherited from a finer world are satisfied, but in exchange there are new events which become satisfiable. To achieve this we extended the syntax and the semantics of L by allowing "family" of function symbols and by adding two new operators.

In our opinion, the logical next step in our work consists in adding a probabilistic dimension to the formalism. Preliminary results (see [11]) confirm that this approach allows a unified framework including the logical formalism and its granular extension, framework in which the property of consistency becomes a consequence of the capacity of a particular stochastic process to obey the strong law of large numbers.

Appendix: Proofs

Proof (of Lemma 2)
Before we start, we introduce the following notation: given two non-empty sets S_1 and S_2 of elements in $\mathcal{A}$, $S_1 \ll S_2$ holds if each number in S_1 is strictly less than each number in S_2 (formally, $S_1 \ll S_2$ if $\forall x \in S_1 \, \forall y \in S_2 \, (x < y)$). Moreover, we say that a set $\mathbb{S}$ of non-empty sets of elements in $\mathcal{A}$ is *monotonic* if for each pair of sets S_1 and S_2 in $\mathbb{S}$ either $S_1 \ll S_2$ or $S_2 \ll S_1$.

The relation $\mu_1 \rightleftharpoons \mu_2$ is equivalent with the existence of a bijection function $h : \mathbb{N} \to \mathbb{N}$ such that $\mu_1(i) = \mu_2(h(i))$, for all i. We will prove by induction that $h(i) = i$, which is equivalent with $\mu_1 = \mu_2$.

- $i = 1$: suppose that $h(1) > 1$. If $a = \min(\mu_1(1))$ – the existence of a is ensured by the condition (3) – then $\mu_1(1) = \mu_2(h(1)) \Rightarrow a \in \mu_2(h(1))$. Because $1 < h(1)$ we have $\mu_2(1) \ll \mu_2(h(1))$ (according to Definition 15) and so there is $b \in \mu_2(1)$ such that $b < a$. The inequality $1 < h(1)$ implies $h^{-1}(1) > 1$, and so $\mu_2(1) = \mu_1(h^{-1}(1)) \gg \mu_1(1)$. But the last inequality ($\gg$) is contradicted by the existence of $b \in \mu_1(h^{-1}(1))$ which is smaller than $a \in \mu_1(1)$. In conclusion, $h(1) = 1$.

- $i = n + 1$: from the induction hypothesis we have $h(i) = i, \forall i \leq n$. Supposing that $h(n+1) \neq n+1$, then the only possibility is $h(n+1) > n+1$. This last relation implies also $h^{-1}(n+1) > n+1$. Using a similar rationing as in the previous case (it's sufficient to replace 1 with $n+1$), we obtain

$$\mu_1(n+1) \ll \mu_1(h^{-1}(n+1)) = \mu_2(n+1) \ll \mu_2(h(n+1)) = \mu_1(n+1)$$

where each of the set from this relation are non-empty, according to (3). The contradiction of the hypothesis, in this case, means that $h(n+1) = n+1$ and, by induction principle, that $h(i) = i, \forall i \in \mathbb{N}$. ■

Proof (of Lemma 3)
Let be $\mu \in \mathcal{G}_2, \nu \in \mathcal{G}_2$.

- $\mu \preccurlyeq \nu$: let $j_0 \in \mathbb{N}$. The relation (4) means that for all $k \in \nu(j_0)$, $\mu^{-1}(k) \neq \emptyset$ and so $S = \bigcup_{k \in \nu(j_0)} \{\mu^{-1}(k)\} \neq \emptyset$. It is obviously, according to Definition 15, that the relation *finer-than* implies that for each $i \in \mathbb{N}$ there is a **unique** $j \in \mathbb{N}$ such that $\mu(i) \subseteq \nu(j)$. Consequently, if $\mu \preccurlyeq \nu$ and $\mu(i) \cap \nu(j) \neq \emptyset$ then $\mu(i) \subseteq \nu(j)$. Therefore, for all $i \in S$, $\mu(i) \subset \nu(j_0)$ which implies $\bigcup_{i \in S} \mu(i) \subset \nu(j_0)$ (a). At the same time, $\forall k \in \nu(j_0)$ we have $k \in \mu\left(\mu^{-1}(k)\right)$ which implies $\nu(j_0) \subseteq \bigcup_{i \in S} \mu(i)$ (b). From (a) and (b) we have $\nu(j_0) = \bigcup_{i \in S} \mu(i)$, which implies $\mu \trianglelefteq \nu$.

- $\mu \trianglelefteq \nu$: let $i_0 \in \mathbb{N}$ and let $k \in \mu(i_0)$. According to (4), there exists $j = \nu^{-1}(k)$. Because $\mu \trianglelefteq \nu$ there is a set S such that $\nu(j) = \bigcup_{i \in S} \mu(i)$. Because the sets $\mu(i), i \in S$ are disjunct and $k \in \nu(j) \cap \mu(i_0)$ we have $i_0 \in S$. Therefore, for each i_0 there is $j \in \mathbb{N}$ such that $\mu(i_0) \subseteq \nu(j)$, which implies $\mu \preccurlyeq \nu$. ■

Proof (of Theorem 2)
The formula p being a temporal free formula, we have $w(p) = 0$. According to Definition 18 and Definition 12, we have

$$\mathbf{I}^{\nu}_{\nu(i)}(p) = \mathrm{supp}(p, \tilde{M}_{\nu(i)}) = \frac{\#\{j \in \nu(i) \,|\, j \models p\}}{\#\nu(i)} \quad (a)$$

On the other hand, because $\nu(i) = \bigcup_{j \in N_i} \mu(j)$, we have also

$$\frac{1}{\#\nu(i)} \sum_{j \in N_i} \#\mu(j) \mathbf{I}^{\mu}_{\mu(j)}(p) = \frac{1}{\#\nu(i)} \sum_{j \in N_i} \#\mu(j) \mathrm{supp}(p, \tilde{M}_{\mu(j)})$$

$$= \frac{1}{\#\nu(i)} \sum_{j \in N_i} \#\{k \in \mu(j) \,|\, k \models p\} = \frac{\#\{j \in \nu(i) \,|\, j \models p\}}{\#\nu(i)} \quad (b)$$

From (a) and (b) we obtain (10). ■

Proof (of Theorem 3)

If $\mu, \nu \in \mathcal{G}_3$ such that $\#\mu(i) = c_\mu$ and $\#\nu(i) = c_\nu$, for all $i \in \mathbb{N}$, it is easy to show that the sets N_i satisfying $\nu(i) = \bigcup_{j \in N_i} \mu(j)$ have all the same cardinality, $\#N_i = c_\nu / c_\mu = k$ and contain successive natural numbers, $N_i = \{j_i, j_i + 1, \ldots, j_i + k - 1\}$. From the relations (7) and (12) we have:

$$
\mathbf{I}^\nu_{\nu(i)}(p \wedge \nabla q) = \frac{1}{2}\left(\mathbf{I}^\nu_{\nu(i)}(p) + \mathbf{I}^\nu_{\nu(i+1)}(q) \right)
$$

$$
= \frac{1}{2}\left(\frac{1}{\#N_i} \sum_{j \in N_i} \mathbf{I}^\mu_{\mu(j)}(p) + \frac{1}{\#N_{i+1}} \sum_{j \in N_{i+1}} \mathbf{I}^\mu_{\mu(j)}(q) \right)
$$

$$
= \frac{1}{2}\left(\frac{1}{k} \sum_{j=j_i}^{j_i+k-1} \mathbf{I}^\mu_{\mu(j)}(p) + \frac{1}{k} \sum_{j=j_i+k}^{j_i+2k-1} \mathbf{I}^\mu_{\mu(j)}(q) \right)
$$

$$
= \frac{1}{2k}\left(\sum_{j=j_i}^{j_i+k-1} \left(\mathbf{I}^\mu_{\mu(j)}(p) + \mathbf{I}^\mu_{\mu(j+k)}(q) \right) \right)
$$

$$
= \frac{1}{2k}\left(\sum_{j=j_i}^{j_i+k-1} 2\mathbf{I}^\mu_{\mu(j)}(p \wedge \nabla_k q) \right) = \frac{1}{k} \sum_{j \in N_i} \mathbf{I}^\mu_{\mu(j)}(p \wedge X_k q) \qquad \blacksquare
$$

Proof (of Theorem 4)

M being a consistent time structure, for each formula p in L the sequence $x(p)_n = n^{-1} \#\{i \le n \,|\, i \models p\}$ has a limit and $\lim_{n \to \infty} x(p)_n = \mathrm{supp}(p, M)$. In the same time, $\mu \in \mathcal{G}_3$ implies $\#\mu(i) = k$ for all $i \in \mathbb{N}$ and $\mu(i) = \{k(i-1)+1, k(i-1)+2, \ldots, ki\}$. Consider the following two cases:

- p *temporal free formula* : We have

$$
\frac{\sum_{i=1}^n \mathbf{I}^\mu_{\mu(i)}(p)}{n} = \frac{\sum_{i=1}^n \mathrm{supp}(p, M_{\mu(i)})}{n}
$$

$$
= \frac{\sum_{i=1}^n \frac{\#\{j \in \mu(i) \,|\, j \models p\}}{\#\mu(i)}}{n} = \frac{\sum_{i=1}^n \#\{j \in \mu(i) \,|\, j \models p\}}{kn}
$$

$$
= \frac{\#\{i \in \bigcup_{i=1}^n \mu(i) \,|\, i \models p\}}{kn} = \frac{\#\{i \le kn \,|\, i \models p\}}{kn} = x(p)_{kn}
$$

Therefore, there exists the limit $\displaystyle\lim_{n \to \infty} \frac{\sum_{i=1}^n \mathbf{I}^\mu_{\mu(i)}(p)}{n} = \lim_{n \to \infty} x(p)_{kn}$ and we have

$$
\mathrm{supp}(p, M_\mu) = \mathrm{supp}(p, M) \text{ for } p \text{ temporal free formula} \qquad (28)
$$

- *temporal formula* $\pi = \nabla_{k_1} p_1 \wedge \ldots \wedge \nabla_{k_m} p_m$: We have

$$\frac{\sum_{i=1}^{n} \mathbf{I}_{\mu(i)}^{\mu}(\nabla_{k_1} p_1 \wedge \ldots \wedge \nabla_{k_m} p_m)}{n} = \frac{\sum_{i=1}^{n} \left(m^{-1} \sum_{j=1}^{m} \mathbf{I}_{\mu(i+k_j)}^{\mu}(p_j) \right)}{n}$$

$$= \frac{1}{m} \frac{\sum_{i=1}^{n} \sum_{j=1}^{m} \mathrm{supp}(p_j, M_{\mu(i+k_j)})}{n} = \frac{1}{mn} \sum_{j=1}^{m} \sum_{i=1}^{n} \mathrm{supp}(p_j, M_{\mu(i+k_j)})$$

$$= \frac{1}{mn} \sum_{j=1}^{m} \sum_{i=1}^{n} \frac{\#\{h \in \mu(k_j + i) \mid h \models p_j\}}{k}$$

$$= \frac{1}{mn} \sum_{j=1}^{m} \frac{\#\{h \in \bigcup_{i=1}^{n} \mu(k_j + i) \mid h \models p_j\}}{k}$$

$$= \frac{1}{mnk} \sum_{j=1}^{m} (\#\{h \leq k(k_j + n) \mid h \models p_j\} - \#\{h \leq kk_j \mid h \models p_j\})$$

$$= \frac{1}{mnk} \sum_{j=1}^{m} \left(k(k_j + n) x(p_j)_{k(k_j+n)} - kk_j x(p_j)_{kk_j} \right)$$

$$= \frac{1}{m} \sum_{j=1}^{m} \frac{k_j + n}{n} x(p_j)_{k(k_j+n)} - \frac{1}{m} \sum_{j=1}^{m} \frac{k_j}{n} x(p_j)_{kk_j}$$

By tacking $n \to \infty$ in the last relation, we obtain

$$\lim_{n \to \infty} \frac{\sum_{i=1}^{n} \mathbf{I}_{\mu(i)}^{\mu}(\nabla_{k_1} p_1 \wedge \ldots \wedge \nabla_{k_m} p_m)}{n}$$

$$= \lim_{n \to \infty} \left(\frac{1}{m} \sum_{j=1}^{m} \frac{k_j + n}{n} x(p_j)_{k(k_j+n)} - \frac{1}{m} \sum_{j=1}^{m} \frac{k_j}{n} x(p_j)_{kk_j} \right)$$

$$= \frac{1}{m} \sum_{j=1}^{m} \lim_{n \to \infty} \frac{k_j + n}{n} x(p_j)_{k(k_j+n)} - \frac{1}{m} \sum_{j=1}^{m} \lim_{n \to \infty} \frac{k_j}{n} x(p_j)_{kk_j}$$

$$= \frac{1}{m} \sum_{j=1}^{m} \lim_{n \to \infty} x(p_j)_{k(k_j+n)} = \frac{1}{m} \sum_{j=1}^{m} \mathrm{supp}(p_j, M)$$

and so we have

$$\mathrm{supp}(\nabla_{k_1} p_1 \wedge \ldots \wedge \nabla_{k_m} p_m, M_\mu) = \frac{1}{m} \sum_{j=1}^{m} \mathrm{supp}(p_j, M) \qquad (29)$$

From 5 and 29 results the conclusion of the theorem   ∎

References

[1] Al-Naemi, S.: A theoretical framework for temporal knowledge discovery. In: Proceedings of International Workshop on Spatio-Temporal Databases, Spain, pp. 23–33 (1994)

[2] Bettini, C., Wang, X.S., Jajodia, S.: Mining temporal relationships with multiple granularities in time sequences. Data Engineering Bulletin 21(1), 32–38 (1998)

[3] Bettini, C., Wang, X.S., Jajodia, S.: A general framework for time granularity and its application to temporal reasoning. Ann. Math. Artif. Intell. 22(1-2), 29–58 (1998)

[4] Bettini, C., Wang, X.S., Jajodia, S., Lin, J.-L.: Discovering frequent event patterns with multiple granularities in time sequences. IEEE Trans. Knowl. Data Eng. 10(2), 222–237 (1998)

[5] Chen, X., Petrounias, I.: A Framework for Temporal Data Mining. In: Quirchmayr, G., Bench-Capon, T.J.M., Schweighofer, E. (eds.) DEXA 1998. LNCS, vol. 1460, pp. 796–805. Springer, Heidelberg (1998)

[6] Chomicki, J., Toman, D.: Temporal Logic in Information Systems. BRICS Lecture Series, vol. LS-97-1, pp. 1–42 (1997)

[7] Ciapessoni, E., Corsetti, E., Montanari, A., Pietro, P.S.: Embedding time granularity in a logical specification language for synchronous real-time systems. Sci. Comput. Program. 20(1-2), 141–171 (1993)

[8] Clifford, J., Rao, A.: A simple general structure for temporal domains. In: Temporal Aspects of Information Systems. Elsevier Science, Amsterdam (1988)

[9] Cotofrei, P., Stoffel, K.: Classification Rules + Time = Temporal Rules. In: Sloot, P.M.A., Tan, C.J.K., Dongarra, J., Hoekstra, A.G. (eds.) ICCS-ComputSci 2002. LNCS, vol. 2329, pp. 572–581. Springer, Heidelberg (2002)

[10] Cotofrei, P., Stoffel, K.: From temporal rules to temporal meta-rules. In: Kambayashi, Y., Mohania, M., Wöß, W. (eds.) DaWaK 2004. LNCS, vol. 3181, pp. 169–178. Springer, Heidelberg (2004)

[11] Cotofrei, P., Stoffel, K.: Stochastic processes and temporal data mining. In: Proceedings of the 13th ACM SIGKDD International Conference on Knowledge Discovery and Data Mining, San Jose, USA, pp. 183–190 (August 2007)

[12] Emerson, E.A.: Temporal and Modal Logic. In: Handbook of Theoretical Computer Science, pp. 995–1072 (1990)

[13] Euzenat, J.: An algebraic approach to granularity in qualitative time and space representation. In: IJCAI (1), pp. 894–900 (1995)

[14] Evans, C.: The macro-event calculus: representing temporal granularity. In: Proceedings of PRICAI, Japan (1990)

[15] Giunchglia, F., Walsh, T.: A theory of abstraction. Artificial Intelligence 56, 323–390 (1992)

[16] Han, J., Cai, Y., Cercone, N.: Data-driven discovery of quantitative rules in databases. IEEE Transactions on Knowledge and Data Engineering 5, 29–40 (1993)

[17] Hobbs, J.: Granularity. In: Proceedings of the IJCAI 1985, pp. 432–435 (1985)

[18] Hornsby, K.: Temporal zooming. Transactions in GIS 5, 255–272 (2001)

[19] Knoblock, C.: Generating Abstraction Hierarchies: an Automated Approach to Reducing Search in Planning. Kluwer Academic Publishers, Dordrecht (1993)

[20] Lin, T.Y., Louie, E.: Data mining using granular computing: fast algorithms for finding association rules. Data mining, rough sets and granular computing, 23–45 (2002)

[21] Malerba, D., Esposito, F., Lisi, F.: A logical framework for frequent pattern discovery in spatial data. In: Proceedings of 5th Conference Knowledge Discovery in Data (2001)

[22] Mani, I.: A theory of granularity and its application to problems of polysemy and underspecification of meaning. In: Proceedings of the Sixth International Conference Principles of Knowledge Representation and Reasoning, pp. 245–255 (1998)

[23] McCalla, G., Greer, J., Barrie, J., Pospisil, P.: Granularity hierarchies. Computers and Mathematics with Applications 23, 363–375 (1992)

[24] Roman, G.-C.: Formal specification of geographic data processing requirements. IEEE Trans. Knowl. Data Eng. 2(4), 370–380 (1990)

[25] Saitta, L., Zucker, J.-D.: Semantic abstraction for concept representation and learning. In: Proceedings of the Symposium on Abstraction, Reformulation and Approximation, pp. 103–120 (1998)

[26] Stell, J., Worboys, M.: Stratified map spaces: a formal basis for multi-resolution spatial databases. In: Proceedings of the 8th International Symposium on Spatial Data Handling, pp. 180–189 (1998)

[27] Yao, Y.: Granular computing: basic issues and possible solutions. In: Wang, P. (ed.) Proceedings of the 5th Joint Conference on Information Sciences, Atlantic City, New Jersey, USA, pp. 186–189 (2000)

[28] Yao, Y., Zhong, N.: Potential applications of granular computing in knowledge discovery and data mining. In: Torres, M., Sanchez, B., Aguilar, J. (eds.) Proceedings of World Multiconference on Systemics, Cybernetics and Informatics, Orlando, Florida, USA, pp. 573–580 (1999) (International Institute of Informatics and Systematics)

[29] Zadeh, L.A.: Information granulation and its centrality in human and machine intelligence. In: Rough Sets and Current Trends in Computing, pp. 35–36 (1998)

[30] Zhang, B., Zhang, L.: Theory and Applications of Problem Solving. North-Holland, Amsterdam (1992)

[31] Zhang, L., Zhang, B.: The quotient space theory of problem solving. In: Proceedings of International Conference on Rough Sets, Fuzzy Set, Data Mining and Granular Computing, pp. 11–15 (2003)

Mining User Preference Model from Utterances

Yasufumi Takama and Yuki Muto

Summary. This chapter introduces a method for mining user preference model from user's behavior including utterances. Information recommendation is one of promising applications in the field of data mining. The key component of information recommendation is generation of user preference model (also called as user profile). Approaches for modeling user preference are divided into explicit and implicit approaches. Although implicit approach is preferable in terms of user's workload, quality of obtained model is usually lower than explicit approach. In order to improve the quality of implicit approach, this chapter proposes to analyze utterance of user. The proposed approach is applied to TV program recommendation, in which the log of watched TV programs as well as utterances while watching TV is collected. First, user's interest in a TV program is estimated based on fuzzy inference, of which inputs are watching time, utterance frequency, and contents of utterances obtained by sentiment analysis. Then, user profile is generated by identifying features common to user's favorite TV programs. Experimental results show the quality of generated profile is improved compared with existing methods.

1 Introduction

This chapter introduces a method for mining user preference model from user's behavior including utterances. Information recommendation is one of promising applications in the field of data mining. One of the key components of information recommendation is generation of user preference model (also called as user profile). Approaches for modeling user preference are divided into explicit and implicit approaches. Although implicit approach is

Yasufumi Takama and Yuki Muto
Tokyo Metropolitan University, 6-6 Asahigaoka, Hino, Tokyo 191-0065, Japan
e-mail: ytakama@sd.tmu.ac.jp

A. Abraham et al. (Eds.): Foundations of Comput. Intel. Vol. 6, SCI 206, pp. 97–123.
springerlink.com   © Springer-Verlag Berlin Heidelberg 2009

preferable in terms of user's workload, quality of obtained model is usually lower than explicit approach.

In order to improve the quality of implicit approach, this chapter proposes to analyze utterance of users. The proposed approach is applied to TV program recommendation, in which the log of watched TV programs as well as utterances while watching TV is collected.

First, user's interest in a TV program is estimated based on fuzzy inference, of which inputs are watching time, utterance frequency, and contents of utterances obtained by sentiment analysis. Then, user profile is generated by identifying attributes common to user's favorite TV programs. Experimental results show the quality of generated profile is improved compared with existing methods.

This chapter is organized as follows. In Sect. 2, information recommendation technologies such as type of recommendation and basic methods are introduced. Section 3 surveys studies on TV program recommendation, which includes methods / systems for TV program recommendation, as well as analysis of user's TV watching behavior. The profile generation algorithm for TV program recommendation is proposed in Sect. 4, in which the main part is the introduction of utterance analysis to profile generation. After the experimental results are shown in Sect. 5, future challenges and promising applications of the proposed method are discussed in Sect. 6.

2   Information Recommendation

As the growth of the Internet and information processing, target of services has been shifting from mass to individuals. For example, web-based advertising targets individuals has been popular recently. The target of conventional advertising such as TV commercial and magazine advertising is ordinary people. Even if the effect of advertising for each person is small, expected sales that are estimated by the product of the advertising effect and number of potential customers will be large by targeting ordinary people. On the other hand, target of Web-based advertising such as affiliate advertising and Pay-per-click advertising is individuals or small groups of customers. By targeting at individuals, the effect of advertising will be improved compared with targeting at ordinary people. Furthermore, by summing up sales from vast number of individuals, expected sales will be large. It is a basic idea behind advertising targeting at individuals.

In the case of manufacturing, producing vast number of products, each of which is customized to small group of customers, is not realistitic in terms of production and inventory cost. On the other hand, those costs of services on the Web are quite low, which makes individual-targeted service possible.

Information recommendation is another important factor for the success of individual-targeted service (including advertising). Its aim is to recommend information, product, etc., that a user might be interested in. Without combining the nature of services on the Web and the power of information recommendation technology, long tail market could not be promising target of business.

Information recommendation can be viewed as one of data mining technologies. In general, the purpose of data mining is to extract useful patterns from data set of huge volume, and utilize the obtained patterns for producing new values or improving existing systems, products, technologies, etc. Recent growth of computer systems including computer networks and data storage has made it possible to collect huge volume of log data such as POS (point-of-sales) data and medical records. That has created the demand for data mining, and various technologies have been studied in various fields [20]. Information recommendation also extracts and utilizes the patterns from given data set. Compared with typical applications of data mining, which aim at extracting patterns that are observed in common among a group such as customer group, information recommendation extracts not only patterns from a group, but also from an individual. For example, collaborative filtering which will be introduced in Sect. 2.2 utilizes patterns found in user / item group, whereas user profile is a preference pattern of an individual. As with other applications of data mining, extraction and utilization of patterns from data is key component of information recommendation.

2.1 Type of Recommendation

Typical application of information recommendation can be found on the Web sites of online shpping, in which various kinds of recommendations are performed.

- Best-selling items
- Best buy items
- Related items

Best-selling items and best buy items are recommended for everyone. That is, these are user-independent recommendation, which is the same as conventional advertising.

On the other hand, recommendation of related items is user dependent. When target user (user to which recommendation is performed) is watching information about an object (target object), a set of items that tend to be bought / checked together with it by past customers is recommended. Another example is that a set of items that relate with the user's purchase / browsing history is recommended.

As for other kinds of user-dependent recommendation, reminder is also a kind of recommendation. Typical application of reminder is e-mail alert of online journal, which notice a user when new documents that related with user's predefined query are found.

Recommendation does not only find related items, but also remove unrelated items. For example, a spam mail filter keeps the condition of unwanted e-mails, based on which removes e-mails satisfying the condition.

2.2 Information Recommendation Algorithm

This chapter focuses on user-dependent recommendation. In order to find related items, relation between items that are useful in terms of recommendation should be considered. In the field of information recommendation, the following assumption is adopted.

- A user tends to like items that are similar to what s/he liked in the past.
- Users having similar preference tend to like similar items.

From these assumptions, it is said that related items can be obtained from similarity between items, as well as similarity between users. Based on employed similarity, recommendation methods can be classified into 2 types: inter-item similarity-based and inter-user similarity-based recommendation.

It is also possible to classify information recommendation methods from viewpoint of evidences used for similarity calculation, such as follows.

- Similarity calculation based on attribute of items / users
- Similarity calculation based on user's behavior

The former approach is usually called content-based approach, whereas typical method of the latter approach is collaborative filtering. The content-based approach calculates similarity between items based on their attributes. For example, attributes of a movie are its genre, performer, director, distributing company, opening date, etc. Based on attributes, each item is represented as a vector, and similarity between items is calculated with cosine or Euclidean distance. When calculating inter-user similarity, demographic information such as job, address, age, and salary, are used as attributes of a user.

Both memory-based approach such as nearest-neighbor method and model-based approach can be available for performing recommendation. For example, when model-based approach is employed together with vector space model, a user profile that represents user preference is represented as a vector as well.

Advantage of content-based approach is that information about other users is not required. On the other hand, it requires information about items (attributes and attribute values), which is often costly and time-consuming. Furthermore, privacy problems should be considered when treating demographic information.

Collaborative filtering [12] is one of typical algorithms for information recommendation. Characteristic of collaborative filtering is that it does not require information about items. That is, it utilizes only information about whether a user likes an item or not. Usually, user's purchase / browsing history is used as the evidence of user's preference. The evidence for user's preference is discussed in more detail in the next subsection.

In collaborative filtering, user-item matrix is employed as data representation, in which each cell $c(i,j)$ represents whether or not a user $u_i (\in U)$ likes an item $i_j (\in I)$. A value of $c(i,j)$ represents user's evaluation on item, which can be binary (e.g., $1 \ldots u_i$ bought i_j, $0 \ldots$ otherwise) or user's rating (e.g., $c(i,j) \in \{1, \cdots, 5\}$). The inter-user similarity between user u_i and u_j, $sim(i,j)$ is often calculated with Pearson-r correlation (1) or cosine similarity.

$$sim(i,j) = \frac{\sum_{k \in I_{ij}} (c(i,k) - \bar{c}'_i)(c(j,k) - \bar{c}'_j)}{\sqrt{\sum_{k \in I_{ij}} (c(i,k) - \bar{c}'_i)^2} \sqrt{\sum_{k \in I_{ij}} (c(j,k) - \bar{c}'_j)^2}}, \quad (1)$$

where $I_{i,j} (\subset I)$ is a set of items evaluated by both users, $\bar{c}'_i$ is average evaluation of u_i over $I_{i,j}$. User u_i's evaluation on unknown item i_j, $\hat{c}(i,j)$ can be predicted based on the obtained inter-user similarity with (2).

$$\hat{c}(i,j) = \bar{c}_i + \frac{\sum_{k \in U_j} sim(i,k)(c(k,j) - \bar{c}'_k)}{\sum_{k \in U_j} |sim(i,k)|}, \quad (2)$$

where U_j is a set of users evaluating i_j, and $\bar{c}_i$ is average evaluation of u_i over items s/he has already evaluated. When using inter-item similarity, calculation can be done in similar way by switching row and column of the matrix [13].

Advantage of collaborative filtering against content-based approach is that it does not require information about items. On the other hand, it is known that collaborative filtering has a cold-start problem. Usually, cold-start problem has two meanings, such as follows.

- Difficulty in providing appropriate recommendation for a new user
- Difficulty in recommend a new item

As most of the cells corresponding to a new user / item inevitably have no value, its similarity to other user / item becomes low. That causes the above problems.

Although typical collaborative filtering as noted above is memory-based approach, its model-based approach has been also studied [13]. As the number of items / users increases, performance (processing time required for recommendation) of memory-based approach get to be low. Introducing model-based approach can keep performance within reasonable processing time.

Privacy problems should be considered in the case of collaborative filtering as well. That is, the scale of database (user-item matrix) affects accuracy and coverage (the numter of itms that can be recommended) of recommendation. If such databases of different organizations (shops, enterprises, etc.) could be combined, both accuracy and coverage of recommendation would be improved to a large extent. However, as such a database is valuable asset to organizations, they would never provide it to others. Furthermore, organizations must keep customers' privacy. Therefore, combining databases is actually impossible except that privacy of customers can be preserved. To solve this problem, privacy-preserving collaborative filtering has been studied, such as secure multiparty computation [1] and perturbation approach [11]. In secure multiparty computation, each participant provide the data concerning privacy in encrypted format, with which calculation required for recommendation is performed without decryption.

On the other hand, perturbation approach adds perturbation to data (cell values) in user-item matrix in order to avoid providing exact data. As recommendation is based on aggregation of data, it is expected added perturbation is cancelled thorough computation and thus decrease in performance is avoided.

2.3 User's Behavior as Evidence for Recommendation

Collaborative filtering is based on user-item matrix, which stores each user's preference for each item. Even in the case of content-based approach, knowing items for which a user prefers is important. Therefore, obtaining evidences for recommendation is one of important processes in information recommendation.

Methods for obtaining such evidences are divided into the following 2 approaches.

Explicit approach: Asking users to express clearly their preference for items
Implicit approach: Estimating users' preference from their behaviors

Typical example of explicit approach is to ask users to rate items with specified scale, such as binary or 5-point scale. Although explicit approach can obtain exact user's preference for items, rating a number of items is burden for users.

In order to reduce user's burden, implicit approach is often employed, which estimates users preference from their behaviors. For example, it is assumed that a user prefers items he bought. The followings are typical user's behaviors that are often used as the evidence of their positive feeling about target item (product, Web page, etc.).

- Buying a product
- Browsing a Web page
- Clicking a link to a Web page
- Printing / saving a Web page

These behaviors are different in terms of preference strength. That is, "buying an item" is usually judged as expressing user's stronger preference for the item than "browsing web page about the item." Other than those examples, various behaviors have been studied in order to improve the accuracy of estimating user preference without increasing user's workload [4, 7]. For example, user's action of picking up items in actual stores would be an evidence for his/her preference (or interests). This kind of action can be recorded with using cameras or RFID tags attached with items.

Time spent on those behaviors are also used for estimating users' preference strength quantitatively. For example, when a user spent longer time browsing a web page A than B, it is supposed that s/he has stronger preference for A than B. However, it should be noted that elapsed time does not always reflect users' preference strength correctly. That is, a user might have left his/her seat when too long browsing time was observed.

3 TV Program Recommendation

3.1 Outline of TV Program Recommendation

As TV is one of the most popular information resources in our daily lives, recommendation of TV programs as well as related information with TV programs has been studied. One of motivations for performing TV program recommendation is to handle increasing choices of TV programs / movies, which is brought by video on demand and shift from analog to digital broadcasting. In particular, regular TV broadcast in Japan is planned to be replaced with digital terrestrial broadcasting by 2011. It is said that finding relevant information from the Web is as difficult as finding a needle in a haystack. Currently, the number of TV channels available is much less than the number of Web pages. However, when the number of TV channels increases or huge number of programs (movies) are available with video on demand service, it is difficult for us to decide which programs to be watched. Therefore, recommendation of TV programs based on users' preference is expected to be useful.

TV program recommendation systems have been studied for realizing personalized TV guides [3, 6, 21, 22, 23]. When the purpose is to recommend TV programs on-air now or in future, collaborative filtering is not suitable because of cold-start problem as noted in Sect. 2.2. That is, as all TV programs that are to be recommended is considered as new items, similarity calculation between such programs and past programs is difficult with using collaborative filtering. Therefore, content-based approach is popular in the field of TV program recommendation

Typical attributes for TV program is such as follows.

- Channel
- Genre
- Cast members
- Keywords
- Parental rating
- Broadcast schedule

These information can be obtained from metadata provided by some standards such as iEPG (internet Electronic Program Guide) and TV-Anytime [3, 22, 23]. The iEPG has been proposed by Sony Corp., aiming at programmed recording of TV programs using TV program listing sites on the Web. Tags that are used in iEPG format includes station (broadcasting station), year, month, date, start / end (start / end time of a program), program-title, program-subtitle, performer, etc. The summary of a TV program can be also described in a format without corresponding tag. TV-Anytime Forum is an association of organizations aiming to develop specifications which enable audio-visual and other services based on consumer platforms. Members of the forum distribute all over the world, which includes major broadcasting companies such as BBC and NHK Japan.

From these metadata, genre and cast members are easily obtained. Keywords that represent the topic of a program are usually extracted from title and summary metadata. As for Broadcast schedule, time slot for broadcasting and day of the week are often used, which are easily calculated from month, date, start / end time metadata. Zhang et al. [21] generate a user profile consisting of 42 subprofiles according to time slot (morning, noon, afternoon, evening, night, midnight) and a day of the week (i.e. 6 slots × 7 days = 42 profiles).

In order to implicitly estimate user preference as noted in Sect. 2.3, watching time is measured by most of TV program recommendation systems [3, 21, 22, 23]. Zhang et al. [21] use percentage of watching time among airtime (duration of a TV program) as a user's degree of interest. This value has been widely used in other systems.

In addition to watching time, AVATAR[3] also focuses on user's reaction to recommendation, which is either accept or reject of the recommendation, for estimating user's degree of interest (DOI) in watched TV program. The time spent on deciding to accept the recommendation is also measured.

AVATAR decides a TV program to be recommended for target user based on both of content-based approach and collaborative filtering. That is, the following two criteria are employed.

- Program should be recommended to target user if semantic similarity between the user profile and the program is high.
- If the neighborhoods have already watched target program, their DOI to target program are used for decision on recommendation.

Former criterion corresponds to content-based approach. The semantic similarity represents the similarity between target program and programs already watched by target user. In order to calculate similarity between TV programs, it uses semantic relationship between attribute values, which is found by an ontology. Resources and relationships used in TV domain are defined with OWL [10], which is used for the similarity calculation.

The latter criterion corresponds to collaborative filtering. User-item matrix, in which each cell stores DOI of a user to a TV program s/he already watched. Based on the matrix, neighborhoods (like-minded users) are found in similar way as noted in Sect. 2.2 and used for recommendation. As already noted, collaborative filtering is basically unsuitable for TV program recommendation. In this system, the drawback of collaborative filtering is covered by combining content-based approach.

3.2 Study on Viewing Behavior

As for related works, users' behaviors of watching TV programs (viewing behaviors) have also been investigated[2, 6, 19]. In the studies, various types of viewing behaviors, such as concerned viewing and diversion viewing, are observed according to period of time (tim slot) in a day [6, 19].

In concerned viewing, a user is concentrating on watching news programs, educational programs, movies, dramas, etc. On the other hand, diversion watching is relaxed watching, which includes users watching TV programs while doing something else. In diversion viewing, a user is not concentrating on TV programs. For example, user is often watching TV during breakfast. In such a situation, a user is watching TV just to know the time, instead of looking at a clock.

Daita et al. have studied the influence of time factor on users' viewing behaviors [2]. They analyzed audience data consisting of 1000 households (2500 viewers) in Portugal with K-means clustering. As the results of analysis, they found the followings.

- There is no big difference in viewing patterns between days of the week.
- Viewing pattern of viewers who seldom watch TV (less than 2 hours) is stable independently of a day of the week.
- Viewing pattern of average viewers who watch TV around 4 hours is dependent on a day of the week.
- viewing pattern of weekend is less stable than weekday.

Taylor and Harper have studied a role of TV in our daily lives based on household interview and ethnographic fieldwork [19]. The followings are the result of their study.

- Viewers tend to establish regular viewing pattern in weekday evening.
- There are three time slots for watching TV: coming home period, mid-evening viewing, and later-evening viewing.
- In coming home period, viewers watch TV to relax. They tend to watch TV while doing something else.
- In mid-evening (8:30-9:00), viewers tend to have planned viewing of certain programs.
- Later-evening correspond to time slot after finishing their chores, until going to bed. In this period, each viewer tends to select TV programs of specific genre. Using TV program guide and channel surfing are often observed in the period.

The first one partially corresponds to the result by Daita et al., and the third one is a kind of diversion viewing. They said that TV is bound up with the ordinary, natural rhythms of daily life in the household, as if it is a part of the furniture. They also claimed that the effort for selecting channels is situation-dependent.

The results of these studies suggest that recommendation strategy should be switched according to time slots as well as weekday / weekend [6, 21]. For example, users will not require recommendation at mid-evening, because they have already decided which program to watch. Furthermore, it is expected detection of user's situation in real-time, such as concerned / diversion viewing will improve effectiveness and applicability of recommendation.

4   Profile Generation Algorithm

4.1   Outline of Method

This chapter focuses on TV program recommendation. Content-based approach is employed because it is more suitable for recommendation of TV programs as noted in Sect. 3.1. This section proposes a method for generating user profile from logs of users' TV watching behavior.

Figure 1 shows the outline of the profile generation algorithm, which consists of the following steps.

1. Collection of TV watching logs for target user
2. Estimation of the user's rating on watched TV programs

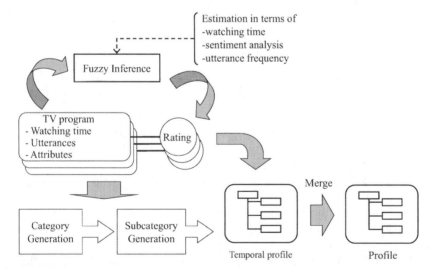

Fig. 1 Overview of profile generation algorithm

3. Generation of temporal user profile from each day log
4. Update user profile by merging temporal user profile into the current profile

The first step collects TV watching logs of target user. The record of each TV program watched by target user has the following data.

- URI
- Program title
- Summary
- airtime
- watching time
- utterance log

The URI is generated from broadcasting date (including year and month), time, and channel so that it can be used to uniquely identify a TV program. The data about year, date, month, and channel are obtained from iEPG metadata. Program title, summary, and airtime are also obtained from iEPG.

One of the contribution of this chapter is that utterances of a user while watching TV are employed as new kind of user's behavior for estimating user preference. It is expected that a user says something about the contents of a TV program while watching it. For example, a user would say "Great! Daisuke won!" when watching baseball game of Boston Red Sox. As such utterances is supposed to contain user's impression on the TV program, estimation accuracy would be improved by analyzing the utterances without forcing users additional workload. The detail of utterance analysis is described in Sect. 4.3.

The second step estimates the user's ratings (i.e., preference strength) of each watched program, which is performed by 2-step fuzzy inference described in Sect. 4.4. After estimation of watched TV programs, temporal profile is generated from logs of specified period. In current setting we generate temporal profile for each day. Considering description capability and general-purpose properties, a format similar to Bookmark is employed for a profile. The profile format has 3 layers: category, subcategory and leaf layers. Generation of temporal profile consists of 2 steps: category generation and subcategory generation. As temporal profile and user profile has the same format, those are merged by category-level and subcategory-level matching. Profile format is described in Sect. 4.2, and generation of profile structure is described in Sect. 4.5.

4.2 Profile Format

This proposed method employs a user profile with the form of bookmark, which is employed by various applications such as ordinary Web browsers. Being used in various applications shows its description capability as well as general-purpose property. Therefore, although this chapter focuses on TV

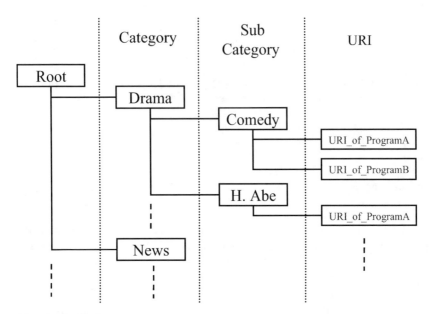

Fig. 2 Profile format

program recommendation, the profile format as well as profile generation method can be applied to other applications.

Figure 2 shows an example of the user profile. The user profile has a 3-layered structure: category layer, subcategory layer, and leaf layer that stores data of TV programs. Each user has two profiles with the same format: positive and negative profile. Positive profile stores TV programs a user is interested in, and negative profile stores TV programs s/he is not interested in.

The category layer has 15 predetermined categories as listed below, which are determined based on genres of TV programs used in iEPG description.

Culture / education, movie, children, drama, news, variety / idol, cartoon / game, music, sports, computer, document, food/travel, science, web, others

Each category in category layer can have multiple subcategories as its childre . Each subcategory represents an arbitrary topic, which characterizes TV programs belonging to it. Examples of subcategories are performers, baseball teams, etc. A TV program must belong to only one category, whereas it can belong to multiple subcategories. In Fig. 2, program A (URI_of_programA) belongs to both of "Comedy" and "H. Abe" (performer's name) subcategories, under "Drama" category. The detail of generating subcategories is described in Sect. 4.5.

When performing recommendation, categories and subcategories are used as the conditions for selecting TV programs to be recommended. For example, a TV program of which genre is contained in positive profile tends to be recommended. Its similarity to user profile (positive or negative) is calculated based on the number of subcategory keywords under the corresponding genre that are contained in summary / title of the program. TV programs with high similarity to positive profile but low similarity to negative one are selected for recommendation.

In the leaf layer, the following data is stored in addition to URI for each TV program.

- estimated rating
- recommendation flag

The estimated rating of a TV program is determined based on the algorithm that is proposed in Sect. 4.3 and 4.4.

The user profile is designed so that it can store not only TV programs that have already watched by target user, but also programs that are to be recommended in future. The recommendation flag is used to discriminate watched programs from those to be recommended. The flag of a TV program is true if it was already recommended by the system, otherwise false.

4.3 Evaluation of Watched TV Program

From the log of a watched TV program as noted in Sect. 4.1, the rating of a TV program by target user is estimated based on the watching time and utterances. In the proposed method, a TV program is estimated in terms of the following 3 scores.

- $Score_w(p)$: a score based on watching time
- $Score_f(p)$: a score based on frequency of utterances
- $Score_{SO}(p)$: a score based on sentiment expressions

The first one is used by most of existing methods. The remaining two scores are newly introduced by the proposed methods, which are obtained by analyzing utterances.

Score based on Watching Time

The score of a TV program p based on watching time is defined by (3), where $wtime(p)$ and $airtime(p)$ indicate the amount of time (minutes) a user watched and its airtime, respectively.

$$Score_w(p) = \frac{wtime(p)}{airtime(p)}. \tag{3}$$

Score based on Utterance Frequency

The score based on utterance frequency, $Score_f(p)$, is defined by (4), which is based on the assumption that a user says something frequently while watching p, if s/he is interested in it.

$$Score_f(p) = \frac{freq(p)}{ceiling(wtime(p)/10)}, \qquad (4)$$

where $freq(p)$ is the number of utterances during watching p, and $ceiling(x)$ is a ceiling function that returns the largest integer less than or equal to x. The score indicates the frequency of utterances per 10 minutes.

Score based on Utterance Analysis

It is often assumed that user's utterances during watching a TV program reflect his/her impression and evaluation on the program. For example, a user would say "I do not like [a performer's name]." if the performer s/he does not like appeared in the program. A user would also say "Great! Daisuke won!" when his / her favorite pitcher won a baseball game. Such utterances contain sentiment expressions [5], which can be used for estimating program's rating as well as for generating subcategories as proposed in Sect. 4.5.

Sentiment analysis [5] has been studied for extracting subjective information such as someone's opinions, impression, and evaluation on something, from documents that are usually written by individuals. The importance of sentiment analysis has increased in the era of Web 2.0, in which consumer generated media such as blogs and BBS have become one of the most important information resources on the Web. Technologies of sentiment analysis include positive /negative judgment of documents in terms of subjective expressions, and extraction of sentences including someone's evaluation / opinion from documents.

The proposed method calculates the score of a TV program p based on utterance analysis, $Score_{SO}(p)$ with the following steps.

1. Extraction of sentiment expressions from utterances
2. Calculation of semantic orientation score for objects o appeared in p
   $(SO(o))$
3. Calculation of $Score_{SO}(p)$ from $SO(o)$

The first step extracts the triple of $<p, o, \text{SO-score}>$ from an utterance containing sentiment expressions, which was issued by target user while watching a program p. The o is an object such as performers, which is a target of user's utterance.

A SO-score (semantic orientation score) represents the degree of positive/ negative sentiment of target user against o in the program. A dictionary of sentiment expression is prepared manually in advance, in which a word (sentiment expression) is stored with its SO-score (ranging from -1.0 to 1.0).

In the second step, a SO-score of an object o, $SO(o)$, is calculated based on extracted triples with (5). Where, $S_e(o)$ indicates a set of sentiment expressions e_i such that a triple $< p, o, e_i >$ is extracted, and $SO(e)$ is the SO-score of e. The range of $SO(e)$ is $[-1, 1]$.

$$SO(o) = \frac{2}{\pi} \arctan \left(\sum_{e_i \in S_e(o)} SO(e_i) \right), \tag{5}$$

In the third step, the score of a program p based on sentiment expressions, $Score_{SO}(p)$, is calculated by (6), where $S_o(p)$ is a set of objects appeared in the utterances during watching p, N_p is the total number of extracted triples, and $N(o)$ is the number of extracted triples regarding o. That is, N_p is the sum of $N(o)$ for all o in $S_o(p)$. The range of $Score_{SO}(p)$ is $[-1, 1]$.

$$Score_{SO}(p) = \sum_{o_i \in S_o(p)} \frac{N(o_i)}{N_p} SO(o_i). \tag{6}$$

4.4 Fuzzy Inference for Rating Estimation

We suppose that watching time and the frequency of utterances do not reflect directly user's interest in a TV program, but his / her attention to it. For example, a user would find a TV program uninteresting after s/he watched it. It would also be possible that a user frequently complain the program or performers in the program. Therefore, this paper proposes to estimate the user's rating of a TV program by 2-step fuzzy inference as shown in Fig. 3. We employ fuzzy inference because of its easiness for describing inference rules.

At the first step, the degree of user's attention to a program p, $Score_a(p)$, is calculated from $Score_w(p)$ (score based on watching time) and $Score_f(p)$ (score based on utterance frequency). Figure 4 shows the membership functions and rule matrix used for calculating $Score_a(p)$. As for fuzzy inference, Min-max composition and center of gravity method for defuzzification are used.

At the second step, the estimated rating of a program, $Score(p)$ is calculated from $Score_a(p)$ and $Score_{SO}(p)$. Figure 5 shows the membership

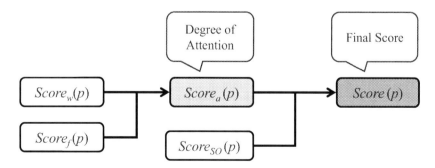

Fig. 3 Process of 2-step fuzzy inference for rating estimation

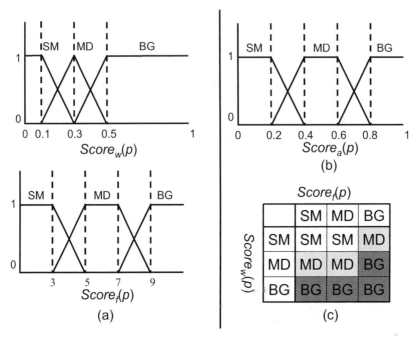

Fig. 4 Fuzzy inference for degree of attention: Membership functions for (a) antecedent and (b) consequent, (c) rule matrix

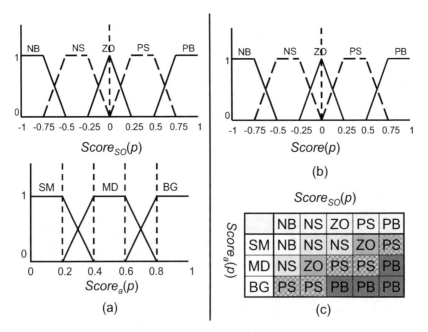

Fig. 5 Fuzzy inference for rating: Membership functions for (a) antecedent and (b) consequent, (c) rule matrix

functions and rule matrix used for the calculation. As $Score_{SO}(p)$ can have negative value, the value of $Score(p)$ ranges from -1.0 to 1.0.

4.5 Generation of Profile Structure

Figure 6 shows the outline of profile structure generation. Given the set of watched TV programs, generation of a user profile (positive / negative) is performed according to the following steps.

1. Classification of watched TV programs
2. Categorization of TV programs into categories
3. Categorization of TV programs into subcategories
4. Storing TV programs in generated profile structure
5. Update of user profile

In step 1, watched TV programs are classified into 2 sets, S_p and S_n. The S_p contains TV programs of which estimated rating $Score(p)$ is positive or 0, while S_n contains those with negative rating.

Profile structure is generated by generating category and subcategory layer. In step 2, TV programs are categorized into 15 categories as noted in Sect. 4.2. The category of a TV program is determined based on the "genre" metadata in its iEPG description.

In step 3, TV programs in each category are further categorized into subcategories. As noted in Sect. 4.2, a TV program can belong to multiple subcategories in a category. The title of a TV program, performers in the program, and keywords that are appeared in the "summary" metadata in iEPG are used as attributes of the TV program. As our system currently focuses on

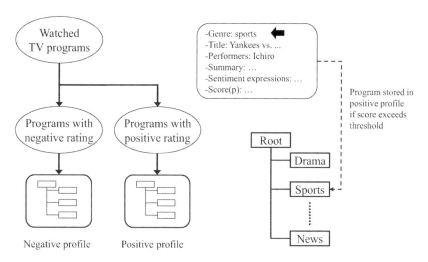

Fig. 6 Outline of profile structure generation

Japanese TV programs, morphological analysis is applied to extract nouns as keywords.

Subcategories are selected from above-mentioned attributes in the following 2 ways.

- Subcategory generation based on speech recognition
- Subcategory generation from common attributes

Subcategory Generation based on Speech Recognition

When a TV program to be stored in a profile has an object o, of which $SO(o)$ exceeds a threshold, the subcategory that corresponds to o is generated.

Subcategory Generation from Common Attributes

It is supposed that TV programs that a user is interested in have some common features such as favorite performers and players. Such common features are expected to be evidence for user's selecting TV programs. Based on this idea, the proposed method extracts an attribute as a subcategory, if the number of TV programs that relate with the attribute exceeds a threshold in recent N-day log of target user. It is also noted all programs in S_p are considered in this step. Figure 7 shows the outline of the subcategory generation process, in which one week log is used for finding common attributes. For example, when generating a temporal profile for Nov. 1, the logs from Oct. 26 to Nov. 1 are used for finding common attributes.

After subcategories are generated, watched TV programs are assigned to the corresponding subcategories in step 4. It is noted that positive profile

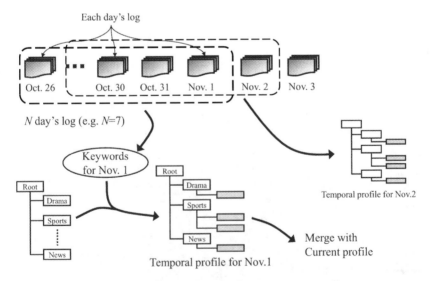

Fig. 7 Outline of subcategory generation from common attributes

contains only TV programs of which the $Score(p)$ exceeds a threshold, while negative profile stores all programs in S_n.

Update of User Profile

By performing Step 1–4, a temporal profile is generated periodically. In our current system, it is generated for each day by collecting the logs of the corresponding day. After a temporal profile is generated, a user profile is updated by merging a temporal profile into current user profile in step 5. Newly-appeared TV programs, categories, and subcategories are merged into current profile, while some existing TV programs are removed. The rating of the stored TV program is updated according to (7), where τ is half period and Δd is elapsed day after being stored in the profile. The program is removed from the profile if its score gets lower than a threshold. The category/subcategory that contains no TV program in it is also removed.

$$Score(p) \leftarrow Score(p) \times \exp(-\frac{\log 2}{\tau}\Delta d). \tag{7}$$

5 Experimental Results

5.1 Outline of Experiments

This section reports the result of experiments that are performed in order to show the performance of proposed methods. The purposes of the experiments are as follows.

- To show the improvement of accuracy of rating estimation by introducing utterance analysis.
- To show user profile that reflects user preference can be generated.

In order to evaluate rating estimation of TV programs, experiments with test subjects are performed, in which utterances issued by each subject during watching TV programs are recorded and analyzed. Because it is difficult even for the state-of-the-art speech recognition modules to recognize unrestricted utterances, we employed Wizard of Oz approach, i.e., human operator was behind an experimental system and responded to test subjects during experiments. The recorded utterances are manually translated into text, to which text processing including sentiment analysis is applied.

As for the latter purpose, long-term experiment, at least more than a week, is required for collecting enough number of utterances, because the proposed profile generation method includes the subcategory generation based on common attributes as noted in Sect. 4.5. Unfortunately, it is quite difficult with our current experimental environments to let us record utterances of test subjects during watching TV for long period. Therefore, we employed diary-based approach, in which test subjects are asked to write down their watched

TV programs along with the watching time, rating, and comments. Instead
of utterances, sentiment analysis is applied to the comments.

5.2   Evaluation of Rating Estimation with Sentiment Analysis

Each of seven test subjects is asked to watch TV for 1 hour alone, during
which 3 TV programs are broadcasted in parallel, and they can freely switch
TV channels to select one of those programs, of which genre is variety shows.
After watching TV, they are asked to rate TV programs they watched with
11 scale (−5 to +5).

We supposed that watching TV alone is not suitable for recording utterances, because we tend to express our opinion in conversation with others.
Therefore, mascot robot[1] is employed in the experiment, which plays a role
to induce test subjects to express their opinion about a TV program while
watching it. That is, when a test subject says nothing for a long time, the
robot attracts his / her attention by its motion and asks a question about
the impression on the TV program being watched. In another case, when a
subject simply said "interesting", the robot ask a question to identify the
object s/he felt interesting. In order to avoid interfering subject's viewing,
the robot ask a question not with voice but with display.

As noted in the beginning of this section, it is difficult to recognize
users' utterances and make above-mentioned question. Therefore, we employ

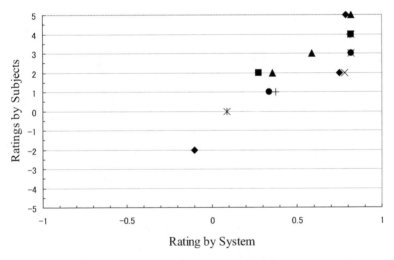

Fig. 8 Comparison of rating by system ($Score(p)$) with rating by test subjects

---

[1] We used nuvo (http://nuvo.jp/nuvo_home_e.html) as the mascot robot.

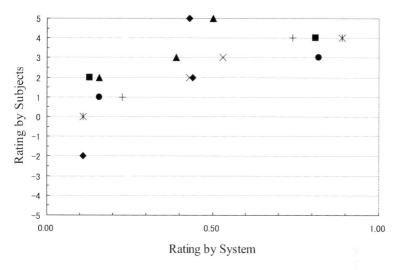

Fig. 9 Comparison of rating by system ($Score_W(p)$) with rating by test subjects

Wizard of Oz approach, in which an operator controls the robot for asking a question about test subjects' impression on the TV programs being watched.

Figure 8 shows the correlation between ratings of TV programs estimated by the proposed method ($Score(p)$) and those by test subjects. Horizontal axis corresponds to ratings by the proposed method (of which range is $[-1, 1]$), and vertical axis corresponds to those by test subjects (of which range is $[-5, 5]$). In Fig. 8, different mark is used for each test subject. The correlation coefficient between $Score(p)$ and subjects' ratings is 0.88.

In order to show the improvement of accuracy by introducing utterance analysis, the correlation between $Score_W(p)$ and ratings by test subjects is shown in Fig. 9. Horizontal axis corresponds to $Score_W(p)$, which is employed by most of existing recommendation systems for estimating user preference. It is noted that the range of $Score_W(p)$ is $[0, 1]$. The correlation coefficient between ratings by test subjects and $Score_W(p)$ is 0.71. Therefore, the results show the proposed method can estimate test subjects' rating on TV programs correctly. In particular, the introduction of sentiment analysis contributes to the improvement of estimation accuracy.

5.3 Evaluation of Profile Generation

Due to the limitation on our current experimental environments as noted in Sect. 5.1, experiments are performed based on the data collected with diary-based approach. We asked 20 test subjects to note TV programs they watched from December 25, 2006 to January 7, 2007. Each subject noted watched TV programs along with the watching time, rating of the program

Table 1 Summary of Profile Generation

Subject	# of programs	# of subcategories / programs (Positive)	# of subcategories / programs (Negative)
A	97	31 / 19	4 / 3
B	65	20 / 15	0 / 0
C	51	21 / 20	0 / 0
D	51	52 / 32	0 / 0
E	49	19 / 17	2 / 2
F	47	29 / 17	0 / 0
G	39	61 / 24	0 / 0
H	31	18 / 9	3 / 1
I	31	5 / 4	1 / 1
Total	461	256 / 157	10 / 7

(11 scale: −5 to +5), and comments on it. Comments are used to extract sentiment expressions, instead of applying speech recognition to utterances.

Table 1 shows summary of the generated profiles: the number of watched TV programs, the numbers of subcategories and programs in positive and negative profiles. It is noted that all watched programs are not contained in profiles, because positive profile contains only the programs of which the $Score(p)$ exceeds a threshold (0.3 in the experiment) as noted in Sect. 4.5. It can be seen from the table that a number of subcategories are generated in positive profiles compared with the number of stored programs. Compared with positive profiles, the number of subcategories / programs in negative profiles is small. It is supposed that test subjects would stop watching a TV program as soon as they found it uninteresting, and they tended to skip writing down TV programs watched only for a short time.

In order to evaluate generated subcategories, 10 subcategories are selected from each positive profile (except subject I) , and evaluated by the corresponding subject. The subjects judge whether or not a subcategory reflects their interests during watching TV. All subcategories in positive profile for subject I and in negative profiles for all subjects are evaluated in the same way.

Table 2 Evaluation of Generated Subcategories

	Positive profile	Negative profile
Reflects interests	61 (0.72)	0 (0.00)
Not reflect interests	24 (0.28)	10 (1.00)
Total	85 (1.00)	10 (1.00)

Table 2 summarizes the evaluation result, in which the subcategories in positive and negative profiles are classified into a set of subcategories that reflect subjects' interests, and those not reflecting their interests.

In Table 2, 72% of subcategories in positive profile are evaluated as reflecting subjects' interests, whereas negative profile contains only subcategories they are not interested in. This result shows that the proposed method can generate user profiles that correctly reflect subjects' interests.

6 Discussion

6.1 Future Challenges

This chapter proposes a method for generating user profile as a basis for TV program recommendation. One of the contributions of the method is introduction of utterance analysis for improving estimation accuracy of users' rating on TV programs.

As for the method of utterance analysis, sentiment analysis is employed. As this analysis is relatively simple, readers would think that applying deeper analysis in the field of natural language processing could further improve the estimation accuracy.

However, we think simple utterance analysis such as used in the proposed method is enough because of the following reasons.

- Capability of speech recognition modules is limited.
- Target utterances are usually short, and often incomplete as a sentence.

In order to apply the method to actual applications, speech recognition module is required for analyzing utterances. As already noted in Sect. 5.1, even the state-of-the-art speech recognition modules cannot recognize various kinds of utterances issued without any restriction. That is, although speech dictation with large vocabulary has been studied, currently typical speech recognition modules that achieve enough recognition accuracy employ word recognition, which can recognize only words in recognition dictionary.

Furthermore, it is supposed that utterances referring to the contents of TV programs, such as players, performers, and places, are usually short and sometime incomplete as a sentence. As a result of word recognition from such utterances, only simple sentences are obtained, to which it is difficult to apply deep natural language processing.

As for other kinds of utterance analysis, it would be also premising to analyze prosody information, which is expected to have similar characteristic to utterance frequency employed in this chapter.

The limitation of the proposed method is that it does not consider genre dependency of users' utterances during watching TV. Currently, proposed method supposes a user says something frequently when s/he is paying attention to a TV program. It is valid to some extent when the genre of TV program is sports, comedy, variety shows, etc, although most of utterances are meaningless such as shouts or laughing voice. On the other hand, a user would say nothing when s/he is enjoying dramas or movies other than comedy.

Another example of genre dependency is that negative utterances to objects appeared in TV programs may have different meanings according to genre. For example, if a user says "It's terrible" while watching some news, the utterance does not mean s/he would not watch news related with the topic. As another example, it would often happen that a user is complaining about judgment calls while watching baseball or football games. Although such utterances might contain sentiment expressions with negative feelings, it does not means s/he would not want to watch baseball games of the same umpire. Therefore, a set of sentiment expressions that are to be extracted for rating estimation should be prepared for each genre. Further study on users' watching behaviors in terms of genre dependency is required.

6.2 Promising Applications

Although the proposed method is useful as a basis for personalized TV guides [3, 6, 21, 22, 23], other applications is also promising. One of the promising applications is to introduce the method into a partner robot in order to improve its communication capability with human partners.

A partner robot is defined as a robot that is able to adapt to human partners, or focuses on interaction with partners. According to a report by working group on robot policy of Ministry of Economy, Trade and Industry [9], market of robots will be expanding within a few decades, because of the growing market of robots for non-industrial use. Robots for non-industrial use cover wide areas, from home robots to robots used in public area, entertainment as well. Home robots are designed for supporting human's activities in his/her home, which include cleaning robots, home security robots, nursing-care robots, and robot information appliance [8]. Robots used in public area include security / cleaning robots in a building, medical robots, and rescue robots. Entertainment robots such as pet robots are also expected to be one of the promising robot applications, some of which will be also used for therapeutic purpose [14]. Partner robots are also for non-industrial use.

Although a partner robot is generally regarded as the achievement of robotics, intelligent information processing such as information recommendation will also contribute to the development of partner robots [15].

In order for partner robots to coexist with humans as a partner, communication between human and a robot when it is not doing a task is important. If communication occurs only when we want to order a robot to do something, the robot is not a partner but one of the tools or home appliances. Therefore, studying communication between humans and robots is very important for realizing a partner robot.

We think personalization technology such as information recommendation and relevance feedback, which are currently hot topic of Web intelligence, is one of the key technologies for realizing a partner robot. Among them, giving a partner robot a capability of TV program recommendation is one of the most promising approaches. As noted in Sect. 3.2, it is said that TV is bound up with the ordinary, natural rhythms of our daily life [19]. If a robot can estimate our interests / preference by observing our watching behavior and have conversation about the topics we might be interested in, which could leave friendly impression on us. Therefore, TV could be a base for establishing friendly relationship between humans and partner robots.

Based on this idea, we are developing a prototype system with using the proposed method, where a mascot robot communicates with human partners while watching TV [16, 17, 18]. The prototype system records user's behavior during watching TV, such as selecting channels and utterances, based on which the mascot robot has a conversation with a user by providing information (including TV programs) that s/he might be interested in. As the purpose of the system is not to simply recommend TV programs, but to communicate with a user, it is also important to decide appropriate timing for recommendation based on user's situation. The development of the system is ongoing project, and we think it is one of the promising future directions of information recommendation.

7 Conclusions

This chapter introduced a method for mining user preference model from user's behavior including utterances. Main target of the chapter is TV program recommendation, and profile generation method is proposed for the purpose.

First, Information recommendation technology was surveyed in Sect. 2, which includes type of recommendation, basic recommendation algorithms, and user's behavior to be analyzed for information recommendation. Studies on TV program recommendation were surveyed in Sect. 3, which includes a method / system for TV program recommendation, as well as analysis of viewing behaviors.

The main contribution of this chapter is to introduce utterance analysis into user profiling. The detail of the profile generation algorithm was described in Sect. 4, in which the method for utterance analysis was proposed. After the experimental results were shown in Sect. 5, future challenges and promising applications of the proposed method were discussed in Sect. 6.

References

1. Canny, J.: Collaborative Filtering with Privacy via Factor Analysis. In: SIGIR 2002, pp. 238–245 (2002)
2. Daita, N., Pires, J.M., Cardoso, M., Pita, H.: Temporal Patterns of TV Watching for Portuguese Viewers. In: 2005 Portuguese Conference on Artificial Intelligence, pp. 151–158 (2005)

3. Fernandez, Y.B., Arias, J.J.P., Nores, M.L., Solla, A.G., Cabrer, M.R.: AVATAR: An Improved Solution for Personalized TV based on Semantic Inference. IEEE Transactions on Consumer Electronics 52(1), 223–231 (2006)

4. Velayathan, G., Yamada, S.: Behavior-Based Web Page Evaluation. In: Proc. 2006 IEEE/WIC/ACM International Conference on Web Intelligence and Intelligent Agent Technology, pp. 409–412 (2006)

5. Inui, T., Okumura, M.: A Survey of Sentiment Analysis. Journal of Natural Language Processing 13(3), 201–241 (2006) (in Japanese)

6. Isobe, T., Fujiwara, M., Kaneta, H., Uratani, N., Morita, T.: Development and Features of a TV Navigation System. IEEE Trans. on Consumer Electronics 49(4), 1035–1042 (2003)

7. Joachims, T., Granka, L., Pan, B., Hembrooke, H., Gay, G.: Accurately Interpreting Clickthrough Data as Implicit Feedback. In: Proc. the 28th annual international ACM SIGIR conference on Research and development in information retrieval, pp. 154–161 (2005)

8. Matsuhira, N., Ogawa, H.: Trends in Development of Home Robots Leading Advanced Technologies. Toshiba Review 59(9) (2004) (in Japanese)

9. Ministry of Economy, Trade and Industry, Report of Working group on robot policy (2006) (in Japanese),
http://www.meti.go.jp/press/20060516002/robot-houkokusho-set.pdf

10. Passin, T.B.: Explorer's Guide to the Semantic Web. Manning Publications Co. (2004)

11. Polat, H., Du, W.: Privacy-preserving Collaborative Filtering Using Randomized Perturbation Techniques. In: Proc. ICDM 2003, pp. 625–628 (2003)

12. Resnick, P., Iacovou, N., Suchak, M., Bergstrom, P., Riedl, J.: GroupLens: An Open Architecture for Collaborative Filtering of Netnews. In: Proc. Conf. on Computer Supported Co-operative Work, pp. 175–186 (1994)

13. Sarwar, B., Karypis, G., Konstan, J., Riedl, J.: Item-Based Collaborative Filtering Recommendation Algorithms. In: Proc. 10th Int'l Conf. on World Wide Web, pp. 285–295 (2001)

14. Shibata, T.: An Overview of Human Interactive Robot for Psychological Enrichment. Proc. of the IEEE 92(11), 1749–1758 (2004)

15. Takama, Y.: Introduction of Humatronics – Towards Integration of Web Intelligence and Robotics. In: Proc. CISIM 2007, pp. 37–44 (2007)

16. Takama, Y., Iwase, Y., Namba, H., Yamaguchi, T.: Information Recommendation Module for Human Robot Communication Support under TV Watching Situation. In: Proc. 4th International Conference on Ubiquitous Robots and Ambient Intelligence (URAI 2007), pp. 395–400 (2007)

17. Takama, Y., Muto, Y.: Profile Generation from TV Watching Behavior Using Sentiment Analysis. In: Proc. IWI2007 (WI-IAT 2007), pp. 191–194 (2007)

18. Takama, Y., Namba, H., Iwase, Y., Muto, Y.: Application of TV Program Recommendation to Communication Support Between Human and Partner Robot. In: Proc. WCCI 2008, pp. 2315–2322 (2008)

19. Taylor, A., Harper, R.: Switching On to Switch Off: a Analysis of Routine TV Watching Habits and Their Implications for Electronic Programme Guide Design, usableiTV, vol. 1, pp. 7–13 (2002)

20. Witten, I.H., Frank, E.: Data Mining: Practical Machine Learning Tools and Techniques, 2nd edn. Morgan Kaufmann, San Francisco (2005)

21. Zhang, H., Zheng, S., Yuan, J.: A Personalized TV Guide System Compliant with MHP. IEEE Transactions on Consumer Electronics 51(2), 731–737 (2005)
22. Zhang, H., Zheng, S.: Personalized TV Program Recommendation based on TV-Anytime Metadata. In: Proc. of the Ninth International Symposium on Consumer Electronics (ISCE 2005), pp. 242–246 (2005)
23. Zhiwen, Y., Zhou, X.: TV3P: An Adaptive Assistant for Personalized TV. IEEE Transactions on Consumer Electronics 50(1), 393–399 (2004)

Part II
Text and Rule Mining

Text Summarization: An Old Challenge and New Approaches

Josef Steinberger and Karel Ježek

Summary. One of the most relevant todays problems called information overloading has increased the necessity of more sophisticated and powerful information compression methods - summarizers. This chapter firstly introduces a taxonomy of summarization methods, an overview of their principles from classical ones, over corpus based, to knowledge rich approaches. We consider various aspects which can affect their categorization. A special attention is devoted to application of recent information reduction methods, based on algebraic transformations. Our own LSA (Latent Semantic Analysis) based approach is included too. The next part is devoted to evaluation measures for assessing quality of a summary. The taxonomy of evaluation measures is presented and their features are discussed. Further, we introduce experiences with the development of our web searching and summarization system. Finally, some new ideas and a conception for the future of this field are mentioned.

1 Introduction

The enormous increase and easy availability of information on the World Wide Web has recently resulted in brushing off the traditional linguistics problem - the condensation of information from text documents. This task is essentially a data reduction process. It has been manually applied since time immemorial and firstly computerized in the late 1950's . The resulting summary must inform by selection and/or by generalization on important content and conclusions in the original text. Recent scientific knowledge and more efficient computers form a new challenge giving the opportunity to resolve *the information overload problem* or at least to decrease its negative impact.

Josef Steinberger and Karel Ježek
University of West Bohemia in Pilsen, Univerzitní 8, Pilsen 306 14, Czech Republic
e-mail: {jstein,jezek_ka}@kiv.zcu.cz

A. Abraham et al. (Eds.): Foundations of Comput. Intel. Vol. 6, SCI 206, pp. 127–149.
springerlink.com   © Springer-Verlag Berlin Heidelberg 2009

There are plenty of definitions of what text summarization actually means. Apart from that mentioned a few lines above, they include:

- 'A brief but accurate representation of the contents of a document',
- 'A distilling of the most important information from a source to produce an abridged version for a particular user/users and task/tasks'.

The quantitative features which can characterize the summary include:

- Semantic informativeness (can be viewed as a measure of the ability to reconstruct from the summary the original text),
- Coherence (expresses the way the parts of the summary together create an integrated sequence),
- Compression ratio.

The history of automatic i.e. computerized summarization began 50 years ago. The oldest publication describing the implementation of an automatic summarizer is often cited as [1]. Luhn's method uses term frequencies to appraise the eligibility of sentences for the summary. It should be mentioned (nowadays it seems strange) that Luhn's motivation was information overload, as well.

The next remarkable step was taken ten years later [2]. Edmundson's work introduced a hypothesis concerning several heuristics (e.g. positional heuristic - a high informational value of sentences at the beginning of an article).

The following years brought further results, but the renaissance in this field and remarkable progress came in the 1990's. We should note [3] or [4]. It was the time of broader use of artificial intelligence methods in this area and the combining of various methods in hybrid systems. The new millennium, due to WWW expansion, shifted the interest of researchers to the summarization of groups of documents, multimedia documents and the application of a new algebraic method for data reduction.

Another related and very ambitious task is the evaluation of the summary's quality. Serious questions remain concerning appropriate methods and types of evaluation. There are a variety of possible bases for the comparison of summarization systems' performance. We can compare a system summary to the source text, to a human-generated summary or to another system summary. Summarization evaluation methods can be broadly classified into two categories [5]. In *extrinsic* evaluation, the summary quality is judged on the basis of how helpful summaries are for a given task, and in *intrinsic* evaluation, it is directly based on an analysis of the summary. The latter can involve a comparison with the source document, measuring how many main ideas of the source document are covered by the summary or a content comparison with an *abstract* written by a human. The problem of matching the system summary to an "ideal summary" is that the ideal summary is hard to establish. The human summary may be supplied by the author of the article, by a judge asked to construct an abstract, or by a judge asked to extract sentences. There can be a large number of abstracts that can summarize a given document. The intrinsic evaluations can then be broadly divided into *content evaluation* and *text quality evaluation*. Whereas content evaluations measure the ability to identify the key topics, text quality evaluations judge the readability, grammar and coherence of automatic summaries.

Searching the web has played an important role in human life in the past couple of years. A user either searches for specific information or just browses topics which interest him/her. Typically, a user enters a query in natural language, or as a set of keywords, and a search engine answers with a set of documents which are relevant to the query. Then, the user needs to go through the documents to find the information that interests him. However, usually just some parts of the documents contain query-relevant information. A benefit to the user would be if the system selected the relevant passages, put them together, made it concise and fluent, and returned the resulting text. Moreover, if the resulting summary is not relevant enough, the user can refine the query. Thus, as a side effect, summarization can be viewed as a technique for improving querying. Our aim is to apply the following step after retrieval of the relevant documents. The set of documents is summarized and the resulting text is returned to the user. So, basically, the key work is done by the summarizer. We briefly describe our SWEeT system (Summarizer of WEb Topics [6]).

This chapter is organized in the following way. The next section describes the basic notions and typology of summarizers. Section 3 is devoted to a short overview of traditional methods. The fourth section is on new approaches with impact on algebraic reduction methods, including our own LSA-based approach. In Section 5, we discuss summarization evaluation methods. The sixth section briefly describes our online searching and summarization system – SWEeT, together with evaluation results and an example summary. The last section concludes the paper, and the further direction of the summarization field is mentioned.

2   Taxonomy of Summarizing Methods

There are several, often orthogonal views which can be used to characterize summarizers. A list and description of the most often cited follows. Comparing the form of summary, we recognize:

- Extracts, summaries completely consisting of word sequences copied from the original document. Phrases, sentences or paragraphs can be used as word sequences. As expected, extracts suffer from inconsistencies, lack of balance, and lack of cohesion. Sentences may be extracted out of context, anaphoric reference may be broken.
- Abstracts, summaries containing word sequences not present in the original. So far, it has been too hard a task for computer research to resolve successfully.

The view coming from the level of processing distinguishes:

- Surface-level approaches, in which case information is represented in notions of shallow features and their combination. Shallow features include e.g. statistically salient terms, positionally salient terms, terms from cue phrases, domain-specific or user's query terms. Results have the form of extracts.
- Deeper-level approaches may produce extracts or abstracts. The latter case uses synthesis involving natural language generation. They need some semantic analysis e.g. they can use entity approaches and build a representation of text entities (text units) and their relationships to determine salient parts. Relationships of entities include thesaural relations, syntactic relations, meaning relations and others. They can also use discourse approaches and model the text structure on the basis of e.g. hypertext markup or rhetorical structure.

Another typology comes from the purpose the summary serves:

- Indicative summaries give abbreviated information on the main topics of a document. They should preserve its most important passages and are often used as the end part of IR systems, being returned by the search system instead of the full document. Their aim should be to help a user to decide whether the original document is worth reading. The typical lengths of indicative summaries range between 5 to 10% of the complete text.
- Informative summaries provide a substitute ('surrogate', 'digest') of the full document, retaining important details, while reducing information volume. An informative summary is typically 20-30 % of the original text.
- Critical or Evaluative summaries capture the point of view of the summary's author on a given subject. Reviews are a typical example, but they are rather outside the scope of today's automatic summarizers. It should be noted, that all three mentioned groups are not mutually exclusive and there are common summaries serving both an indicative and informative function. It is quite usual to hold informative summarizers as a subset of indicative ones.

When distinguished by the audience, we can recognize:

- Generic summaries, when the result is aimed at a broad community of readers, all major topics are equally important.
- Query-based summaries, when the result is based on a question e.g. 'what are the causes of high inflation?'
- User focused or Topic focused summaries, which are tailored to the interest of the particular user or emphasize only particular topics.

There are some other views we can use for the taxonomy of summarizers e.g.: Span of processed text:

- Single document or multi-document summarization.

Language:

- Monolingual versus multilingual.

Genre:

- Scientific article or report or news.

3   Overview of Methods Based on Traditional Principles

3.1   Pioneering Works

The first approaches of automatic text summarization used only simple (surface level) indicators to decide what parts of a text to include in the summary. The oldest sentence extraction algorithm was developed in 1958 [1]. It used frequencies of terms as the sentence relevance criterion. The basic idea was that a writer will repeat certain words when writing about a given topic. The importance of terms is considered proportional to their frequency in the summarized documents. The frequencies are used in the next step to score and select sentences for the extract. Other indicators of relevance used in [7] are the position of a sentence within the document and the presence of certain cue-words (i.e., words like "important" or "relevant") or words contained in the title. The combination of cue-words, title words and the position of a sentence was used in [2] to produce extracts and their similarity with human written abstracts was demonstrated.

3.2   Statistical Methods

In [4] it was proved that the relevance of document terms is inversely proportional to the number of documents in the corpus containing the term. The formula for term relevance evaluation is given by $tf_i \times idf_i$, where tf_i is the frequency of term i in the document and idf_i is the inverted frequency of documents containing this term. Sentences can be subsequently scored for instance by summing the relevance of terms in the sentence.

The implementation of a more ingenious statistical method was described in [3]. It uses a Bayesian classifier to compute the probability that a sentence in a source document should be included in a summary. To train the classifier, the authors used a corpus of 188 pairs of full documents/summaries. The characteristic features used in the Bayesian formula include, apart from word frequency, also uppercase words, sentence length, phrase structure, and in-paragraph position.

An alternative way of measuring term relevance was proposed in [8]. Instead of rough term counting, the authors used concept relevance, which can be determined using WordNet. E.g. the occurrence of the concept 'car' is counted when the word 'auto' is found, as well as when, for instance, 'auto-car', 'tires', or 'brake' are found.

3.3   Methods Based on Text Connectivity

Anaphoric expressions[1] that refer to previously mentioned parts of the text
need to know their antecedents in order to be understood. Extractive methods can fail to capture the relations between concepts in a text. If a sentence
containing an anaphoric link is extracted without the previous context, the
summary can become difficult to understand. Cohesive properties comprise
relations between expressions of the text. They have been explored by different summarization approaches.

Let us mention a method called Lexical chains, which was introduced in [9].
It uses the WordNet thesaurus to determine cohesive relations between terms
(i.e., repetition, synonymy, antonymy, hypernymy, and holonymy) and composes the chains by related terms. Their scores are determined on the basis
of the number and type of relations in the chain. Only those sentences where
the strongest chains are highly concentrated are selected for the summary.
A similar method where sentences are scored according to the objects they
mention was presented in [10]. The objects are identified by a co-reference resolution system. Co-reference resolution is the process of determining whether
two expressions in natural language refer to the same entity. The sentences
where the occurrence of frequently mentioned objects exceeds the given limit
are included in the summary.

In the group of methods based on text connectivity we can include the
methods utilizing Rhetorical Structure Theory (RST). RST is a theory about
text organization. It consists of a number of rhetorical relations that connect
text units. The relations tie together a nucleus - which is central to the writer's
goal, and a satellite - less central or marginal parts. From relations, a tree-like
representation is composed which is used for extraction of the text unit into
the summary. In [11] sentences are penalized according to their rhetorical role
in the tree. A weight of 1 is given to satellite units and a weight of 0 is given
to nuclei units. The final score of a sentence is given by the sum of weights
from the root of the tree to the sentence. In [12], each parent node identifies
its nuclear children as salient. The children are promoted to the parent level.
The process is recursive down the tree. The score of a unit is given by the
level it obtained after promotion.

3.4   Iterative Graph Methods

Iterative graph algorithms, such as HITS [13] or Google's PageRank [14], were
originally developed as exploring tools of the link-structure to rank Web pages.
Later on, they were successfully used in other areas e.g. citation analysis, social
networks, etc. In graph ranking algorithms, the importance of a vertex within
the graph is iteratively computed from the entire graph. In [15] the graph-based

---

[1] Anaphoric expression is a word or phrase which refers back to some previously
    expressed word or phrase or meaning (typically, pronouns such as herself, himself,
    he, she).

model was applied to natural language processing, resulting in an algorithm named TextRank. The same graph-based ranking principles were applied in summarization [16]. A graph is constructed by adding a vertex for each sentence in the text. Edges between vertices are established using sentence interconnections. These connections are defined using a similarity relation, where similarity is measured as a function of content overlap. The overlap of two sentences can be determined as the number of common tokens between lexical representations of two sentences. The iterative part of the algorithm is consequently applied to the graph of sentences. When its processing is finished, vertices (sentences) are sorted by their scores. The top-ranked sentences are included in the result.

3.5   Coming Close to Human Abstracts

There is a qualitative difference between the summaries produced by current automatic summarizers and abstracts written by human abstractors. Computer systems can identify the important topics of an article with only limited accuracy. Another factor is that most summarizers rely on extracting key sentences or paragraphs. However, if the extracted sentences are disconnected in the original article and they are strung together in the extract, the result can be incoherent and sometimes even misleading.

Lately, some non-sentence-extractive summarization methods have started to appear. Instead of reproducing full sentences from the summarized text, these methods either compress the sentences [17, 18, 19, 20], or re-generate new sentences from scratch [21]. In [22] a Cut-and-paste strategy was proposed. The authors have identified six editing operations in human abstracting: (i) sentence reduction; (ii) sentence combination; (iii) syntactic transformation; (iv) lexical paraphrasing; (v) generalization and specification; and (vi) reordering. Summaries produced this way resemble the human summarization process more than extraction does. However, if large quantities of text need to be summarized, sentence extraction is a more efficient method. Extraction is robust towards all irregularities of input text. It is failure-proof and less language dependent.

4   New Approaches Based on Algebraic Reduction

Several approaches based on algebraic reduction methods have appeared in the last couple of years. The most widely used is latent semantic analysis (LSA) [23], however other methods, like non-negative matrix factorization (NMF) [24] and semi-discrete matrix decomposition (SDD) [25] look promising as well.

4.1   LSA in Summarization Background

LSA is a fully automatic algebraic-statistical technique for extracting and representing the contextual usage of words' meanings in passages of discourse.

The basic idea is that the aggregate of all the word contexts in which a given word does or does not appear provides mutual constraints that determine the similarity of meanings of words and sets of words to each other. LSA has been used in a variety of applications (e.g., information retrieval, document categorization, information filtering, and text summarization).

The heart of the analysis in the summarization background is a document representation developed in two steps. The first step is the creation of a term by sentences matrix $A = [A_1, A_2, \ldots, A_n]$, where each column A_i represents the weighted term-frequency vector of sentence i in the document under consideration. If there are m terms and n sentences in the document, then we will obtain an $m \times n$ matrix A. The next step is to apply Singular Value Decomposition (SVD) to matrix A. The SVD of an $m \times n$ matrix A is defined as:

$$A = U \Sigma V^T \tag{1}$$

where $U = [u_{ij}]$ is an $m \times n$ column-orthonormal matrix whose columns are called left singular vectors.

$\Sigma = diag(\sigma_1, \sigma_2, \ldots, \sigma_n)$ is an $n \times n$ diagonal matrix, whose diagonal elements are non-negative singular values sorted in descending order. $V = [v_{ij}]$ is an $n \times n$ orthonormal matrix, whose columns are called right singular vectors. The dimensionality of the matrices is reduced to r most important dimensions and thus, U' is $m \times r$, Σ' is $r \times r$ and V'^T is $r \times n$ matrix[2].

From a mathematical point of view, SVD derives a mapping between the m-dimensional space specified by the weighted term-frequency vectors and the r-dimensional singular vector space. From an NLP perspective, what SVD does is to derive the latent semantic structure of the document represented by matrix A, i.e. a breakdown of the original document into r linearly-independent base vectors which express the main 'topics' of the document. SVD can capture interrelationships among terms, so that terms and sentences can be clustered on a 'semantic' basis rather than on the basis of words only. Furthermore, as demonstrated in [26], if a word combination pattern is salient and recurring in a document, this pattern will be captured and represented by one of the left singular vectors. The magnitude of the corresponding singular value indicates the importance degree of this pattern within the document. Any sentences containing this word combination pattern will be projected along this singular vector, and the sentence that best represents this pattern will have the largest value with this vector. Assuming that each particular word combination pattern describes a certain topic in the document, each left singular vector can be viewed as representing such a topic [27], the magnitude of its singular value representing the importance degree of this topic[3].

---

[2] U', resp. Σ', V'^T, denotes matrix U, resp. Σ, V^T, reduced to r dimensions.

[3] In [27] it was shown that the dependency of the significance of each particular topic on the magnitude of its corresponding singular value is quadratic.

4.2   LSA-Based Single-Document Approaches

The summarization method proposed by Gong and Liu in [28] uses the representation of a document thus obtained to choose the sentences to go in the summary on the basis of the relative importance of the 'topics' they mention, described by the matrix V^T. The summarization algorithm simply chooses for each 'topic' the most important sentence for that topic: i.e., the kth sentence chosen is the one with the largest index value in the kth right singular vector in matrix V^T.

The main drawback of Gong and Liu's method is that when l sentences are extracted the top l topics are treated as equally important. As a result, a summary may include sentences about 'topics' which are not particularly important. In order to fix the problem, we changed the selection criterion to include in the summary sentences whose vectorial representation in the matrix $\Sigma^2 V^T$ has the greatest 'length', instead of the sentences containing the highest index value for each 'topic'. Intuitively, the idea is to choose the sentences with the greatest combined weight across all important topics, possibly including more than one sentence about an important topic, rather than one sentence for each topic. More formally: after computing the SVD of a term by sentences matrix, we compute the length of each sentence vector in $\Sigma^2 V^T$, which represents its summarization score as well (for details see [29]).

In [30] an LSA-based summarization of meeting recordings was presented. The authors followed the Gong and Liu approach, but rather than extracting the best sentence for each topic, n best sentences were extracted, with n determined by the corresponding singular values from matrix Σ. The number of sentences in the summary that will come from the first topic is determined by the ratio of the largest singular value and the sum of all singular values, and so on for each topic. Thus, dimensionality reduction is no longer tied to summary length and more than one sentence per topic can be chosen.

Another summarization method that uses LSA was proposed in [31]. It is a mixture of graph-based and LSA-based approaches. After performing SVD on the word-by-sentence matrix and reducing the dimensionality of the latent space, they reconstruct the corresponding matrix $A' = U'\Sigma'V'^T$. Each column of A' denotes the semantic sentence representation. These sentence representations are then used, instead of a keyword-based frequency vector, for the creation of a text relationship map to represent the structure of a document. A ranking algorithm is then applied in the resulting map (see section 3.4).

4.3   LSA-Based Multi-document Approaches

In [32] we proposed the extension of the method to process a cluster of documents written about the same topic. Multi-document summarization is a one step more complex task than single-document summarization. It brings new problems we have to deal with. The first step is again to create a term by sentence matrix. In this case, we include in the matrix all sentences from the cluster of documents. (In the case of single-document summarization, we included the sentences from the one document.) Then we run sentence ranking. Each sentence gets a score, which is computed in the same way as when we summarize a single document - vector length in the matrix $\Sigma^2 V^T$. Now we are ready to select the best sentences (the ones with the greatest score) for the summary.

However, two documents written about the same topic/event can contain similar sentences and thus we need to resolve redundancy. We propose the following process: before adding a sentence to the summary, see whether there is a similar sentence already in the summary. The similarity is measured by the cosine similarity in the original term space. We determine a threshold here. The extracted sentence should be close to the user query. To satisfy this, query terms get a higher weight in the input matrix.

Another problem of this approach is that it favours long sentences. This is natural because a longer sentence probably contains more significant terms than a shorter one. We resolve this by dividing the sentence score by $number\_of\_terms^{lk}$, where lk is the length coefficient.

Experiments showed good results with low dimensionality. It is sufficient to use up to 10 dimensions (topics). However, the topics are not equally important. The magnitude of each singular value holds the topic importance. To make it more general, we experimented with different power functions in the computation of the final matrix used for determination of the sentence score: $\Sigma^{power} V^T$.

In [33], an interesting multi-document summarization approach based on LSA and maximal marginal relevance (MMR) was proposed. In the MMR framework [34], a single extraction score is derived by combining measures of relevance and redundancy of candidate sentences. The sentences are represented as weighted term-frequency vectors which can thus be compared to query vectors to gauge similarity and already extracted sentence vectors to gauge redundancy, via the cosine of the vector pairs. While this has proved successful to a degree, the sentences are represented merely according to weighted term frequency in the document, and so two similar sentences stand a chance of not being considered similar if they don't share the same terms. One way to rectify this is to do LSA on the matrix first before proceeding to implement MMR, but this still only exploits term co-occurrence within the documents at hand. In contrast, the system described in [33] attempts to derive more robust representations of sentences by building a large semantic space using LSA on a very large corpus.

5   Summarization Evaluation

The summary quality can be evaluated both manually and automatically. Manual evaluation gives more precise results, however, it is a subjective task

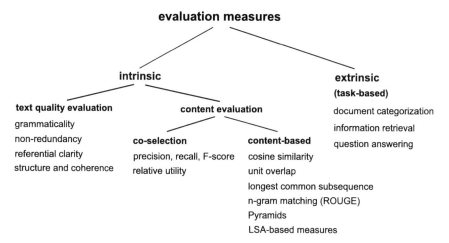

Fig. 1 The taxonomy of summary evaluation measures

and thus needs to be done on a large scale and it is very expensive. On the other hand, automatic evaluation methods give just rough image of summary quality, however, they usually need just some reference documents, like human-written abstracts.

The taxonomy of summary evaluation measures can be found in Figure 1. *Text quality* is often assessed by human annotators. They assign a value from a predefined scale to each summary. The main approach for summary quality determination is the intrinsic *content evaluation* which is often done by comparison with an ideal summary. For *sentence extracts*, it is often measured by *co-selection*. It ascertains how many ideal sentences the automatic summary contains. *Content-based measures* compare the actual words in a sentence, rather than the entire sentence. Their advantage is that they can compare both human and automatic extracts with human abstracts that contain newly written sentences. Another significant group is *extrinsic (task-based)* methods. They measure the performance of using the summaries for a certain task.

5.1 Text Quality Measures

There are several aspects of text (linguistic) quality:

- **Grammaticality** - the text should not contain non-textual items (i.e., markers) or punctuation errors or incorrect words.
- **Non-redundancy** - the text should not contain redundant information.
- **Reference clarity** - the nouns and pronouns should be clearly referred to in the summary. For example, the pronoun *he* has to mean somebody in the context of the summary. The proposed solution can be found in [35].

- **Coherence and structure** - the summary should have good structure and the sentences should be coherent.

This cannot be done automatically. The annotators mostly assign marks (i.e., from A - very good - to E - very poor) to each summary.

5.2   Co-selection Measures

Precision, Recall and F-score

The main evaluation metrics of co-selection are precision, recall and F-score. *Precision* (P) is the number of sentences occurring in both system and ideal summaries divided by the number of sentences in the system summary. *Recall* (R) is the number of sentences occurring in both system and ideal summaries divided by the number of sentences in the ideal summary. *F-score* is a composite measure that combines precision and recall. The basic way to compute the F-score is to count the harmonic average of precision and recall:

$$F = \frac{2 * P * R}{P + R}. \tag{2}$$

Below is a more complex formula for measuring the F-score:

$$F = \frac{(\beta^2 + 1) * P * R}{\beta^2 * P + R}, \tag{3}$$

where β is a weighting factor that favours precision when $\beta > 1$ and favours recall when $\beta < 1$.

Relative Utility

The main problem with P&R is that human judges often disagree on what the top p% most important sentences are in a document. Using P&R creates the possibility that two equally good extracts are judged very differently. Suppose that a manual summary contains sentences [1 2] from a document. Suppose also that two systems, A and B, produce summaries consisting of sentences [1 2] and [1 3], respectively. Using P&R, system A will be ranked much higher than system B. It is quite possible that sentences 2 and 3 are equally important, in which case the two systems should get the same score.

To address the problem with precision and recall, the *relative utility* (RU) measure was introduced [36]. With RU, the model summary represents all sentences of the input document with confidence values for their inclusion in the summary. For example, a document with five sentences [1 2 3 4 5] is represented as [1/5 2/4 3/4 4/1 5/2]. The second number in each pair indicates the degree to which the given sentence should be part of the summary according to a human judge. This number is called the *utility* of the sentence. It depends on the input document, the summary length, and the judge. In the example, the system that selects sentences [1 2] will not get a higher score than a system that chooses sentences [1 3] because both summaries [1 2] and [1 3] carry the same number of utility points (5+4). Given that no other combination of two sentences carries a higher utility, both systems [1 2] and [1 3] produce optimal extracts. To compute relative utility, a number of judges, $(N \geq 1)$ are asked to assign utility scores to all n sentences in a document. The top e sentences according to utility score[4] are then called a sentence extract of size e. We can then define the following system performance metric:

$$RU = \frac{\sum_{j=1}^{n} \delta_j \sum_{i=1}^{N} u_{ij}}{\sum_{j=1}^{n} \epsilon_j \sum_{i=1}^{N} u_{ij}}, \qquad (4)$$

where u_{ij} is a utility score of sentence j from annotator i, ϵ_j is 1 for the top e sentences according to the sum of utility scores from all judges, otherwise its value is 0, and δ_j is equal to 1 for the top e sentences extracted by the system, otherwise its value is 0. For details, see [36].

5.3 Content-Based Measures

Co-selection measures can count in as a match only exactly the same sentences. This ignores the fact that two sentences can contain the same information even if they are written differently. Furthermore, summaries written by two different annotators do not in general share identical sentences. In the following example, it is obvious that both headlines, H_1 and H_2, carry the same meaning and they should somehow count as a match.

H_1: "The visit by the president of the Czech Republic to Slovakia"

H_2: "The Czech president visited Slovakia"

Whereas co-selection measures cannot do this, content-based similarity measures can.

Cosine Similarity

A basic content-based similarity measure is Cosine Similarity [4]:

$$cos(X, Y) = \frac{\sum_i x_i \cdot y_i}{\sqrt{\sum_i (x_i)^2} \cdot \sqrt{\sum_i (y_i)^2}}, \qquad (5)$$

where X and Y are representations of a system summary and its reference document based on the vector space model.

Unit Overlap

Another similarity measure is Unit Overlap [37]:

---

[4] In the case of ties, an arbitrary but consistent mechanism is used to decide which sentences should be included in the summary.

$$overlap(X, Y) = \frac{|X \cap Y|}{|X| + |Y| - |X \cap Y|}, \tag{6}$$

where X and Y are representations based on sets of words or lemmas. $|X|$ is the size of set X.

Longest Common Subsequence

The third content-based measure is called Longest Common Subsequence (*lcs*) [38]:

$$lcs(X, Y) = \frac{length(X) + length(Y) - edit_{di}(X, Y)}{2}, \tag{7}$$

where X and Y are representations based on sequences of words or lemmas, lcs(X,Y) is the length of the longest common subsequence between X and Y, $length(X)$ is the length of the string X, and $edit_{di}(X, Y)$ is the edit distance of X and Y [38].

N-gram Co-occurrence Statistics - ROUGE

In the last edition of DUC conferences, ROUGE (Recall-Oriented Understudy for Gisting Evaluation) was used as an automatic evaluation method. The ROUGE family of measures, which are based on the similarity of n-grams[5], was first introduced in 2003 [39].

Suppose a number of annotators created reference summaries - reference summary set (*RSS*). The ROUGE-*n* score of a candidate summary is computed as follows:

$$\text{ROUGE-}n = \frac{\sum_{C \in RSS} \sum_{gram_n \in C} Count_{match}(gram_n)}{\sum_{C \in RSS} \sum_{gram_n \in C} Count(gram_n)}, \tag{8}$$

where $Count_{match}(gram_n)$ is the maximum number of n-grams co-occurring in a candidate summary and a reference summary and $Count(gram_n)$ is the number of n-grams in the reference summary. Notice that the average n-gram ROUGE score, ROUGE-*n*, is a recall metric. There are other ROUGE scores, such as ROUGE-L - a longest common subsequence measure (see the previous section) - and ROUGE-SU4 - a bigram measure that enables at most 4 unigrams inside of bigram components to be skipped [40].

Pyramids

The Pyramid method is a novel semi-automatic evaluation method [41]. Its basic idea is to identify summarization content units (SCUs) that are used for comparison of information in summaries. SCUs emerge from annotation of a corpus of summaries and are no bigger than a clause. The annotation starts with identifying similar sentences and then proceeds with a more detailed

---

[5] An n-gram is a subsequence of n words from a given text.

inspection that can lead to identifying more tightly related subparts. SCUs that appear in more manual summaries will get greater weights, so a pyramid will be formed after SCU annotation of manual summaries. At the top of the pyramid there are SCUs that appear in most of the summaries and thus have the greatest weight. The lower in the pyramid the SCU appears, the lower its weight is because it is contained in fewer summaries. The SCUs in peer summary are then compared against an existing pyramid to evaluate how much information is agreed between the peer summary and manual summary. However, this promising method still requires some annotation work.

5.4 Task-Based Measures

Task-based evaluation methods do not analyze sentences in the summary. They try to measure the prospect of using summaries for a certain task. Various approaches to task-based summarization evaluation can be found in literature. We mention the three most important tasks - document categorization, information retrieval and question answering.

Document Categorization

The quality of automatic summaries can be measured by their suitability in surrogating full documents for *categorization*. Here the evaluation seeks to determine whether the generic summary is effective in capturing whatever information in the document is needed to correctly categorize the document. A corpus of documents together with the topics they belong to is needed for this task. Results obtained by categorizing summaries are usually compared to ones obtained by categorizing full documents (an upper bound) or random sentence extracts (lower bound). Categorization can be performed either manually [42] or by a machine classifier [43]. If we use automatic categorization, we must keep in mind that the classifier demonstrates some inherent errors. It is therefore necessary to differentiate between the error generated by a classifier and one caused by a summarizer. This is often done only by comparing the system performance with the upper and lower bounds.

In SUMMAC evaluation [42], apart from other tasks, 16 participating summarization systems were compared by a manual categorization task. Given a document, which could be a generic summary or a full text source (the subject was not told which), the human subject chose a single category (from five categories, each of which had an associated topic description) to which the document is relevant, or else chose "none of the above".

Precision and recall of categorization are the main evaluation metrics. *Precision* in this context is the number of correct topics assigned to a document divided by the total number of topics assigned to the document. *Recall* is the number of correct topics assigned to a document divided by the total number of topics that should be assigned to the document. The measures go against each other and therefore a composite measure - the F-score - can be used (see the section 5.2).

Information Retrieval

Information Retrieval (IR) is another task appropriate for the task-based evaluation of summary quality. *Relevance correlation* [38] is an IR-based measure for assessing the relative reduction in retrieval performance when moving from full documents to summaries. If a summary captures the main points of a document, then an IR machine indexed on a set of such summaries (instead of a set of full documents) should produce (almost) as good a result. Moreover, the difference between how well the summaries do and how well the full documents do should serve as a possible measure for the quality of summaries.

Suppose that given query Q and corpus of documents D, a search engine ranks all documents in D according to their relevance to query Q. If instead of corpus D, the corresponding summaries of all documents are substituted for the full documents and the resulting corpus of summaries S is ranked by the same retrieval engine for relevance to the query, a different ranking will be obtained. If the summaries are good surrogates for the full documents, then it can be expected that the ranking will be similar. There exist several methods for measuring the similarity of rankings. One such method is Kendall's tau and another is Spearman's rank correlation [44]. However, since search engines produce relevance scores in addition to rankings, we can use a stronger similarity test - linear correlation.

Relevance correlation (RC) is defined as the linear correlation of the relevance scores assigned by the same IR algorithm in different data sets (for details see [38]).

Question Answering

An extrinsic evaluation of the impact of summarization in a task of *question answering* was carried out in [45]. The authors picked four Graduate Management Admission Test (GMAT) reading comprehension exercises. The exercises were multiple-choice, with a single answer to be selected from answers shown alongside each question. The authors measured how many of the questions the subjects answered correctly under different conditions. Firstly, they were shown the original passages, then an automatically generated summary, furthermore a human abstract created by a professional abstractor instructed to create informative abstracts, and finally, the subjects had to pick the correct answer just from seeing the questions without seeing anything else. The results of answering in the different conditions were then compared.

6 The SWEeT System

In this section, we will describe our experiences with the development of our summarizer and its using in the web searching and summarization system – SWEeT (Summarizer of WEb Topics) [46, 6].

A user enters a query in the system. That query should describe the topic he would like to read about (e.g. "George Bush Iraq War"). The system passes the query to a search engine. It answers with a set of relevant documents sorted by relevance to the query. Top n documents, where n is a parameter of the system, are then passed to our summarizer, the core of the system. The created summary is returned to the user, together with references to the searched documents that can help him to get more details about the topic.

The system has a modular architecture. We can easily change the search engine or the summarizer or any of its modules. Summarization modules, e.g. a sentence compression module, can be easily plugged, unplugged, or changed. And thus the system output will improve with improvements in the modules.

There are two crucial parts that affect the performance of the system: the quality of searching and the quality of summarization. As for searching, we will present figures showing its accuracy (how many retrieved documents were relevant to the user query and how many were not). We use manual annotations. The quality of the summarization is assessed by the widely-used ROUGE measure [39, 40]. At the end of the section, we present an example of an output summary.

6.1 Searching Quality

We use well-known external search engines to guarantee the highest searching quality. The first one is Google, whose performance cannot be doubted. However, we need to search just a single domain and thus we use the modifier "site:domain". For searching in the Czech news site novinky.cz, we directly use their search engine. It is based on the Seznam engine, one of the most widely used engines on the Czech Web. When we evaluated the searching quality we obtained mostly relevant documents. Just a couple of documents were classified as marginally relevant (i.e., the query terms are mentioned there in the right sense, but the main document's topic is different from the query topic). A few documents were irrelevant (e.g., when we submitted a query about a huge accident on Czech highway D1, the system returned a document about an accident on an Austrian highway). Proper names can increase the accuracy of searching. We analyzed a maximum of the top ten retrieved documents for each topic/query.

6.2 Summarization Quality

Assessing the quality of a summary is much more problematic. The DUC[6] series of annual conferences controls the direction of the evaluation.

---

[6] The National Institute of Standards and Technology (NIST) initiated the Document Understanding Conference (DUC) [47] series to evaluate automatic text summarization. Its goal is to further the progress in summarization and enable researchers to participate in large-scale experiments. Since the year 2008 DUC has moved to the Text Analysis Conference (TAC) [48].

Table 1 Multiple comparisons of all systems based on ANOVA of ROUGE-2 recall

Summarizer	ROUGE-2	95% significance groups for ROUGE scores.
15	0.0725	A
17	0.0717	A
10	0.0698	A B
8	0.0696	A B
4	0.0686	A B C
SWEeT	0.0679	A B C
5	0.0675	A B C
11	0.0643	A B C D
14	0.0635	A B C D E
16	0.0633	A B C D E
19	0.0632	A B C D E
7	0.0628	A B C D E F
9	0.0625	A B C D E F
29	0.0609	A B C D E F G
25	0.0609	A B C D E F G
6	0.0609	A B C D E F G
24	0.0597	A B C D E F G
28	0.0594	A B C D E F G
3	0.0594	A B C D E F G
21	0.0573	A B C D E F G
12	0.0563	B C D E F G
18	0.0553	B C D E F G H
26	0.0547	B C D E F G H
27	0.0546	B C D E F G H
32	0.0534	C D E F G H
20	0.0515	D E F G H
13	0.0497	D E F G H
30	0.0496	D E F G H
31	0.0487	E F G H
2	0.0478	F G H
22	0.0462	G H
1	0.0403	H I
23	0.0256	I

However, the only fully automatic and widely used method so far is ROUGE (see Section 5.3). We present a comparison of our summarizer with those that participated at DUC 2005 - Tables 1 and 2. Not all of the differences are statistically significant. Therefore, we show by the letters the multiple systems' comparison - the differences between systems that share the same letter (in the last column) are NOT statistically significant. To summarize these tables: in ROUGE-2, our summarizer performs worse than 5 systems

Table 2 Multiple comparisons of all systems based on ANOVA of ROUGE-SU4 recall

Summarizer	ROUGE-SU4	95% significance groups for ROUGE scores.
15	0.1316	A
17	0.1297	A B
8	0.1279	A B
4	0.1277	A B C
10	0.1253	A B C D
SWEeT	0.1239	A B C D
5	0.1232	A B C D E
11	0.1225	A B C D E
19	0.1218	A B C D E
16	0.1190	A B C D E F
7	0.1190	A B C D E F
6	0.1188	A B C D E F G
25	0.1187	A B C D E F G
14	0.1176	A B C D E F G
9	0.1174	A B C D E F G
24	0.1168	A B C D E F G
3	0.1167	A B C D E F G
28	0.1146	B C D E F G H
29	0.1139	B C D E F G H
21	0.1112	C D E F G H I
12	0.1107	D E F G H I
18	0.1095	D E F G H I J
27	0.1085	E F G H I J
32	0.1041	F G H I J
13	0.1041	F G H I J
26	0.1023	G H I J K
30	0.0995	H I J K
2	0.0981	H I J K
22	0.0970	I J K
31	0.0967	I J K
20	0.0940	J K
1	0.0872	K
23	0.0557	L

and better than 27 systems; however, when we count in significance, none of the systems performs significantly better than ours and 8 of them perform significantly worse. And similarly in ROUGE-SU4, our summarizer performs worse than 5 systems and better than 27 systems; however, when we count in significance, none of the systems performs significantly better than ours and 11 of them perform significantly worse.

6.3 Example Summary

To demonstrate the system output, we show an example summary. Its desired length is 250 words. The query was "Al Qaeda and Osama bin Laden" - see Figure 2.

Even as American officials portrayed the case as mainly a Canadian operation, the arrests so close to the United States border jangled the nerves of intelligence officials who have been warning of the continuing danger posed by small "homegrown" extremist groups, who appeared to operate without any direct control by known leaders of Al Qaeda. These fighters include Afghans and seasoned Taliban leaders, Uzbek and other Central Asian militants, and what intelligence officials estimate to be 80 to 90 Arab terrorist operatives and fugitives, possibly including the Qaeda leaders Osama bin Laden and his second in command, Ayman al-Zawahri. In recent weeks, Pakistani intelligence officials said the number of foreign fighters in the tribal areas was far higher than the official estimate of 500, perhaps as high as 2,000 today. The area is becoming a magnet for an influx of foreign fighters, who not only challenge government authority in the area, but are even wresting control from local tribes and spreading their influence to neighboring areas, according to several American and NATO officials and Pakistani and Afghan intelligence officials. Some American officials and politicians maintain that Sunni insurgents have deep ties with Qaeda networks loyal to Osama bin Laden in other countries. Hussein's government, one senior refinery official confided to American soldiers. In fact, money, far more than jihadist ideology, is a crucial motivation for a majority of Sunni insurgents, according to American officers in some Sunni provinces and other military officials in Iraq who have reviewed detainee surveys and other intelligence on the insurgency.

Fig. 2 Resulting summary for the query: "Al Qaeda and Osama bin Laden"

7 Conclusion

We presented the history and the state of the art in the automatic text summarization research area. We paid most attention to the approaches based on algebraic reduction methods, including our LSA-based method. Their strong property is that they work only with the context of terms and thus they do not depend on a particular language. The evaluation of summarization methods has the same importance as own summarizing. The annual summarization evaluation conference TAC (formerly DUC) sets the direction in evaluation processes. The summarization method we proposed forms the heart of our searching and summarization system, SWEeT. Its searching performance is not problematic and the summarizer is comparable with state-of-the-art systems. A couple of years ago, the summarization field moved from single-document summarization to multi-document summarization. At present, it is again moving forward to update summarization. The task is to create a summary of a set of newer documents where the reader has some knowledge of the topic from a set of older documents. Thus, solving redundancy will be more important.

Acknowledgements. This research was partly supported by National Research Programme II, project 2C06009 (COT-SEWing).

References

1. Luhn, H.P.: The Automatic Creation of Literature Abstracts. IBM Journal of Research Development 2(2), 159–165 (1958)
2. Edmundson, H.P.: New Methods in Automatic Extracting. Journal of the Association for Computing Machinery 16(2), 264–285 (1969)
3. Kupiec, J., Pedersen, J.O., Chen, F.: A Trainable Document Summarizer. In: Proceedings of the 18th annual international ACM SIGIR conference on research and development in information retrieval (1995)
4. Salton, G.: Automatic Text Processing. Addison-Wesley Publishing Company, Reading (1988)
5. Sparck Jones, K., Galliers, J.R.: Evaluating Natural Language Processing Systems. LNCS (LNAI), vol. 1083. Springer, Heidelberg (1996)
6. The SWEeT System (2008), http://tmrg.kiv.zcu.cz:8080/sweet
7. Baxendale, P.B.: Man-made Index for Technical Literature - an experiment. IBM Journal of Research Development 2(4), 354–361 (1958)
8. Hovy, E., Lin, C.Y.: Automated Text Summarization in SUMMARIST. In: Mani, I., Maybury, M.T. (eds.) Advances in Automatic Text Summarization. MIT Press, Cambridge (1999)
9. Barzilay, R., Elhadad, M.: Using Lexical Chains for Text Summarization. In: Proceedings of the ACL/EACL1997 Workshop on Intelligent Scalable Text Summarization, Madrid, Spain (1997)
10. Boguraev, B., Kennedy, C.: Salience-based content characterization of text documents. In: Mani, I., Maybury, M.T. (eds.) Advances in Automatic Text Summarization. MIT Press, Cambridge (1999)
11. Ono, K., Sumita, K., Miike, S.: Abstract Generation Based on Rhetorical Structure Extraction. In: Proceedings of the International Conference on Computational Linguistics, Kyoto, Japan (1994)
12. Marcu, D.: From Discourse Structures to Text Summaries. In: Proceedings of the ACL 1997/EACL 1997 Workshop on Intelligent Scalable Text Summarization, Madrid, Spain (1997)
13. Kleinberg, J.M.: Authoritative sources in a hyperlinked environment. Journal of the ACM 46(5), 604–632 (1999)
14. Brin, S., Page, L.: The anatomy of a large-scale hypertextual Web search engine. Computer Networks and ISDN Systems 30, 1–7 (1998)
15. Mihalcea, R., Tarau, P.: Text-rank - bringing order into texts. In: Proceeding of the Comference on Empirical Methods in Natural Language Processing, Barcelona, Spain (2004)
16. Mihalcea, R., Tarau, P.: An Algorithm for Language Independent Single and Multiple Document Summarization. In: Proceedings of the International Joint Conference on Natural Language Processing, Korea (2005)

17. Jing, H.: Sentence Reduction for Automatic Text Summarization. In: Proceedings of the 6th Applied Natural Language Processing Conference, Seattle, USA, pp. 310–315 (2000)
18. Knight, K., Marcu, D.: Statistics-Based Summarization – Step One: Sentence Compression. In: Proceeding of The 17th National Conference of the American Association for Artificial Intelligence (2000)
19. Sporleder, C., Lapata, M.: Discourse chunking and its application to sentence compression. In: Proceedings of HLT/EMNLP, Vancouver, Canada (2005)
20. Steinberger, J., Tesař, R.: Knowledge-poor Multilingual Sentence Compression. In: Proceedings of 7th Conference on Language Engineering, Cairo, Egypt, The Egyptian Society of Language Engineering (2007)
21. McKeown, K., Klavans, J., Hatzivassiloglou, V., Barzilay, R., Eskin, E.: From Discourse Structures to Text Summaries. In: Towards Multidocument Summarization by Reformulation: Progress and Prospects. AAAI/IAAI (1999)
22. Jing, H., McKeown, K.: Cut and Paste Based Text Summarization. In: Proceedings of the 1st Meeting of the North American Chapter of the Association for Computational Linguistics, Seattle, USA (2000)
23. Landauer, T.K., Dumais, S.T.: A solution to plato's problem: The latent semantic analysis theory of the acquisition, induction, and representation of knowledge. Psychological Review 104, 211–240 (1997)
24. Lee, D.D., Seung, H.S.: Learning the parts of objects by non-negative matrix factorization. Nature 401(6755), 788–791 (1999)
25. Kolda, T.G., O'Leary, D.P.: A semidiscrete matrix decomposition for latent semantic in-dexing information retrieval. ACM Transactions on Information Systems 16(4), 322–346 (1998)
26. Berry, M.W., Dumais, S.T., O'Brien, G.W.: Using linear algebra for intelligent IR. SIAM Review 37(4) (1995)
27. Ding, C.: A probabilistic model for latent semantic indexing. Journal of the American Society for Information Science and Technology 56(6), 597–608 (2005)
28. Gong, Y., Liu, X.: Generic text summarization using relevance measure and latent semantic analysis. In: Proceedings of ACM SIGIR, New Orleans, USA (2002)
29. Steinberger, J., Ježek, K.: Text Summarization and Singular Value Decomposition. In: Yakhno, T. (ed.) ADVIS 2002. LNCS, vol. 2457, pp. 245–254. Springer, Heidelberg (2002)
30. Murray, G., Renals, S., Carletta, J.: Extractive Summarization of Meeting Recordings. In: Proceedings of Interspeech, Lisboa, Portugal (2005)
31. Yeh, J.Y., Ke, H.R., Yang, W.P., Meng, I.H.: Text summarization using a trainable summarizer and latent semantic analysis. Special issue of Information Processing and Management on An Asian digital libraries perspective 41(1), 75–95 (2005)
32. Steinberger, J., Křišťan, M.: LSA-Based Multi-Document Summarization. In: Proceedings of 8th International PhD Workshop on Systems and Control, a Young Generation Viewpoint, Balatonfured, Hungary (2007)
33. Hachey, B., Murray, G., Reitter, D.: The embra system at duc 2005: Query-oriented multi-document summarization with a very large latent semantic space. In: Proceedings of the Document Understanding Conference (DUC), Vancouver, Canada (2005)

34. Carbonell, J.G., Goldstein, J.: The use of mmr, diversity-based reranking for reordering documents and producing summaries. In: Proceedings of the 21st Annual International ACM SIGIR Conference on Research and Development in Information Retrieval, Melbourne, Australia (1998)
35. Steinberger, J., Poesio, M., Kabadjov, M.A., Ježek, K.: Two Uses of Anaphora Resolution in Summarization. Special Issue of Information Processing & Management on Summarization 43(6), 1663–1680 (2007)
36. Radev, D., Jing, H., Budzikowska, M.: Centroid-based summarization of multiple documents. In: ANLP/NAACL Workshop on Automatic Summarization, Seattle, USA (2000)
37. Saggion, H., Radev, D., Teufel, S., Lam, W., Strassel, S.: Developing infrastructure for the evaluation of single and multi-document summarization systems in a cross-lingual environment. In: Proceedings of LREC, Las Palmas, Spain (2002)
38. Radev, D., Teufel, S., Saggion, H., Lam, W., Blitzer, J., Qi, H., Celebi, A., Liu, D., Drabek, E.: Evaluation Challenges in Large-scale Document Summarization. In: Proceeding of the 41st meeting of the Association for Computational Linguistics, Sapporo, Japan (2003)
39. Lin, Ch., Hovy, E.: Automatic evaluation of summaries using n-gram co-occurrence statistics. In: Proceedings of HLT-NAACL, Edmonton, Canada (2003)
40. Lin, C.: Rouge: a package for automatic evaluation of summaries. In: Proceedings of the Workshop on Text Summarization Branches Out, Barcelona, Spain (2004)
41. Nenkova, A., Passonneau, R.: Evaluating Content Selection in Summarization: The Pyramid Method. In: Document Understanding Conference, Vancouver, Canada (2005)
42. Mani, I., Firmin, T., House, D., Klein, G., Sundheim, B., Hirschman, L.: The TIPSTER Summac Text Summarization Evaluation. In: Proceedings of the 9th Meeting of the European Chapter of the Association for Computational Linguistics (1999)
43. Hynek, J., Ježek, K.: Practical Approach to Automatic Text Summarization. In: Proceedings of the ELPUB 2003 conference, Guimaraes, Portugal (2003)
44. Siegel, S., Castellan, N.J.: Nonparametric Statistics for the Behavioral Sciences, 2nd edn. McGraw-Hill, Berkeley (1988)
45. Morris, A., Kasper, G., Adams, D.: The Effects and Limitations of Automatic Text Condensing on Reading Comprehension Performance. Information Systems Research 3(1), 17–35 (1992)
46. Steinberger, J., Ježek, K., Sloup, M.: Web Topic Summarization. In: Proceedings of the 12th International Conference on Electronic Publishing, Toronto, Canada (2008)
47. Document Understanding Conference, NIST (2001-2007), http://duc.nist.gov
48. Text Analysis Conference, NIST (2008), http://nist.gov/tac

From Faceted Classification to Knowledge Discovery of Semi-structured Text Records

Yee Mey Goh, Matt Giess, Chris McMahon, and Ying Liu

Abstract. The maintenance and service records collected and maintained by the aerospace companies are a useful resource to the in-service engineers in providing their ongoing support of their aircrafts. Such records are typically semi-structured and contain useful information such as a description of the issue and references to correspondences and documentation generated during its resolution. The information in the database is frequently retrieved to aid resolution of newly reported issues. At present, engineers may rely on a keyword search in conjunction with a number field filters to retrieve relevant records from the database. It is believed that further values can be realised from the collection of these records for indicating recurrent and systemic issues which may not have been apparent previously. A faceted classification approach was implemented to enhance the retrieval and knowledge discovery from extensive aerospace in-service records. The retrieval mechanism afforded by faceted classification can expedite responses to urgent in-service issues as well as enable knowledge discovery that could potentially lead to root-cause findings and continuous improvement. The approach can be described as a structured text mining involving records preparation, construction of the classification schemes and data mining.

1 Introduction

Observation from in-service events can often indicate how well the design of an engineering artefact satisfies the performance, operational and maintenance requirements. Such knowledge is useful to the engineering company in providing maintenance support, repair and upgrade of their existing products as well as in improving the design of their future products. Therefore, information collected from in-service is a useful resource to the engineering companies, particularly aerospace because of the high value and extended life involved. The use of in-service information for decision-making is most evident in condition-based and predictive maintenance using sensors-based data logging and monitoring systems. These systems allow the detection, isolation, identification and prediction of fault condition(s) based on data processing and analysis. Statistical, artificial intelligence, model-based and rule-based approaches are typically employed to algorithmically relate the patterns in data to the expected outcomes in-service (Jardine et al., 2006). As an example, the Distributed Aircraft Maintenance Environment (DAME) is an e-Science project demonstrating the use of GRID

A. Abraham et al. (Eds.): Foundations of Comput. Intel. Vol. 6, SCI 206, pp. 151–169.
springerlink.com   © Springer-Verlag Berlin Heidelberg 2009

infrastructure to implement online health monitoring and fault diagnostic systems for the maintenance of aircraft engines in distributed environments (Ong, 2004).

Besides supporting the ongoing maintenance activities, in-service information about the functional or operational performance of the product through life is often useful in providing validation and feedback of knowledge assumed at the design stage. Experience in service allows the assumptions made in the design process to be refined and corrected, and this new information in turn informs the next design process. However, knowledge and information gained from in-service experience is found to be consistently under-utilised in decision-making in design and development (Fundin, 2003). Increasing pressure for *design right first time* due to outsourcing and global competition means that there are greater incentives for companies to learn about their products in service. Increasingly common practice in aerospace industry, non-trivial experiences may also be recorded in best practice and lessons learned databases for future reference (Lowe et al., 2003, NASA, 2007). The main aim of lessons learned systems is to support sharing and distribution of knowledge (in the form of lessons learned) to the wider organisation in order to avoid repeating the same mistakes. They are usually generic and are not just limited to supporting in-service feedback but also other activities within the organisation.

A common means of codifying maintenance and service events is through the use of semi-structured documents, usually generated from standard forms or based on templates. These records may be generated from manual or digital form entries that are subsequently stored. Each instance of the records is usually referenced using unique identification systems such as serial numbering. Information contained in these records generally is a combination of mandatory fields such as service engineer's name, date, equipment attended, customer's name, site and free-text describing the problem and solution. Alternatively, electronic and online reporting systems such as Computerised Maintenance Management System (CMMS) or proprietary e-maintenance systems may be used with associated workflow management and events logging capabilities. For example, when a customer reports a problem through a phone call to a service centre, this issue is logged automatically and becomes a live issue that is managed electronically. These systems allow companies to manage and systematically respond to issues thus potentially improving customer relationships. Some systems also allow automatic user information capture, correspondence, reminders and archiving. The information within these systems may be structured[1] (with predefined fields such as date, issue, customer, location etc.), but it is fundamentally different to the numerical data provided by sensors.

As semi-structured records the terminology and content of the maintenance and in-service information is subject to significant variance. There are opportunities for improving the information organisation aspect related to the service and maintenance records in the electronic form by abstracting this information into broader, aggregated and consistent viewpoints. In the immediate term, the

---

[1] 'Structured' data or information is that which is represented electronically in a way that makes it interpretable to a computer, such that a computer can understand what to do with it when it is encountered.

abstracted viewpoint serves as a browsable organisational scheme for the retrieval of similar past issues and their resolution, thus reducing response time to urgent in-service issues. This gives a significant advantage to the in-service engineers as demands for decisions to be reached in much less time, such as stipulated in modern aircraft contracts. In the longer term, the organisation scheme may allow for repetitive systemic issues to be highlighted using statistical analysis and data mining to reveal emergent patterns or correlations in the collective data set that may not have been apparent previously. Such indications may help engineers to prioritise root cause finding and development of solutions to rectify them. The paper describes a faceted classification approach for organising semi-structured records to enhance retrieval and also to facilitate knowledge discovery from the records. The approach can be described as a structured text mining involving records preparation, construction of the classification schemes and data mining from the records using OLAP-like approach. The method is demonstrated with an application in an aerospace case study.

2 Literature Review

Two distinctive strategies are typically adopted in searching and subsequent retrieval of information. The first approach involves term search techniques with the user entering a query or keywords that initiates a search for documents in the collection which are likely to be relevant to the user. Methods for matching the query to the document content (or the document index) are then carried out at search time. A variety of methods can be used to indicate relevance of document to query including Boolean, vector and probabilistic models (Baeza-Yates and Ribeiro-Neto, 1999). The biggest weakness in term search approach is that it relies greatly on the user formulating the suitable query to the targeted document corpus. Technique such as search expansion using thesauri has been used to improve query-based retrieval (Tikk et al., 2002). The second approach involves browsing documents that are organised, a priori, according to some predefined structure or classification schemes. The user browses the categories when searching for information. This approach requires more effort up-front in deciding and creating meaningful classification schemes, which may be difficult sometimes due to viewpoint dependency (which is addressed to some extent in faceted classification as discussed later). Ontologies have also been applied for improving information organisation and retrieval in various domains. An ontology, which is a specification of a conceptualisation, defines all the entities (objects or concepts) that are of interest in a domain and the relationships that connect these entities together, usually in some formal and preferably machine-readable manner (Hendler, 2001). Both classification and ontological approaches rely on the identification and abstraction of key concepts and their relationships from the underlying document corpus, thus requiring more efforts in construction and maintenance. However, they provide added advantages over the term search strategy in information retrieval as the abstracted schemes provide a browsable structure to facilitate retrieval as well as the analysis to deduce common patterns across the assigned concepts (or categorisations). This method is particularly useful when a user is unfamiliar with the domain terminology.

2.1 Information Classification

Classification is achieved by arranging objects into classes which can further be divided into subclasses according to some kind of *principles of division*. A class or subclass is a group of objects which share a particular set of attributes, with no other objects having this particular set of attributes. The purpose of classification may be analytical, i.e. to provide a systematic understanding of physical phenomena such as the biological classification for species of organisms. The famous Linnaean taxonomy uses five hierarchical ranks: class, order, genus, species, and variety. Another purpose of classification is for management of documents or information. The origins of modern approaches towards document classification are founded on principles originally developed by library scientists (e.g. Foskett, 1996, Rowley and Farrow, 2000, Taylor, 1992).

Information classification involves the development and use of schemes for the systematic organisation of knowledge (represented as information) to facilitate retrieval. There are three general types of classification schemes (Taylor, 1992):

- **Enumerative** – a scheme based on the concept of a universe of knowledge that is divided into successively narrower and more specific subjects.
- **Synthetic** – is a scheme in which new classes can be developed for new topics that are not already listed and establishes logical rules for dividing topics into classes, divisions and sub-divisions.
- **Analytico-synthetic** (or **faceted**) – assigns terms to individual concepts and provides rules for the local cataloguer to use in constructing classification headings for composite subjects.

Enumerative classification seeks to list exhaustively all possible subclasses of interest in a particular class. The Library of Congress Classification (LCC) and the Universal Decimal Classification (UDC) are primarily enumerative. The major weakness with the enumerative schemes is that it imposes some subjectivity to the principle of division used, and introduces issues with viewpoint dependency. The number of specific subjects that have to be related in a classification (i.e. a field of knowledge) can potentially be infinite. For instance, a compound subject of "statistics for scientists and engineers" can be ambiguously classified under subject "statistics" and/or target audience "scientists" or "engineers". A further weakness of the enumerative scheme is the need for regular revision to deal with new and emerging subject that has not been considered during the construction of the scheme. This method requires significant maintenance efforts if the subject of classification is dynamic and changes regularly. These weaknesses are, to varying degrees, addressed by faceted classification approaches.

Although not the inventor of facet analysis, Ranganathan is credited as the first to systematise and formalise the theory (Foskett, 1996, Rowley and Farrow, 2000, Vickery, 1975). According to (Taylor, 1992), facets are clearly defined, mutually exclusive, and collectively exhaustive aspects, properties or characteristics of a class or specific subject. Facet analysis, applied to the presentation and physical layout of *thesauri*, involves (Rowley and Farrow, 2000):

- Identifying sets of terms representing concepts. This involves the identification of multiple 'simple' concepts from the compound subjects (s) that describe a document.
- The grouping of the terms representing the simple concepts into a number of mutually exclusive categories (called facets).
- Organising the facets into a limited number of fundamental categories – these fundamental categories can be viewed a being different types of classification schemes. Thus the process of organising facets is essentially analogous to a process of classification scheme construction.

When a new subject is needed, a classifier using an enumerative scheme will have to wait until the scheme provides a term for that particular subject. In the case of a faceted scheme it is much easier to combine already-existing terms to form a new subject. In addition, because the concepts or facets for each classification scheme are compiled independently of each other, this ensures that the terms chosen can be developed and maintained independently. A key strength of the faceted classification is that it allows for multiple viewpoints to be represented through concurrent schemes.

2.2 Data and Text Mining

Data mining has been used for extracting new and potentially interesting patterns from highly structured data. Methods in machine learning, pattern recognition, and statistics are used extensively in data mining to discover features of interest from databases. Two types of analysis are generally considered in data mining application. The first approach creates models or predictive patterns for making an educated guess about the value of an unknown attribute given the values of known attributes (Witten and Frank, 2000). This modelling method is also known as the supervised learning, using techniques and algorithms like classification (decision tree) and numerical prediction (regression). On the other hand, the second approach aims to create models that do not solve a specific problem but simply to provide some description of the domain of interest. This modelling method is also known as the unsupervised learning, where techniques and algorithms such as clustering and association are employed in the exploratory exercise. Decision tree seeks to generate a series of logical rules that prescribe the performance of a series of instances or cases given their characteristics. The algorithm divides the set of instances into separate segments, each of similar performance, based upon the values of certain characteristics as seen within the full range of instances. By dividing these segments into smaller and smaller partitions, ultimately to the point where each member of a group has identical performance, a number of logical rules are generated which relate the value of the characteristics to performance. Association rule induction can be useful to describe interesting relationships between variables (or dimensions) from the data. The boundary between the two types of analysis is, however, not sharp as depending on the degree of the model interpretability some of the predictive models can be descriptive, and vice versa (Fayyad et al., 2006).

Text mining is about looking for patterns in natural language text, and may be defined as the process of analysing text to extract information from it for particular purposes. Related research in Natural Language Processing (NLP) seeks to use computers to automatically understand human languages. To date some progress has been made in this area, particularly within the subjects of bioinformatics where samples of human language are converted into more formal representations that are easier for computer programs to manipulate. As oppose to NLP, text mining focuses on extracting a small amount of information from text algorithmically and discovering relationships from the extracted information in a similar manner to data mining. The wider data mining process also includes data preparation, data selection, data cleaning, incorporation of appropriate prior knowledge, and proper interpretation of the results of mining, to ensure that useful knowledge is derived from the data (Fayyad et al., 2006). Similarly, text mining process generally consists of document collection, text preparation and mining operations which may include feature extraction, classification and clustering. These activities are particularly useful, for example, for the identification of key phrases in the absence of any thesaurus or controlled vocabulary in unstructured text (Witten and Frank, 2000). The overall process of turning unstructured textual data into high level information and knowledge is sometimes known as knowledge discovery in text.

3 Faceted Classification Using Waypoint

Faceted schemes have two notable weaknesses, both related to the manner in which a user may interact with a constructed faceted classification. In cases where a set of documents are retrieved via concurrent selections from multiple facets, it is difficult to rank the documents in order of relevance as it is unclear how to assign suitable relevance across these different facets – it is possible that a given document will have great relevance to one selection, and a less strong (but still significant) relevance to a separate selection. It is also possible to make concurrent selections across different facets which would be infeasible in practice, where the set of returned documents under a selection for one facet might not intersect with the set of returned documents from a concurrent selection under a different facet. This second weakness, and to a lesser extent the first weakness, may be addressed using a system entitled Adaptive Concept Matching (McMahon et al., 2002) which has been implemented within the Waypoint faceted classification environment.

3.1 Adaptive Concept Matching^{TM} (Adiuri Systems)

Adaptive Concept Matching (ACM) describes a method in which a series of concurrent selections from different facets may be built up incrementally to avoid infeasible combinations of concepts. A faceted classification scheme is populated by a document corpus, and users are invited to make a series of selections from each facet. As the first selection is made, all documents that are not relevant to that selection are removed from the returned document set. This process is repeated, where the returned document set is gradually pruned of non-relevant documents. It is perhaps conceptually simpler to think of each selection as being a query in its own right, and that a complex query can be incrementally constructed via a graphical interface (McMahon et al., 2002). The benefit of the graphical interface is that the size of returned document set given at each incremental step can be evaluated and displayed, thus providing the user with an impression of the discrimination of their search. This display is extended to include the number of surviving documents contained under each class, making it possible to remove all unpopulated classes from the visual interface, preventing the user from making selections that would return no documents thus preventing infeasible queries.

As opposed to Document Management systems, the Waypoint environment does not collect and store documents, instead it is an environment within which a user may explore document corpora and retrieve relevant documents from their original source and in their native format. As the system operates separately from the document management, and thus has no mechanism through which to enforce manual classification at the point of document creation or archive, it is essential that documents may be automatically classified within the faceted scheme. Such an approach also extends the applicability of the system to legacy documents. The automated classification is conducted via the use of constraints, syntactic rules that identify key terms within a document which indicate its relevance to a category within the classification scheme (Figure 1). A hypothetical example of such a constraint is given below.

IF document contains Term 'Landing Gear' AND Term 'Bush' OR Term 'product SN XXXXX' THEN document is relevant to concept 'Landing Gear Bush'

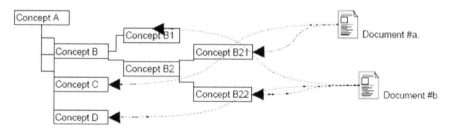

Fig. 1 Faceted classification associating documents to categories under each facet (Waypoint, 2006)

These constraints may be constructed for each *concept node* within the *concept map*, and is typically carried out concurrently with the construction of the faceted scheme – the scheme is itself a distillation of terms that appear in the documents (as concepts). The association of a document to a concept node through constraints is explained in the following section. This construction does not have to be conducted upon all documents, a subset of documents may be interrogated for this purpose (called the training set). It is also possible to automatically parse the documents to extract all significant terms (for example, by comparing the prevalence of certain terms within a document as compared to the prevalence within the overall corpus, referred to as "term frequency – inverse document frequency" (Baeza-Yates and Ribeiro-Neto, 1999)) and aggregate those key terms to form concepts. Methods for assisting concept identification will be discussed for the case study in later section although controlled vocabulary can greatly assist in this task.

3.2 Exploiting Document Structure

'Structured' data or information is that which is represented electronically in a way that makes it interpretable to a computer, such that a computer can understand what to do with it when it is encountered. Structured data can be found in such things as relational databases and in files which are tab or location delineated. Here the data are organised into a data structure according to the relationships and data type definitions prescribed by a data model. Unfortunately, much of the information generated in the course of the engineering process (in such things as reports, communications, procedures, catalogues, etc) is not only unstructured – making it difficult for electronic information systems to handle – but by its nature difficult to structure in its entirety. Nonetheless, it is possible to bring some order and machine-interpretability to unstructured information by making it semi-structured. This is done conventionally using such things as headings, paragraphs and sections (physical structure) and more recently by tagging or marking up interesting elements of the content explicitly using purpose-built formal languages such as HTML or XML (WC3). For instance, it may be possible to identify commonly occurring information elements, and develop a scheme or model by which these information elements in a document can be partially structured (and indeed, new documents constructed).

Particularly useful are the semantic (that is, meaning-bearing) information-bearing elements of content. One document may have many semantic dimensions, each of which is represented by a different structure. Revealing the semantic content of information allows information to be searched for based not on conventional pattern-matching techniques, but on meaning of the content. The process of making semantic structure explicit by marking up information content effectively means that documents can be decomposed into smaller and meaningful chunks. Waypoint specifically makes use of the explicit structure of XML document when making an association between a document and a concept. This is done by specifying the part of structure to which the constraints apply. Figure 2 illustrates the following rule in associating a structured document to a *concept node* in the *concept map* with the following rule.

Document is relevant to concept 'Damage' IF AND ONLY IF document contains Term 'Damage' within the 'Description' field.

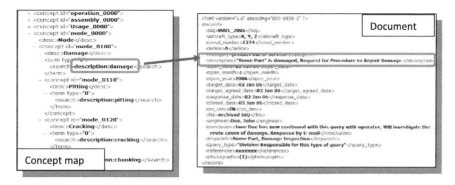

Fig. 2 Association of a structured document to a concept node in the concept map

4 Aerospace In-Service Records

In face of ever reducing time to respond to unplanned in-service events, the In-Service Support (ISS) teams rely greatly on past experiences to expedite their responses to the aircraft operators. Previously, they depend greatly on knowledge of the work context and company's organisation structures, along with a keyword search on the text descriptions to retrieve and reuse similar past cases. Although the keyword search approach may be efficient to the users familiar with the domain vocabulary, it is problematic to users outside of the ISS teams (such as the designers). By classifying the In-Service Queries (ISQ) according to additional facets such as assembly/component, failure mode, operational phase etc. retrieval by users who are unfamiliar with the context can be facilitated. For instance, the databases become more useful for the designers to interrogate for component-related failure modes, operational-induced issues etc. This added advantage was also recognised by the ISS engineers.

4.1 Construction of the Classification Schemes

In practice, the creation of classification schemes or schedules involves a great deal of intellectual effort. Two approaches, the 'top-down' or the 'bottom-up' approach may be used. The 'top-down' approach is usually simpler where the scope of the domain is considered, and these are divided and further sub-divided according to a consideration of specific distinguishing characteristics. Such an approach presupposes that the classifier's view of the world is both complete and accurate and that they can clearly identify which characteristics of a document to use as points of division. The more complex approach is referred to as a 'bottom-up' approach, which may be considered as an analytico-synthetic approach. This does not require an accurate or complete view of the domain in question, rather it depends upon adequate possession of the documents available within that domain. The view of the domain is obtained by distilling the essential characteristics or concepts of each document, which define the scope of the domain, and defining the classification scheme according to a natural grouping of these distilled concepts. Faceted schemes may be constructed via the use of both approaches, however the bottom-up approach ensures independence between concepts across facets, that there is no duplication within the scheme, and each given document is described by a combination of concepts that are adequate to completely describe and identify that document.

The method of identifying key concepts, terms and relationships within documentation and structuring these into a classification scheme can also be influenced to some extent by *warrant* (Beghtol, 1986). Many types of warrant have been suggested, including user warrant, scientific warrant, educational warrant and cultural warrant. For example, domain experts could create classification schemes based upon the *scientific warrant* whereas the *user warrant* aims at supporting the end user. The *literary warrant* describes the practice of constructing a classification scheme based upon the specific content of literature (Hulme, 1911). In terms of applicability, analytico-synthetic schemes rely upon literary warrant where the concepts which are contained within the document corpus are identified beforehand and the scheme arranged to fit these concepts. In this case study, the faceted schemes were constructed bottom-up based on *literary warrant*, which was then compared, refined and verified by domain experts.

From the ISQ database with many thousands of instances, the records for a selected number of years were scripted into structured XML documents with each piece of information tagged (shown in Figure 2 "Document"). An XML document can be viewed as the meta-information about an ISQ instance, where hyperlinks to the original documents such as reports, photographs and e-mails may be embedded for future retrieval. As mentioned previously, the structure improves the effectiveness in the Waypoint classification, such that rules can be coded under each tag/field. Some of these tags were inherited from the current ISQ database, particularly those referring to the airline, aircraft type and the ATA chapters which naturally become facets by which to organise the documents. These facets however, mainly relate contextually to the ISQ instance (except the ATA chapters which reflect the subsystem and functions to a limited extent). They do not sufficiently describe the technical detail of the issues. Some fields may contain numerical information such as the dates, the flight cycles and hours, which can be classified according to ranges specified in the rules. Nevertheless, information being described in the brief descriptions (free text and usually in an incomplete sentence) was found to be useful in deriving additional facets by which to classify the documents. The descriptions provide content-based information to the issue such as the failure mode, assembly, operational phase, flight type, topological location on the aircraft[2] (RH/LH, rib #, leading/trailing edge) etc. Table 1 summarises the concepts and facets that are distilled from the description in addition to the inherited ones.

Central to the effectiveness of the information organisation system depend on the information quality and the classification schemes used. By performing the bottom-up approach from a subset of the document corpus a number of key

---

[2] This information is more pertinent in the Wing Structures ISQ.

Table 1 Concepts and facets distilled from the "description" of Fuel Systems and Landing Gear ISQ records in addition to the inherited facets

Facets	Categories	Concepts (examples of)
Assembly:	Landing gear	shock absorber, downlock actuator, uplock roller, grounding stud hole, sliding piston, sliding tube, lower bearing, lower harness, tyres, axle, seal, aft pintle pin, bogie harness wire, retraction actuator, wheel, bushes, disconnect boxes, freefall system, bell crank bush, primary seal, steering feedback sensor, chrome plate, wing bracket
	Fuel system	transfer valves, dry bay drain hole, trim tank harness, centre tank pump, inner tank pump, refuel panel plug, outer LHS tank, outer RHS tank, pipes, vapour seal, probes, engine feed pipe, standby pump, collector cell, water scavenging system
	Steering system	turnbuckle
	Software	FCMC, BSCU
	Air conditioning system	N/A
	Engine	N/A
	Brake system	N/A
Failure mode:	Damage	Corrosion Mechanical: cracking, wear, pitting, dent, rupture, delamination, chafing Heat: overheat, high temperature Chemical Other: manufacturing, leakage, spillage, debris contamination, water contamination
	System	fault, complete failure, partial failure
	Event	collision, bird strike, lightning
	General query	agreement to defer, agreement to modify, update procedure, clarification, resolution
Operational phase:	In-flight	cruise, take-off, landing
	Ground operation	refuelling
	Maintenance	jacked, cleaning, overhaul, lubrication, fuel leak, test, inspection, brake bleeding, installation
Flight type:	Revenue flight Test flight Ferry	

concepts were found that are consistently used by the ISS engineers, which provides the domain vocabulary for evolving taxonomy. Since the quality of legacy records and the classification schemes may not be adequate in the first attempt, continuous evolution and refinement of the classification schemes and the reporting of newly identified key information is necessary to improve utility of the system. For instance, some common concepts may not be readily identifiable from the records due to variation in the way of reporting. Some of the issues include:

- Inconsistency in referring to the product, failure mode, operational environment etc. in the description
    o Information may be reported at different level of granularity e.g. damage/cracking
    o Ambiguous terms and synonyms e.g. damage/broken, cracking/chunking
    o Acronyms and abbreviation e.g. MLG/Main Landing Gear; assembly/assy
    o Spelling variation and errors, e.g. centre/center

- Missing or implied information in the description field
  o "MLG Pin Assembly" – the issue affecting the product is not explicitly reported
  o "Lower Bearing Corrosion" – the main assembly of the part is not reported.

From the information contained in the Fuel Systems and Landing Gear database for the selected number of years, the full classification show sparse records in some of the facets, inevitably for those derived from the description, as summarised in Table 2. The free-text description allows flexibility for engineers to report issues according to facets of information which are deemed relevant by the engineers. The flight type is the least frequently quoted information in the records with only 1.03 %. The operational phase and the failure mode have 30.14 % and 25.79 % respectively of the whole document set successfully classified. However, 58.79 % of the records contain information about the product assembly and components that is being queried. These statistics reflect the inconsistency in the facets of information that are considered important to each ISQ instance as well as the efficiency of the constraints used in the classification. It should be noted that the statistics only apply to a fraction of records in the database. The system has subsequently been upgraded with more structured information entry. Therefore, the statistics is expected to improve if more recent records are being classified.

In order to facilitate the construction of classification schemes from text description, methods in text mining were considered in order to extract key phrases that appear frequently in the text to represent concepts (and the corresponding rules/constraints). As previously mentioned, the information in the description field is critically important to the engineers as it provides a summary to help retrieve more complete information from the original correspondences and technical reports that are linked to the ISQ instances. These documents are often in scanned PDF format, albeit for legislative reasons, are not readily treatable.

4.2 Text Mining Methods

As the classification schemes were to be constructed based on the literary warrant, i.e. evolved from the underlying content of corpus, text mining methods are

Table 2 Statistics of the Classification

Facets	Inherit/From Description	% of Classified Documents
ISQ Date	Inherit	99.75
Category	Inherit	99.07
ATA Chapter	Inherit	90.27
Aircraft Type	Inherit	90.00
Assembly	From description	58.79
Query Type	Inherit	44.88
Operational Phase	From description	30.14
Failure Mode	From description	25.79
Flight Type	From description	1.03

suitable to facilitate the automated identification of key phrases from text. In information retrieval, methods such as stemming are often used to deal with common lemma of words. Additionally, spelling variation, abbreviation and acronyms can be dealt with using dictionary and domain knowledge to link between similar concepts (Wren et al., 2005). The Waypoint system includes such functionality by incorporating the Open-Source Lucene indexing and search engine library (Apache, 2006). This library has a number of different document analysers which cater for different forms of such treatment. For the Wing Structures database, facets related to the topological location of the fault were particularly important. Although the facet was inherited from the original company database, the records in these fields were only sparsely populated. Improving the records will improve the classification as the information is valuable for learning about the characteristics of the in-service issues. Information was extracted from the unstructured description field to populate the classification schemes using a dictionary of terms that are expected in each concept node. The records with missing information are matched against those terms and are filled-in. Pre-processing operations such as this can improve the accuracy of the association of the documents to the rules/constraints in the concept map.

As the domain is quite restricted, many concepts are repeated in the records. Although methods proposed in Natural Language Processing (NLP) such as Part-of-Speech (POS) tagging appear to be relevant in the first instance, they are found to be less useful in dealing with technical reporting which does not conform to standard grammatical rules (Ciravegna, 1995). As shown in Table 1, the concepts need to be grouped into classes to provide a logical classification structure (either enumerative or faceted schemes). For instance, aircraft types can be grouped according to their families of variants which share similar characteristics e.g. single/twin aisle, medium/long range, twin/four engine. By doing so, behaviour displayed by the families of aircrafts could be discriminated in the classification. As such, the means of aggregating the concepts into classes and the granularity of the classification have significant influence on ability to make inference from the patterns and trends in the records. Although methods like WordNet and Formal Concept Analysis have been used with some success in a number of areas (Stoica and Hearst, 2006, Cimiano et al., 2004), it was found that domain knowledge is still essential to provide meaningful abstractions at present.

In this case study, statistical methods were used to analyse the frequency of collocated words to extract most frequent word clusters. High frequency word clusters are a good indication of common phrases referring to some meaningful concepts. The concepts extracted from the ISQ records reflect the domain vocabulary used by the in-service engineers. Table 3 shows the results of word clusters with a minimum frequency of 10. The clusters shown are generated from indices in concordance with the keyword "tank". The number of words in a cluster (known as window) and the minimum frequency were arbitrarily set depending on the characteristics of the corpus. The significance of the word clusters as concepts can be improved by breaking the sentence at punctuations and common stop words. Also as can be seen from the table, automatically extracted concepts can be hierarchically arranged into the main assembly, sub-assembly and component of

Table 3 2-5 word clusters in concordance with "Tank", minimum frequency = 10

Main assembly (freq)	Sub-assembly (freq)	Components (freq)
Fuel centre tank (10)	Centre tank (110) Center tank (41)	Tank FQI (12)
Fuel trim tank (12)	Fuel tank (69)	Tank fuel (44)
LH inner tank (12)	Inner tank (111)	Tank harness (24)
RH inner tank (10)	Surge tank (10)	Tank indication (12)
RH outer tank (10)	Trim tank (220)	Tank inlet (21) Tank inlet valve (19)
	Outer tank (75)	Tank low (16)
	Wing tank (55)	Tank NTO (12)
		Tank overflow (12)
		Tank pump (33) Tank pumps (12)
		Tank transfer (35)
		Tank vent (10)

tank with sufficient level of accuracy. This is similar to the concept pairs heuristics proposed by Yang (2004).

4.3 Retrieval and Knowledge Discovery

In the Waypoint implementation, the ISQ instances (XML documents) were classified according to the faceted schemes that were constructed as previously described. Through the interface, a number of concepts in the faceted schemes can be selected to form a compound query. For instance, if the user is interested in corrosion on aircraft type X, he/she will select the concept "X" under the facet Aircraft Type and the concept "Corrosion" under the facet Failure Mode. The returned documents are pruned from the system leaving only relevant records that satisfy the conjunction of the selected concepts. At this point, a much smaller set but highly relevant documents can be retrieved as a list by clicking on the 'Results' button on the top right hand corner as illustrated in Figure 3(a). A list of results that are only relevant to both concepts will be displayed as shown in Figure 3(b). The display can be customised as well as hyperlinked to other sources of documents that are relevant to those ISQ instances. Due to confidentiality, the actual numbers of documents and some fields are not shown in the figures. The use of Dynamic HTML (DHTML) allows the output to be easily formatted or inserted into other applications or web pages. If the returned document set is still too large, the user can introduce further constraints (for example by selecting the concept "Main Landing Gear" under Assembly) to reduce the set further and increase precision. This way, the faceted classification schemes help to quickly identify and retrieve experiences and resolution to past issues based on a combination of the concepts selected.

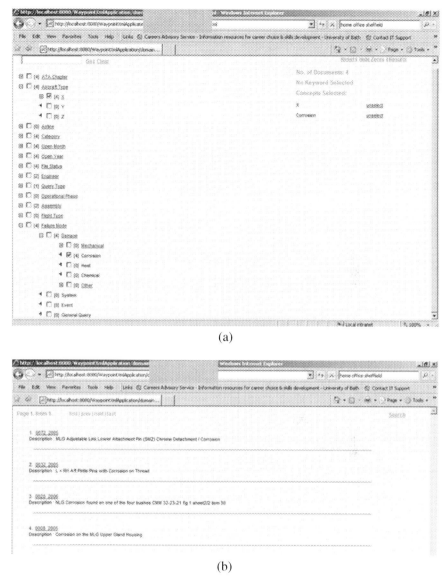

(a)

(b)

Fig. 3 (a) A Waypoint faceted classification interface (numbers do not reflect actual ISQ records) (b) List of documents relevant

As the users interactively browse through the classification tree and introduce constraints (by selecting more concepts) in Waypoint, the document set is pruned and the remaining relevant document set is updated dynamically. The document count is reflected in the numbers next to each node (at all levels of the tree), where only those documents classified under that node and are also relevant to the current selected concepts are displayed (Figure 3 (a)). For example, if the user wants to find out about the failure modes that most frequently affect the main landing gear of aircraft X, with these two nodes selected, he/she can navigate to the Failure Mode facet. The ISQ documents will be distributed across the four classes (Damage [a], System [b], Event [c] and General Query [d] as shown in Figure 3(a)) with the number of relevant documents displayed next to each node in brackets. This allows the user to infer dominant class(es) of failure mode reported about the main landing gear of aircraft X, say Damage [a]. By expanding the Damage class, all the active (still relevant) concept nodes are displayed with numbers next to each one of them (Figure 3(a)). Then, the user might also want to limit the results to those reported in a particular year or month. At any stage, the concepts with no document count against them can be hidden or shown depending on the users' preference. The number of documents at a parent node may be equal to or greater than the sum of documents for all the children nodes as some instances may be classified under more than one concept nodes in the same facet. For example, an ISQ instance that reports both mechanical and heat issues will be classified and counted against both the nodes but only counted once for the Damage class (Figure 3(a)). This flexibility allows for documents to be associated to concept nodes at any level of the tree to cope with the varying level of abstraction in the reporting of ISQ.

By browsing through the faceted schemes in Waypoint, one might be able to discover classes of in-service issues that are systemic such as design-induced (if there is apparent correlation between a component/topological location and a failure mode), operation-induced (if there is apparent correlation between an operational phase/event and a failure mode) and use-induced (if there is apparent correlation between a flight type/route/aircraft operator and a failure mode) issues. Correlation can be indicated by higher than average number of instances between two or more concept nodes. Although the correlations do not necessarily indicate an underlying problem, by indicating that it is a frequent occurrence, it acts as a prompt for engineers to determine root causes, which may lead to understanding of the operating conditions practiced by different airlines (e.g. maintenance procedure, type of operation – short or long haul). Furthermore, it becomes possible to infer less apparent correlation such as between a time period and failure mode to suggest potential inherent reliability issues. The capability will help towards highlighting recurrent patterns and correlations between different facets across a large set of ISQ instances. Such effort could have been achieved in the past through manual tracking and compilation of data or through experience but it can be achieved interactively in the Waypoint environment. This mechanism can be used to provide evidence to the design team when formally requesting engineering change or in prioritising continuous development efforts as well as for analysing reliability of components and operational interruptions. Some of these insights may also be usefully reported as lessons learnt which might benefit future aircraft development programs. A unified knowledge management solution for learning from experience can potentially be realised given that the company has also implemented a software tool for capturing and sharing lessons learned.

5 Discussions and Conclusions

Information collected from service that are commonly of semi-structured form can be more effectively reused if subject to some organisation. The faceted classification approach described in this paper not only provides an improved retrieval mechanism to aid speedy resolution of ongoing in-service issues but also provides a potentially useful mechanism for learning from the collective instances. The faceted schemes allow the user to arrive quickly at a highly relevant set of results by selecting concepts relevant to the query. In addition, the facets derived from the content of the ISQ records are especially useful to a non-familiar user to browse for information. As a result, the system becomes meaningful to others to interrogate, such as designers looking for typical issues raised on a particular component of the aircraft. Perhaps more importantly, the faceted schemes also allow for patterns and trends in the records to be highlighted, either by browsing the classification tree or automatically detecting patterns using data mining algorithms. The patterns and correlations may suggest for systemic and repetitive issues observed in service, which can be used as objective evidence to prioritise root-cause finding and continuous development efforts. Although the role of learning and discovering knowledge still lies with the domain experts, this is in contrast to the present way of learning that depends on subjective and manual assimilation.

The construction of faceted classification schemes may involve significant intellectual efforts in defining the levels of granularity and the dimensions of information relating to the records that are subject to organisation. In doing so, faceted classification provides a mechanism for viewing information at different levels of granularity using the concept hierarchy similar to On-Line Analytical Processing (OLAP) (Hand et al., 2001). Waypoint in particular allows the user to interrogate for patterns and trends through its dynamic interface. Aggregations can be built by changing the granularity on specific dimensions (facet) and aggregating up data along these dimensions. For instance, the user can choose to look for patterns that are characteristics of a family of aircraft or of a specific aircraft model, of a sub-system or of a specific component etc. Further functionality of data mining can also be easily integrated with Waypoint to automate pattern identification from the semi-structured records. It can be anticipated that the construction of the hierarchical structure of the classification schemes will affect inferences that might be drawn because the number of instances in a class is determined by the lower level nodes that are associated with it. Therefore, selecting the suitable level of abstraction is not trivial and may require trade-off to be made between retrieval efficiency and usefulness for knowledge discovery.

To be most effective, the classification schemes and records have to be evolved and refined iteratively. With current records as the training set, a taxonomy that is based on the literary warrant can be constructed which will in turn guide the creation of high quality records through prescribed information entry. The use of a controlled vocabulary or taxonomy such as through prescribed data entry (e.g. through selection lists) would significantly enhance ability to classify, retrieve and mine the in-service information. The constraints and rules for the classification may also be refined to improve the association of the records to those abstracted concepts. Additionally, more value can be attained from in-service information if the records can be linked to manufacturing-related information, such as the batch number, supplier, etc. For example, using the part and serial number to allow for cross referencing the ISQ records to the Bill of Materials and manufacturing information can enhance prospects of knowledge discovery. Patterns related to the design and manufacture to the in use performance and events (such as the frequency of maintenance) can be investigated that may result in insights and learning. The previously unsuspected correlations can be made explicit so that proactive steps can be taken to incorporate learning and avoiding repeating similar mistakes in the future. This will also help designers to understand the performance limits of a design, by understanding how the product is being used. Designers in due course may want to adapt the design to changes in use that may be more costly to design but may be more economical to maintain in the extended lifecycle.

Acknowledgement. The authors gratefully acknowledge the funding provided by the Engineering and Physical Science Research Council (EPSRC) for the KIM Project (http://www-edc.eng.cam.ac.uk/kim/) under Grant No. EP/C534220/1 and the I-d/MRC under Grant No. GR/R67507/01 for the research reported in this paper. The contribution of our industrial collaborator and Mr. Joe Cloonan is also gratefully acknowledged.

References

The Apache Software Foundation, Apache Lucene – Overview (2006), http://lucene.apache.org/java/docs/index.pdf

Baeza-Yates, R., Ribeiro-Neto, B.: Modern Information Retrieval. ACM Press Books, Harlow (1999)

Beghtol, C.: Semantic validity: Concepts of warrant in bibliographic classification systems. Library Resources & Technical Services 30, 109–125 (1986)

Cimiano, P., Hotho, A., Staab, S.: Clustering ontologies from text. In: Proceedings of the 4th International Conference on Language Resources and Evaluation, Lisbon, Portugal (2004)

Ciravegna, F.: Understanding messages in a diagnostic domain. Information Processing and Management 31, 687–701 (1995)

Fayyad, U., Piatetsky-Shapiro, G., Smyth, P.: From Data Mining to Knowledge Discovery in Databases. American Association for Artificial Intelligence, vol. 17, pp. 37–54 (Fall, 2006)

Foskett, A.C.: The subject approach to information. Library Association Publishing, London (1996)

Fundin, A.P., Bergman, B.L.S.: Exploring the customer feedback process. Measuring Business Excellence 7, 56–65 (2003)

Hand, D., Mannila, H., Smyth, P.: Principles of Data Mining. MIT Press, Cambridge (2001)

Hendler, J.: Agents and the Semantic Web. IEEE Intelligent Systems 16, 30–37 (2001)

Hulme, E.W.: Principles of Book Classification. Library Association Record 13, 354–358 (1911)

Jardine, A.K.S., Lin, D., Banjevic, D.: A review on machinery diagnostics and prognostics implementing condition-based maintenance. Mechanical Systems and Signal Processing 20, 1483–1510 (2006)

Lowe, A., McMahon, C.A., Culley, S.J., Coleman, P., Dotter, M.: A Novel Approach towards Design Information Management within Airbus. In: ICED 2003, Stockholm, Sweden (2003)

McMahon, C., Crossland, R., Lowe, A., Shah, T., Williams, J.S., Culley, S.: No zero match browsing of hierarchically categorized information entities. Artificial Intelligence for Engineering Design, Analysis and Manufacturing 16 (2002)

NASA, The NASA Engineering Network. Lessons Learned Information Systems (LLIS) (2007)

Ong, M., Ren, X., Allan, G., Kadirkamanathan, V., Thompson, H.A., Fleming, P.J.: Decision Support System on The Grid. In: Int'l Conference on Knowledge-Based Intelligent Information & Engineering Systems, New Zealand (2004)

Rowley, J., Farrow, J.: Organising Knowledge, Aldershot, Hants, UK, Gower (2000)

Stoica, E., Hearst, M.: Demonstration: Using WordNet to Build Hierarchical Facet Categories. In: ACM SIGIR Workshop on Faceted Search (2006)

Taylor, A.: Introduction to Cataloguing and Classification, Westport, CT, Libraries Unlimited (1992)

Tikk, D., Yang, J.D., Baranyi, P., Szakal, A.: Fuzzy relational thesauri in information retrieval: automatic knowledge base expansion by means of classiffied textual data. In: 6th International Conference on Intelligent Engineering Systems, Opatija, Croatia (2002)

Vickery, B.C.: Classification and Indexing in Science, London, Butterworth (1975)

Waypoint, Faceted Classification and Adaptive Concept Matching. Gemstone Business Intelligence Ltd. (2006)

Witten, I.H., Frank, E.: Data Mining: Practical machine learning tools with Java implementations. Morgan Kaufmann, San Francisco (2000)

Wren, J.D., Chang, J.T., Pustejovsky, J., Adar, E., Garner, H.R., Altman, R.B.: Biomedical term mapping databases. Nucl. Acids Res. 33, D289–D293 (2005)

Yang, K., Jacob, E., Loehrlein, A., Lee, S., Yu, N.: Organizing the Web: Semi-automatic construction of a faceted scheme. In: IADIS International Conference WWW/Internet, Madrid, Spain (2004)

Multi-value Association Patterns and Data Mining

Thomas W.H. Lui and David K.Y. Chiu

Summary. Mining patterns involving multiple values that are significantly relevant is a difficult but very important problem that crosses many disciplines. Multi-value association patterns, which generalize sequentially ordered patterns, are sets of associated values extracted from sampling outcomes of a random N-tuple. Because they are value patterns from multiple variables, they are more specifically defined than their corresponding variable patterns. They are also easier to interpret. Normally, they can be detected by statistical testing if the occurrence of a pattern event is significantly deviated from the expected according to a prior model or null hypothesis. When the null hypothesis presumes the values of a pattern to be independent, the alternative hypothesis asserts that the values as a whole are associated, allowing some values to be independent within the detected set. Recently, a special type of multi-value association pattern is proposed which we called nested high-order pattern (NHOP), which is a subtype of the high-order pattern (HOP). We discuss here these patterns together with a related one called consigned pattern (CP). Evaluations using relevant experiments of synthetic and biomolecular data are also included.

1 Introduction

Identifying significant relationships in data when multiple sources are involved is extremely important in bioinformatics and systems biology. For example, linking genotype and phenotype attributes plays a vital role in

Thomas W.H. Lui
Department of Computing and Information Science, University of Guelph
e-mail: tlui27@yahoo.com

David K.Y. Chiu
Department of Computing and Information Science, University of Guelph
e-mail: david.chiu@cis.uoguelph.ca

A. Abraham et al. (Eds.): Foundations of Comput. Intel. Vol. 6, SCI 206, pp. 171–191.
springerlink.com   © Springer-Verlag Berlin Heidelberg 2009

understanding biological processes, including cellular and organismal responses under normal or novel conditions such as in diseases or genetic modifications. At the level of biomolecules, identifying relationships between certain sequence segments or molecular sites and their effects on structure and biological functionality is a very active area of research.

The study of sequential patterns is recognized as an active research area in data mining with many unsolved challenges [19]. Some problems include unexpected variations due to perturbations in the form of insertions, deletions and substitutions. Other problems can be due to skewed distributions. On the other hand, there are important applications in signal recognition, image analysis, biomolecular analysis, and others where such discovery can enhance understanding.

High-order pattern (or HOP), introduced in [20], generalizes sequential patterns and multi-variable patterns. HOP is a pattern that reflects a complex relationship among its composed values (Fig. 1a). It is based on sampling outcomes of a random N-tuple, sometimes identified as a multiple sequence alignment [5]. It is a set of jointly occurring interdependent (or associated) values that are evaluated to be statistically significant. The order of a high-order pattern is defined as the number of values detected. For example, given sample outcomes of $(X_1, X_2, \ldots, X_{20})$, an example of 3-order pattern can be $(X_3 = A, X_7 = B, X_{12} = D)$. Based on an analysis of statistical significance, a high-order pattern can be defined as a set of values whose observed occurrence is significantly deviated from the prior model assumption, such as the independence, uniform or any known distribution assumption [7].

In this paper, we consider the interrelationship of three related types of multi-value association patterns identified in [5] [6] [14] [15] [20] [21]. All of them are composed from multiple values, and that the values involved in the patterns are built from statistically significant associations. Thus, they are newly discovered patterns as a deviation from the prior model, and may not be previously known. In addition to the general type of *high-order pattern* (HOP) [20] [21], we consider the *consigned pattern* (CP) [5] [6], and *nested high-order pattern* (NHOP) [14] [15]. CP is identified as a single value, but interdependent with other multiple values (Fig. 1b). NHOP has a nested associative structure among its detected values (Fig. 1c).

In [20], the significance of a high-order pattern is evaluated using a statistical test. An overview of the test can be described as follows. First, the observed frequency of a pattern candidate is obtained from the data. When prior knowledge is not available, the expected frequency of the pattern candidate is estimated based on a null hypothesis which normally assumes that all the values are mutually independent. It is rejected if the actual observed frequency of the pattern candidate statistically deviates from the expected frequency. In this case, the candidate is accepted as a significant pattern.

The null hypothesis assumption of independence has been used extensively in many pattern discovery tasks. When the independence hypothesis is rejected, the alternative hypothesis is accepted. The alternative hypothesis

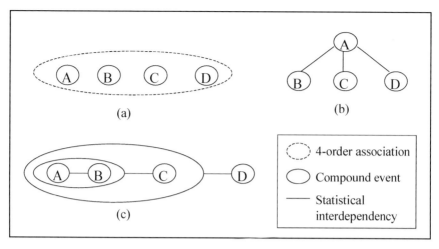

Fig. 1 Three related types of multi-value association patterns: (a) High-order pattern (b) Consigned pattern (c) Nested high-order pattern

suggests that not all of the values in the pattern are independent. In other words, it suggests that at least some values are interdependent, without specifying which values are interdependent when more than two values are involved. A comprehensive interpretation of the internal interdependency is required later.

In general, an n-order pattern ($2 < n \leq N$), discovered from the data set of a random N-tuple, may still contain independent disjoint parts. For example, in a 4-order pattern (A,B,C,D), the value subset {A,B} may be independent of the other subset {C,D}, even though the whole set {A,B,C,D} deviates from its null hypothesis of the prior model.

The second type referred to as *consigned pattern* (CP) in [5] [6], is identified as single value with multi-value pattern relationships such that the interdependencies associated with that value is large (Fig. 1a). This pattern is important because if associations form in such a way, the identified value is likely a highly significant but in a compact form indicative of the complex underlying interdependencies of the data. This compact form is found to be useful to facilitate the difficult mining tasks [5] [6] [8] [18]. The single-value, multiple-association pattern is also easier to interpret for further mining.

Recently, we identify a third type of multiple-value association patterns referred to as *nested high-order pattern*, or NHOP [14] [15]. By NHOP, the pattern has nested associated values, identified progressively when the pattern is expanded (Fig. 1c). This pattern satisfies a consistent statistical criterion when the pattern is iteratively identified. NHOP is a sub-type of HOP where each value is interdependently connected with other values in the pattern. Hence, it is a pattern that avoids independent disjoint values.

To construct a NHOP, an identified interdependent value is added iteratively to a previous lower order pattern. The first pair is initiated by identifying the strongest association. The motivation is that we aim to add statistically significant values into the constructed pattern. As a result, NHOP is
a pattern that has strongly interrelated associations, and is more consistent
internally for the pattern discovery task. In this chapter, we focus on this
type of pattern and further evaluate its relevance for pattern discovery using
experiments.

The rest of the paper is organized as follows. Section 2 provides a brief
review of the related work. Section 3 describes the notation and basic concepts
used. Section 4 presents an algorithm in extracting NHOP patterns. Section
5 presents the relationships among three types of multi-value association
patterns and their proofs. Section 6 presents the experimental results.

2   Related Works

In data mining, pattern association such as the extraction of association rules
is an important problem. In the past, many approaches to discover association
patterns have been proposed, including the reports in [1] [3] [4] [12] [13] [16]
[17] [22]. Below, we review them according to three main aspects: (1) the
types of input data in the analysis, (2) the target of the discovery process,
and (3) the main features of the method.

In this paper, we will focus on the analysis of N discrete-valued variable
data, where each variable has a finite number of possible non-overlapping
discrete-value outcomes. Using these variables as data input, a number of
analyses can be performed. For example, contingency table analysis can be
used to evaluate dependency between the variables involved. However, it is
worth to point out that dependency between values is more specific and
descriptive than dependency between variables. Dependency between value
outcomes of variables allows us to obtain more descriptive understanding
about the association of the domain, and hence is more desirable for the
purpose of data mining.

The data input used in many existing methods such as in market basket
data analysis [1] [4] [22] is N binary variables, such that each variable has
two possible outcomes. They may not be readily applicable to multiple valued
variables. For example, to analyze M valued variables, one way is to transform the M discrete-valued variables into a large number of binary variables.
For instance, suppose the data consists of three variables and each variable
has ten value outcomes. By considering each value as a binary variable, we
transform such data into 30 binary variables. Clearly, the constraints relating
interdependency have been changed in the transformed binary variables, and
may not be the same when interpreting the original set of variables.

One of the first approaches for market basket data analysis is described
in [1]. The goal is to discover customer buying habits in supermarket. For

example, if customers are buying milk, how likely are they going to buy cereal as well? Discover such relationship can help the retailers in arranging their products in the shelf that may increase the sales. The basket data can be explained as follow. Let $I=\{i_1, i_2,\ldots,i_N\}$ be a set of N variables, called items. Let X and Y be two disjoint subset of items (i.e. $X \in I$, $Y \in I$ and $X \cap Y = \phi$). In [1], their objective is to discover the association rule in this form: $X \Rightarrow Y$, which can be interpreted as "A customer purchases set of items X is likely to purchase set of items Y". The Apriori method [1] generates a complete set of rules based on two predefined criteria: minimum support (i.e. $P(X, Y) > \epsilon_1$) and minimum confidence (i.e. $P(Y \mid X) > \epsilon_2$). Based on these two criteria, more efficient methods for mining association rules are developed such as FP-growth method [11] and Eclat method [23]. However, we argue that an evaluation on statistical significance of the discovered interdependency between itemsets is a very important aspect of the analysis, especially in problems in the natural sciences.

By considering the statistical interdependence between items, Brin et al. [4] aims to identify N-order associated itemsets (or we called N-order association patterns) from market basket data. Their method first generates itemsets with minimum support (i.e. $P(X) > \epsilon$). Next, for each itemset, a high dimensional (or multi-way) contingency table is constructed. Then, based on the chi-squared value (χ^2) of the table, they evaluate statistical significance of the interdependency of each itemset. However, a limitation of the method is that the computational complexity can be high in constructing the high-dimensional contingency tables.

To prevent constructing very high dimensional contingency tables, Wu et al. [22] suggests to first group all item variables into smaller variable groupings. In each group, they aim to discover minimal sets of high-order associations with highest interestingness that fits the loglinear model. An outline of their method can be described as follows. First, they partition the N items into groups using a graph representation. The graph is constructed based on pairwise interdependency between the items. Then, for each group of items, they construct contingency table and then apply loglinear modeling to each table. Finally, they measure the interestingness of the itemsets by examining the loglinear model's parameters. However, the computational complexity can still be very high if the group sizes are large.

Jaroszewicz and Simovici [12] aims to discover N-order associated itemsets that is unexpected from a difference calculation. In their method, they first construct a Bayesian network from background knowledge. The interestingness of an itemset is then measured based on the absolute difference between the observed probability estimated from data and the expected probability estimated from the network. Finally, they extract the N-order patterns with interestingness above predefined threshold or in [13], select a predefined number of the most interesting N-order patterns. However, no evaluation of statistical significance of the interestingness measure is used and thus the patterns detected may be arbitrary. Thus, the reliability of the N-order patterns generated in the detection of inherent interdependency of the domain is weakened.Another concern of this method is that user of the method may need to be involved in the updating of the network. Jaroszewicz and Scheffer suggest to iteratively applying the identified patterns to modify the network [13]. However, there is no reason to believe that transforming the variable dependency into causal relationship is appropriate.

Bazzi and Glass [3] aims to discover N-gram from N discrete valued data inputs converted from speech signals. They use an iterative and bottom-up detection method to generate patterns based on maximum weighted mutual information between word pairs, without an evaluation on their statistical significance between words. Since the generated N-order patterns can be arbitrary, their interpretation is also weakened.

Sy [16] [17] aims to discover N-order association pattern from N discrete-valued variables. They evaluate the significance of interdependency based on a statistical test of adjusted residual, similar to [20]. They have also considered different criteria, using a simple probability ratio in [16] and a mutual information measure in [17].

3   Notations and Basic Concepts

3.1   Notations and Data Representation

Our data are represented as sampling outcomes of a random N-tuple. Alternatively, we can describe them as an ensemble of a dataset represented as a relation in a relational database. The relation can then be represented as $X=(X_1,X_2,\ldots,X_N)$ with N variables. An instance of X is a realization that can be denoted as $x=(x_1,x_2,\ldots,x_N)$. Each x_i $(1\leq i\leq N)$ can take up an attribute value denoted as $x_i=a_{ip}$.

A attribute value, $x_i=a_{ip}$, is taken from the attribute value set, $\Gamma_i=\{\,a_{ip}|\,p=1,2,\ldots,L_i\}$ where L_i is the number of possible values for the attribute X_i. A value a_{ip} in Γ_i is then defined as a primitive. Let E be a variable subset of X. E then has at least one variable, but less than or equal to N variables. The realization of E is denoted as e. A compound pattern event (or just event) associated with the variable set E is a set of attribute values instantiated by a realization e. The event can be represented by $e=\{X_j=a_{jq}|X_j\in E, q\in 1,2,\ldots,L_j\}$. The order of the event is $n=|E|$ which is usually less than the whole size of the random tuple ($<N$).

3.2   Evaluation of Statistical Significance for NHOP

Statistically significant patterns are considered in order to differentiate from those that are known from prior model. We evaluate the statistical interdependency between an event $E=e$ and an attribute value $X_i=a_{ip}$ based on residual analysis. The notation E refers to the joint variable composed of the corresponding variable subset that makes up the event e. The standardized residual $z(e, a_{ip})$ [10] [20] is defined as:

$$z(e, a_{ip}) = \frac{obs(e, a_{ip}) - exp(e, a_{ip})}{\sqrt{exp(e, a_{ip})}} , \tag{1}$$

where $obs(e, a_{ip})$ is the observed frequency and $exp(e, a_{ip})$ is the expected frequency for (e, a_{ip}). The expected frequency of (e, a_{ip}) can be estimated based on the independence assumption where the null hypothesis assumes e and a_{ip} are independent. If the deviation is significantly large, then the pattern is found to be previously unexpected. Thus, a new unexpected pattern is discovered. The standardized residual is considered to be normally distributed only when the asymptotic variance of $z(e, a_{ip})$ is close to one, otherwise, standardized residual has to be adjusted by its variance using the formula below [20]. The adjusted residual is expressed as:

$$d(e, a_{ip}) = \frac{z(e, a_{ip})}{\sqrt{v(e, a_{ip})}} , \tag{2}$$

where $v(e, a_{ip})$ estimates the maximum likelihood of the variance of $z(e, a_{ip})$. It is expressed as:

$$v(e, a_{ip}) = (1 - P(E = e))(1 - P(X_i = a_{ip})) . \tag{3}$$

The adjusted residual $d(e, a_{ip})$ has an asymptotic normal distribution. Thus the significance level can be chosen to be 99%, or another acceptable level. The test for statistical significance is based on the following inequality:

$$d(e, a_{ip}) \geq N_a , \tag{4}$$

where N_a is the tabulated threshold with α being the confidence level.

3.3   Sample Size for Significance Test

When calculating the adjusted residual in the statistical test of interdependence, it is reliable only when the sample size is large. When the sample size is not large enough, no reliable conclusion can be drawn. For the test of interdependency between a compound event $E=e$ and an attribute value $X_i=a_{ip}$, a safe estimation of the sample size, M, is described in [20]:

$$M = w \, |X_i| \, |E| , \tag{5}$$

where $|X_i|$ is the cardinality of attribute X_i and $|E|$ is the cardinality of the attribute set E. w is the minimum sample size required for a test of a particular cell in a contingency table. Based on the common practice in hypothesis testing for contingency table, w can be selected as three or five [20]. The range of $|E|$ is:

$$arg \max_{X_j \in E}(|X_j|) \leq |E| \leq \prod_{X_j \in E} (|X_j|) , \qquad (6)$$

where $|X_j|$ is the cardinality of attribute X_j. The lower bound is the cardinality of an attribute X_j in E, where X_j has the highest cardinality among the attributes in set E. The upper bound is the product of the cardinalities of all attributes in E.

In detecting pattern from second to a higher order, the required sample size M for a valid test increases when $|X_i|\,|E|$ increases. When the required sample size is higher than the actual sample size, the detection is terminated. In other words, the sample size can limit the order of a high-order pattern to be detected.

3.4 Measure of Interdependency of NHOP

The amount of interdependency between a previously identified pattern $E=e$ and an additional interdependent value $X_i=a_{ip}$ is estimated by weighted self mutual information (MI), which is calculated as:

$$MI(e, a_{ip}) = P(e, a_{ip})log\frac{P(e, a_{ip})}{P(e)P(a_{ip})} . \qquad (7)$$

To calculate the overall relevance of a detected NHOP, combining the interdependency among all the iterations, we use a product of the weighted self mutual information (Ω). For a n-order pattern, $(((X_i=a_{ip},\ X_j=a_{jq}),\ X_m=a_{ml}),...,\ X_h=a_{hL})$, Ω is defined as:

$$\begin{aligned}\Omega&(((X_i = a_{ip}, X_j = a_{jq}), X_m = a_{ml}), ..., X_h = a_{hL})\\ &= MI(X_i = a_{ip}, X_j = a_{jq})\\ &\times MI((X_i = a_{ip}, X_j = a_{jq}), X_m = a_{ml})\\ &\times MI((X_i = a_{ip}, X_j = a_{jq}, X_m = a_{ml}, ...), X_h = a_{hL}) . \qquad (8)\end{aligned}$$

This measure calculates the total interdependency as the joint information of that identified for all iterations. Notice that, Ω is high when the calculated self mutual information for all the added values in the pattern is consistently high. Furthermore, the kth value is added only when it is significantly interdependent with the (k-1)-order NHOP.

4 Best-k Algorithm in Detecting NHOP

To detect a set of NHOP patterns with high interdependency, an algorithm known as Best-k algorithm is proposed in [14] . In the algorithm, a desired upper bound (k) of the detected patterns is selected a priori in order to speed up the search and even out the search space between lower and higher iteration levels. We select the best k patterns at each iteration based on the objective function and only expand on these best k patterns for higher order search. The search is speeded up to explore the best k patterns, without the need to evaluate the rest of the patterns. In addition, the search allows overall best patterns to be selected at each iteration, which is analogous to a refinement of the detected patterns.

Here is an outline of the search algorithm in detecting the best k NHOP patterns:

1. (Initialization) Identify all primitive-pairs (S) from two distinct variables that are statistically significant from the data;
2. Identify the best k 2-order patterns from the set S based on the weighted self mutual information (Equation 7), denoted the set of best k patterns as S_n, where $n=2$;
3. (Iteration) Define the set of $(n+1)$-order patterns as R_{n+1}. Initially, $R_{n+1}=\phi$;
4. Consider a compound event e in S_n. Identify a set of primitives A such that each primitive $X_i=a_{ip}$ in A is statistically interdependent with e. (i.e. $d(e,X_i=a_{ip}) \geq N_\alpha$, where X_i is not a variable in the variable set of e);
5. Add the set of identified patterns to R_{n+1}. (i.e. $R_{n+1} = R_{n+1}\cup\{(e, X_i = a_{ip})\}$);
6. Go back to Step 4 if there is another compound event e in S_n for consideration;
7. Select a set of best k patterns from R_{n+1} based on the objective function Ω (Equation 8). Denote this set of best k $(n+1)$-order patterns as S_{n+1};
8. (Termination) Consider a pattern $(e,X_i=a_{ip})$ in S_{n+1} as one compound event (i.e. $e=e\cup(X_i=a_{ip})$). Set $n=n+1$ and go back to Step 3 for the next iteration. The search is terminated either when no pattern is identified in the current iteration (i.e. $S_{n+1}=\phi$) or when there are not enough samples to conduct a reliable statistical test.

5   Relationships between Three Related Types of Multi-value Association Patterns

HOP, NHOP, and CP are three different but related types of patterns. In HOP, the null hypothesis assumes all pattern values are mutually independent. The pattern may be disjoint with independent parts. On the other hand, in NHOP and CP patterns, each value is interdependent to some other values in the pattern. NHOP is detected by adding values dependent to the previously constructed lower-order patterns. NHOP then emphasizes the consistently associated structure among its nested value subgroups. CP focuses on a single value with multiple association relationships to other values.

The mathematical definition for HOP, CP and NHOP are presented as follows. Consider a pattern that consists of a set of n values, denoted as $\{a_i\}$, where $i=1,2,\ldots,n$.

Definition 1. *Consider the case of the independent assumption, a high-order pattern (HOP), denoted as $e=(a_1,a_2,\ldots,a_n)$, satisfies the following properties:*

$$P(e = (a_1, a_2, \ldots, a_n)) = hP(a_1)P(a_2)\ldots P(a_n),$$

where h represents the ratio of the joint probability of an event e to the expected probability by assuming all the pattern values independent; taking $h>1$ and h is significantly large.

Definition 2. *A consign pattern (CP), denoted as $(e=a_1,\{a_2,\ldots,a_n\})$, such that $e=a_1$ is the consigned value, satisfies the following properties:*

$$P(e = a_1, a_i) = c_i P(e = a_1)P(a_i), \quad for\ i=2,\ldots,n,$$

where c_i represents the interdependency (or weight) between $e=a_1$ and a_i; taking $c_i>1$ and c_i is significantly large.

Definition 3. *A nested high-order pattern (NHOP), denoted as $e=(((a_1,a_2), a_3), \ldots, a_{n-1}), a_n)$, satisfies the following properties for the ith case:*

$$P((a_1, \ldots, a_{i-1}), a_i) = k_i P(a_1, \ldots, a_{i-1})P(a_i), \quad for\ i=2,\ldots,n,$$

where k_i represents the interdependency (or weight) between $(a_1,\ldots,a_{i-1})$ and a_i; taking $k_i>1$ and k_i is significantly large. Note that the probability $P((a_1,\ldots,a_{i-1}), a_i)$ is the same as $P(a_1,\ldots,a_{i-1},a_i)$.

With these definitions, we describe their relationships with proofs.

Lemma 1. *Consider the following properties:*

$$P((a_1, \ldots, a_{i-1}), a_i) = k_i' P(a_1, \ldots, a_{i-1})P(a_i), \quad for\ i=2,\ldots,n, \quad (L1.1)$$

where k_i' represents the interdependency (or weight) between $(a_1,\ldots,a_{i-1})$ and a_i. Note that i starts from 2. Then the joint probability of $(a_1,\ldots,a_n)$ can be expressed as:

$$P(a_1, a_2, \ldots, a_n) = k_2' \ldots k_n' P(a_1) \ldots P(a_n). \quad (L1.2)$$

Proof. From (L1.1), consider $i=2$,

$$P(a_1, a_2) = k_2' P(a_1)P(a_2). \quad (L1.3)$$

Consider $i=3$,

$$P(a_1, a_2, a_3) = k_2' P(a_1, a_2)P(a_3). \quad (L1.4)$$

Substituting (L1.3) to (L1.4) gives

$$P(a_1, a_2, a_3) = k_2' k_3' P(a_1) P(a_2) P(a_3). \qquad \text{(L1.5)}$$

Consider the general case, substituting $P(a_1, a_2, \ldots, a_j)$ to (L1.1) (considering $i=j+1$) for $j=3,\ldots,n\text{-}1$, it can be shown that,

$$P(a_1, a_2, \ldots, a_n) = k_2' \ldots k_n' P(a_1) \ldots P(a_n). \qquad \square$$

Theorem 1. *NHOP is Always HOP*
That is, if NHOP is true (from definition 3), HOP is true (from definition 1).

Proof. From Lemma 1, the joint probability of $(a_1,\ldots,a_n)$ in definition 3 can then be expressed as:

$$P(a_1, a_2, \ldots, a_n) = k_2 \ldots k_n P(a_1) \ldots P(a_n). \qquad \text{(T1.1)}$$

Substituting (T1.1) to the formula of HOP (from definition 1) gives,

$$k_2 \ldots k_n P(a_1) \ldots P(a_n) = h P(a_1) \ldots P(a_n)$$

$$\Rightarrow h = k_2 \ldots k_n.$$

Hence, in order to satisfy HOP (from definition 1), $(k_2 k_3 \ldots k_n)$ needs to be significantly large. From definition 2, since each of $k_2, k_3, \ldots, k_n$ is greater than one and significantly large, $(k_2 k_3 \ldots k_n)$ is also significantly large. HOP is satisfied. $\qquad \square$

Theorem 2. *HOP is Not Necessary NHOP*
That is, if HOP is true (from definition 1), it does not guarantee NHOP is true (from definition 3).

Proof. Consider the following properties:

$$P((a_1, \ldots, a_{i-1}), a_i) = k_i' P(a_1, \ldots, a_{i-1}) P(a_i), \quad for\ i=2,\ldots,n, \qquad \text{(T2.1)}$$

where k_i' represents the interdependency between $(a_1,\ldots,a_{i-1})$ and a_i.
From Lemma 1, the joint probability of $(a_1,\ldots,a_n)$ can then be expressed as:

$$P(a_1, a_2, \ldots, a_n) = k_2' \ldots k_n' P(a_1) \ldots P(a_n). \qquad \text{(T2.2)}$$

Substituting (T2.2) to definition 1 gives,

$$k_2' \ldots k_n' P(a_1) \ldots P(a_n) = h P(a_1) \ldots P(a_n)$$

$$\Rightarrow h = k_2' \ldots k_n'.$$

To satisfy definition 3, each of $k_2', k_3', \ldots, k_n'$ needs to be significantly large. However, given h is significantly large, it does not guarantee that each of $k_2', k_3', \ldots, k_n'$ is significantly large. Thus, HOP is NHOP only if each of $k_2', k_3', \ldots, k_n'$ is significantly large. In other words, HOP is NHOP if the constructed pattern at each iteration is significantly interdependent with each added value. In conclusion, HOP may not be NHOP.   □

Theorem 3. *HOP is Not Necessary CP*
That is, even if HOP is true (from definition 1), it does not guarantee CP is true (from definition 2).

Proof. Consider HOP (from definition 1),

$$P(a_1, a_2, \ldots, a_n) = hP(a_1)P(a_2)\ldots P(a_n), \qquad\qquad \text{(T3.1)}$$

where *h>1* and *h* is significantly large.
Consider the following properties:

$$P((a_1, \ldots, a_{i-1}), a_i) = k_i'P(a_1, \ldots, a_{i-1})P(a_i), \quad \text{for } i{=}2, \ldots, n, \qquad \text{(T3.2)}$$

where k_i' represents the interdependency (or weight) between $(a_1, \ldots, a_{i-1})$ and a_i.
From Lemma 1, the joint probability of $(a_1, \ldots, a_n)$ can then be expressed as:

$$P(a_1, a_2, \ldots, a_n) = k_2'k_3' \ldots k_n'P(a_1)P(a_2)P(a_3)\ldots P(a_n). \qquad \text{(T3.3)}$$

Here, we express the interdependency (or weight) between a_1 and a_2 in terms of *h* in definition 1:
Consider (T3.2) where *i=2*,

$$P(a_1, a_2) = k_2'P(a_1)P(a_2)$$

$$\Rightarrow k_2' = P(a_1, a_2)/(P(a_1)P(a_2)). \qquad\qquad \text{(T3.4)}$$

Substituting (T3.4) to (T3.3) gives,

$$P(a_1, a_2, \ldots, a_n) = k_3' \ldots k_n'P(a_1)\ldots P(a_n)P(a_1, a_2)/(P(a_1)P(a_2)) \quad \text{(T3.5)}$$

Substituting (T3.1) to (T3.5) gives,

$$hP(a_1)\ldots P(a_n) = k_3' \ldots k_n'P(a_1)\ldots P(a_n)P(a_1, a_2)/(P(a_1)P(a_2))$$

$$\Rightarrow h = k_3' \ldots k_n'P(a_1, a_2)/(P(a_1)P(a_2))$$

$$\Rightarrow P(a_1, a_2) = P(a_1)P(a_2)h/(k_3' \ldots k_n'), \qquad\qquad \text{(T3.6)}$$

where $h/(k_3' \ldots k_n')$ represents the interdependency between a_1 and a_2.
    To satisfy CP (definition 2), at least a_1 needs to be interdependent with a_2. In other words, $h/(k_3' \ldots k_n')$ in (T3.6) needs to be significantly large. However, given *h* is significantly large (from definition 1), it does not guarantee that $h/(k_3' \ldots k_n')$ is significantly large. Thus, definition 2 may not be satisfied. HOP is CP only if at least a value (e.g. a_1) in the pattern is interdependent with all other values. In conclusion, HOP may not be CP.   □

Theorem 4. *CP is Not Necessary HOP*
That is, if CP is true (from definition 2), it does not guarantee HOP is true (from definition 1).

Proof. Consider CP (from definition 2), $(e=a_1,\{a_2,\ldots,a_n\})$, with pairwise relationships between $e=a_1$ and a_i, for $i=2,\ldots,n$.

$$P(e = a_1, a_i) = c_i P(e = a_1) P(a_i), \tag{T4.1}$$

where c_i represents the interdependency (or weight) between a_1 and a_i; taking $c_i>1$ and c_i is significantly large.
From (T4.1), consider a case where $i=2$,

$$P(a_1, a_2) = c_2 P(a_1) P(a_2), \tag{T4.2}$$

where c_2 represents the interdependency between a_1 and a_2; taking $c_2>1$ and c_2 is significantly large.
Consider the following properties:

$$P((a_1,\ldots,a_{i-1}), a_i) = k_i' P(a_1,\ldots,a_{i-1}) P(a_i), \quad \text{for } i=2,\ldots,n, \tag{T4.3}$$

where k_i' represents the interdependency between $(a_1,\ldots,a_{i-1})$ and a_i.
From Lemma 1, the joint probability of $(a_1,\ldots,a_n)$ can then be expressed as:

$$P(a_1, a_2,\ldots, a_n) = k_2'\ldots k_n' P(a_1)\ldots P(a_n). \tag{T4.4}$$

From (T4.3), consider a case where $i=2$,

$$P(a_1, a_2) = k_2' P(a_1) P(a_2), \tag{T4.5}$$

where k_2 represents the interdependency between a_1 and a_2.
Substituting (T4.5) to CP (T4.2) gives,

$$k_2' = c_2. \tag{T4.6}$$

Since c_2 is significantly large from definition 2, k_2' is also significantly large.
Consider HOP (from definition 1),

$$P(a_1, a_2,\ldots, a_n) = h P(a_1)\ldots P(a_n), \tag{T4.7}$$

where $h>1$ and h is significantly large.
Substituting (T4.4) to HOP (T4.7) gives,

$$k_2'\ldots k_n' P(a_1)\ldots P(a_n) = h P(a_1)\ldots P(a_n)$$

$$\Rightarrow h = k_2'\ldots k_n'. \tag{T4.8}$$

To satisfy HOP (from definition 1), $(k_2'\ldots k_n')$ in (T4.8) needs to be significantly large. Given CP, we know that k_2' is significantly large (T4.6). However, only given k_2' is significantly large does not guarantee that $(k_2'k_3' \dots k_n')$ is significantly large. Thus, definition 1 may not be satisfied. CP can be HOP if each value in CP is jointly interdependent. In conclusion, CP may not be HOP. □

Theorem 5. *CP is Not Necessary NHOP*
That is, if CP is true (from definition 2), it does not guarantee NHOP is true (from definition 3).

Proof. Consider CP (from definition 2), $(e=a_1,\{a_2,\dots,a_n\})$, with pairwise relationships between $e=a_1$ and a_i, for $i=2,\dots,n$.

$$P(e = a_1, a_i) = c_i P(e = a_1) P(a_i), \qquad (T5.1)$$

where c_i represents the interdependency (or weight) between a_1 and a_i; taking $c_i>1$ and c_i is significantly large.
Without loss of generality for all $i=2,\dots,n$. Consider the case where $i=2$,

$$P(a_1, a_2) = c_2 P(a_1) P(a_2), \qquad (T5.2)$$

where c_2 represents the interdependency between a_1 and a_2; taking $c_2>1$ and c_2 is significantly large.
Consider NHOP (from definition 3), denoted as $e=(((a_1,a_2),a_3)\dots,a_{n-1}),a_n)$, satisfies the following properties:

$$P((a_1,\dots,a_{i-1}),a_i) = k_i P(a_1,\dots,a_{i-1}) P(a_i), \quad \text{for } i=2,\dots,n, \qquad (T5.3)$$

where k_i represents the interdependency between $(a_1,\dots,a_{i-1})$ and a_i; taking $k_i>1$ and k_i is significantly large.
From Lemma 1, the joint probability of $(a_1,\dots,a_n)$ can then be expressed as:

$$P(a_1, a_2, \dots, a_n) = k_2 \dots k_n P(a_1) \dots P(a_n). \qquad (T5.4)$$

Consider $i=2$ in NHOP (T5.3),

$$P(a_1, a_2) = k_2 P(a_1) P(a_2). \qquad (T5.5)$$

Substituting (T5.2) to (T5.5) gives,

$$k_2 = c_2. \qquad (T5.6)$$

Given CP, c_2 is significantly large, so k_2 is significantly large (T5.6). To satisfy NHOP (from definition 3), each of k_2, $k_3,\dots$, and k_n in (T5.4) needs to be significantly large. However, given only k_2 is significantly large, it does not guarantee that each of $k_3,\dots$, k_n is significantly large. Thus, definition 3 of NHOP may not be satisfied. CP is NHOP only if the constructed pattern at each iteration is significantly interdependent with each added value. In conclusion, CP may not be NHOP.   □

Theorem 6. *NHOP is Not Necessary CP*
That is, if NHOP is true (from definition 3), it does not guarantee CP is true (from definition 2).

Proof. Consider NHOP (from definition 3), denoted as $e=((((a_1,a_2), a_3), \ldots, a_{n-1}),a_n)$, satisfies the following properties:

$$P((a_1,\ldots,a_{i-1}),a_i) = k_i P(a_1,\ldots,a_{i-1})P(a_i), \quad \text{for } i=2,\ldots,n, \qquad \text{(T6.1)}$$

where k_i represents the interdependency between $(a_1,\ldots,a_{i-1})$ and a_i; taking $k_i>1$ and k_i is significantly large.
Substituting $i=2$ in (T6.1) to $i=3$ in (T6.1) gives,

$$P(a_1,a_2,a_3) = k_2 k_3 P(a_1)P(a_2)P(a_3), \qquad \text{(T6.2)}$$

where k_2 and k_3 are significantly large.
Here, we express the interdependency (or weight) between a_1 and a_3 in terms of k_2 and k_3 in NHOP. Consider the following the following properties:

$$P(a_1,a_2,a_3) = k_2' P(a_1,a_3)P(a_2), \qquad \text{(T6.3)}$$

where k_2' represents the interdependency between (a_1,a_3) and a_2.
Substituting (T6.3) to (T6.2) gives,

$$k_2' P(a_1,a_3)P(a_2) = k_2 k_3 P(a_1)P(a_2)P(a_3)$$

$$\Rightarrow P(a_1,a_3) = P(a_1)P(a_3)k_2 k_3/k_2', \qquad \text{(T6.4)}$$

where $(k_2 k_3/k_2')$ represents the interdependency between a_1 and a_3.

   To satisfy CP (definition 2), at least a_1 and a_3 need to be significantly interdependent. In other word, $(k_2 k_3/k_2')$ needs to be significantly large in (T6.4). However, given k_2 and k_3 are significantly large in (T6.2), it does not guarantee that $(k_2 k_3/k_2')$ are significantly large when k_2' is sufficiently large. Thus, definition 2 of CP may not be satisfied. NHOP is CP only if the initial value is interdependent with the added value at each iteration. In conclusion, NHOP may not be CP.   □

In summary, since NHOP is always HOP (Theorem 1) and HOP may not be NHOP (Theorem 2), we conclude that NHOP is a special case of HOP. In other word, NHOP is a sub-type of HOP where each value is interdependent to a nested subgroup of values in the pattern. For the relationship between HOP and CP, HOP may not be CP and vice versa (Theorem 3 and Theorem 4). Similarly, for the relationship between NHOP and CP, NHOP may not be CP and vice versa (Theorem 5 and Theorem 6).

6   Results and Discussion

6.1   Experiment 1: Comparison between NHOP and HOP Using Synthetic Data

This experiment evaluates the patterns detected as NHOP and compares them to HOP using synthetic data. To compare the two types of patterns, first, the complete set of high order patterns is detected using the statistics of adjusted residual as described in [20]. We use a high level of confidence of 99.9%. All significant NHOP patterns are then identified using the same confidence level. After obtaining these two sets of patterns, a comparison between them is then performed.

The synthetic data set generated is a data set with seven variables of attributes, as $(X_1, X_2,...,X_7)$. Each attribute can take one of five possible attributed values, or $X_i=1,2,..,5$. A total of 1000 samples are randomly generated. Next, two strictly independent 3-order patterns, A$=(X_1=1, X_2=1, X_3=1)$ and B$=(X_4=2, X_5=2, X_6=2)$ are designed and imposed into the randomly generated data set. They are independent since the occurrence of each pattern has no bearing on each other. To impose one of these two independent patterns, 50 samples are randomly selected. The values of the selected samples are then modified to that of the pattern and return back to the sample set. The process is done for the two patterns. After that the resulting data set is used in the evaluation. Note that A and B separately conforms to NHOP, but HOP may be composed of some sub-patterns, or even a mixture of them.

Experimental Results

A total of 21 high-order patterns (HOP) are detected (Fig. 2). The two embedded 3-order patterns and their six 2-order sub-patterns are successfully extracted (pattern 1 to 8 in Fig. 2). However, 13 additional 3-order HOP patterns are extracted (pattern 9 to 21). Each of them composes of a significant 2-order pattern and an independent value. For example in pattern 9, $(X_1=1, X_5=2, X_6=2)$, composes of a significant 2-order pattern $(X_5=2, X_6=2)$ and an independent value $X_1=1$. The other 13 patterns are detected because the whole set of values is significant, even though the two original patterns A and B are independent.

As comparison, a total of eight NHOP patterns are identified (Fig. 3) correctly. They include the two embedded 3-order patterns and their sub-patterns. Note that the number of the detected NHOP patterns is less than that of HOP, the analysis of the NHOP patterns is also easier.

This experiment demonstrates two different types of patterns detected in the synthetic data set. We conclude that the NHOP pattern is an important type of multi-value association patterns that can facilitate interpretation. In other words, there are tightly connected relationships between the composed values in NHOP. On the other hand, when the internal interdependency may

Fig. 2 Identified HOP patterns in synthetic data

Pattern ID	Order	X_1	X_2	X_3	X_4	X_5	X_6	X_7
1	2					2	2	
2	2				2	2		
3	2				2		2	
4	2	1		1				
5	2		1	1				
6	2	1	1					
7	3				2	2	2	
8	3	1	1	1				
9	3	1				2	2	
10	3					2	2	4
11	3			4		2	2	
12	3				2	2		1
13	3			1		2	2	
14	3			2		2	2	
15	3					2	2	1
16	3				2		2	1
17	3				4	2	2	
18	3	4				2	2	
19	3	1		1			2	
20	3	1		1	1			
21	3		1	1				4

Fig. 3 Identified NHOP patterns in synthetic data

Pattern ID	Order	X_1	X_2	X_3	X_4	X_5	X_6	X_7
1	2					2	2	
2	2				2	2		
3	2				2		2	
4	2	1		1				
5	2		1	1				
6	2	1	1					
7	3				2	2	2	
8	3	1	1	1				

not be clear or very weak, then HOP may provide a more flexible indication for interpretation.

6.2 Experiment 2: Evaluate NHOP Using Biomolecular Data

In this experiment, we consider a protein family known as SH3 domains. They are identified in a wide range of organisms such as yeast and human. Their length is around sixty amino acids among different species. They are found in different types of proteins such as kinases, lipases, GTPases, adopter proteins, and structural proteins. These proteins act in diverse processes including signal transduction, cell cycle regulation, and actin organization. The function of the SH3 domain is to mediate protein-protein interactions by binding to specific sequence motifs in target proteins. Here, we evaluate the correspondence between the patterns detected to what is known about SH3 domains.

The aligned SH3 domain sequences can be obtained from the Pfam protein family database [2]. The alignment has 3373 sequences with the alignment length of 57 units. A goal is to identify crucial functional and structural locations of the molecule even when the 3-dimensional structure may not be known.

From a previous study of SH3 domains [9], two important types of sites in the protein structure are defined. One type is known as the core sites, which are located at the hydrophobic core region in the structure. They are crucial in maintaining the stability of the molecular structure. Another type is known as the binding sites, which are located on the binding surface. The binding sites are critical for peptide binding. Among a total of 57 residues in the sequence of SH3 domain, 10 and 15 residues are defined as the core sites and the binding sites respectively. The remaining 32 residues are without specific functions in their study. We can compare the sites of the identified patterns to these types of structural sites. For example, given an identified 4-order pattern, $(((X_{39}=G, X_{50}=F), X_{26}=I), X_{54}=F)$, we found that site 39, site 50, and site 26 are core sites. Site 54 is a binding site.

Experimental Results

When the *Best-k* algorithm is applied by setting k to a large value (k=1000), a total of 1000 4-order NHOP patterns are generated. To facilitate interpretation, we group the detected patterns based on the nearest-neighbor clustering method. The distance measures is defined as the number of differences between corresponding values of the two patterns to be compared. When a threshold is set on the nearest-neighbor distance, four meaningful pattern groups form.

Table 1 shows the best three patterns for each pattern group. The first pattern group found to be related to the protein core, where most pattern sites (sites 39, 50, 55, 26, and 20) are located at the molecular core. The second pattern group overlaps to the SH3 domain core at site 39 and site 50 with different pattern values, $X_{39}=A$ (Alanine) and $X_{50}=I$ (Isoleucine), which indicate the importance of these two sites in the SH3 domain. The third group is highly related to the binding surface, where most of the pattern sites (sites 10, 16, 17, 34, 35) are binding sites. The fourth pattern group found to be related to both core and binding sites.

In summary, the extracted patterns are found to be important locations of the molecule, including the core positions and those related to the binding surface of the protein. They are important because SH3 has function in mediating protein-protein interactions. The core position is important for maintaining the molecular structure and stability. The binding surface is crucial for binding activities when the molecule interacts. It is clear that identifying

Table 1 Groupings of 4-order NHOP patterns detected by *Best-k* algorithm (k=1000)

Pattern Group	Best 3 patterns in group	Core	Bind	Relationship to functional and structural locations
	$(((X_{39}{=}G, X_{50}{=}F), X_{55}{=}V), X_{26}{=}I)$	4	0	Related to core
1	$(((X_{39}{=}G, X_{50}{=}F), X_{26}{=}I), X_{20}{=}F)$	4	0	
	$(((X_{39}{=}G, X_{50}{=}F), X_{26}{=}I), X_{54}{=}F)$	3	1	
	$(((X_{39}{=}A, X_{50}{=}I), X_{17}{=}D), X_{56}{=}A)$	2	1	Related to core
2	$(((X_{39}{=}A, X_{50}{=}I), X_{56}{=}A), X_{41}{=}S)$	2	0	
	$(((X_{39}{=}A, X_{50}{=}I), X_{17}{=}D), X_{41}{=}S)$	2	1	
	$(((X_{16}{=}D, X_{17}{=}E), X_{35}{=}G), X_{10}{=}Y)$	0	4	Related to binding
3	$(((X_{16}{=}D, X_{17}{=}E), X_{35}{=}G), X_{39}{=}G)$	1	3	surface
	$(((X_{16}{=}D, X_{17}{=}E), X_{35}{=}G), X_{34}{=}D)$	0	4	
	$(((X_{6}{=}A, X_{8}{=}Y), X_{54}{=}Y), X_{53}{=}N)$	1	3	Related to core and
4	$(((X_{6}{=}A, X_{8}{=}Y), X_{53}{=}N), X_{20}{=}F)$	2	2	binding surface
	$(((X_{8}{=}Y, X_{54}{=}Y), X_{53}{=}N), X_{20}{=}F)$	1	3	

Note: 'Core' and 'Bind' are the number of core sites and binding sites in the pattern respectively.

these positions is not trivial when only the sequence data are given. However, when found and in relation to the 3-dimensional molecular structure as binding sites or core sites, are important in understanding the molecule.

7 Conclusion

In this chapter, we have discussed three related multi-value association patterns and their relationships. All of them have shown to be very important for data mining involving discrete valued data. Furthermore, they generalize sequential data and have a more specific interpretation than multi-variable patterns. Here, further evaluations are made with respect to their conceptual mathematical properties. On the more recently identified nested high-order pattern (NHOP), the experiments further highlight its unique characteristics as well as its usefulness in biomolecular analysis using multiple sequences of protein family. The experiment on SH3 domain shows that the identified patterns can be interpreted as important molecular sites as a group that forms together with important biological functions of binding interactions or core sites in preserving structural stability. Hence we conclude that multi-value association patterns are extremely useful for data mining.

References

1. Agrawal, R., Imielinski, T., Swami, A.N.: Mining association rules between sets of items in large databases. In: SIGMOD Conference 1993, pp. 207–216 (1993)
2. Bateman, A., Coin, L., Durbin, R., Finn, R.D., Hollich, V., Griffiths-Jones, S., Khanna, A., Marshall, M., Moxon, S.: The Pfam protein families database. Nucleic Acids Research 32, D138–D141 (2004)
3. Bazzi, I., Glass, J.: Learning units for domain-independent out-of-vocabulary word modelling. In: Proceedings of European Conference on Speech Communication and Technology, Aalborg, pp. 61–64 (September 2001)
4. Brin, S., Motwani, R., Silverstein, C.: Beyond market baskets: Generalizing association rules to correlations. In: SIGMOD Conference 1997, pp. 265–276 (1997)
5. Chiu, D.K.Y., Lui, T.W.H.: Integrated use of multiple interdependent patterns for biomolecular sequence analysis. International Journal of Fuzzy Systems, Special Issue on Intelligent Computation for Data Mining and Knowledge Discovery 4(3), 766–775 (2002)
6. Chiu, D.K.Y., Lui, T.W.H.: A multiple-pattern biosequence analysis method for diverse source association mining. Applied Bioinformatics 4(2), 85–92 (2005)
7. Chiu, D.K.Y., Wong, A.K.C., Cheung, B.: Information discovery through hierarchical maximum entropy discretization and synthesis. In: Piatetsky-Shapiro, G., Frawley, W.J. (eds.) Knowledge Discovery in Databases, pp. 125–140. MIT/AAAI Press (1991)
8. Chiu, D.K.Y., Wong, A.K.C.: Multiple pattern associations for interpreting structural and functional characteristics of biomolecules. Information Science, An International Journal 167, 23–39 (2004)
9. Di Nardo, A.A., Larson, S.M., Davidson, A.R.: The relationship between conservation, thermodynamic stability, and function in the SH3 domain hydrophobic core. Journal of Molecular Biology 333(3), 641–655 (2003)
10. Haberman, S.J.: The analysis of residuals in cross-classified tables. Biometrics 29, 205–220 (1973)
11. Han, J., Pei, J., Yin, Y.: Mining frequent patterns without candidate generation. In: Proceedings of the 2000 ACM-SIGMOD international conference on management of data (SIGMOD 2000), Dallas, TX, pp. 1–12 (2000)
12. Jaroszewicz, S., Simovici, D.A.: Interestingness of frequent itemsets using Bayesian networks as background knowledge. In: KDD 2004, pp. 178–186 (2004)
13. Jaroszewicz, S., Scheffer, T.: Fast discovery of unexpected patterns in data, relative to a Bayesian network. In: KDD 2005, pp. 118–127 (2005)
14. Lui, T.W.H., Chiu, D.K.Y.: Discovering maximized progressive high-order patterns in biosequences. In: Cao, P.Y., et al. (eds.) Proceedings of the 10th Joint Conference on Information Sciences, pp. 110–115 (2007)
15. Lui, T.W.H., Chiu, D.K.Y.: Complementary Analysis of High-Order Association Patterns and Classification. In: Proceedings of the 21st Florida Artificial Intelligence Research Society Conference (FLAIRS), Florida, USA, pp. 294–299 (2008)
16. Sy, B.K.: Information-statistical pattern based approach for data mining. Journal of Statistical Computing and Simulation 69(2), 1–31 (2001)
17. Sy, B.K.: Discovering association patterns based on mutual information. In: Perner, P., Rosenfeld, A. (eds.) MLDM 2003. LNCS, vol. 2734, pp. 369–378. Springer, Heidelberg (2003)

18. Tillier, E.R., Lui, T.W.H.: Using multiple interdependency to separate functional from phylogenetic correlations in protein alignments. Bioinformatics 19, 750–755 (2003)
19. Wang, W., Yang, J.: Mining sequential patterns from large data sets. In: Elmagarmid, A.K. (ed.) Advances in Database Systems. Springer, Heidelberg (2005)
20. Wong, A.K.C., Wang, Y.: High-order pattern discovery from discrete-valued data. IEEE Transactions on Knowledge and Data Engineering 8(6), 877–892 (1997)
21. Wong, A.K.C., Wang, Y.: Pattern discovery: A data driven approach to decision support. IEEE Transactions on Knowledge and Data Engineering 15(3), 914–925 (2003)
22. Wu, X., Barbara, D., Ye, Y.: Screening and interpreting multi-item associations based on log-linear modeling. In: KDD 2003, pp. 276–285 (2003)
23. Zaki, M.J.: Scalable algorithms for association mining. IEEE Transactions on Knowlegde and Data Engineering 12, 372–390 (2000)

Clustering Time Series Data: An Evolutionary Approach

Monica Chiş, Soumya Banerjee, and Aboul Ella Hassanien

Abstract. Time series clustering is an important topic, particularly for similarity search amongst long time series such as those arising in bioinformatics, in marketing research, software engineering and management. This chapter discusses the state-of-the-art methodology for some mining time series databases and presents a new evolutionary algorithm for times series clustering an input time series data set. The data mining methods presented include techniques for efficient segmentation, indexing, and clustering time series.

Keywords: Time series, clustering, evolutionary computation, evolutionary clustering.

1 Introduction

Time series are an ordered sequence of values of a variable at equally spaced time intervals. Time series analysis is often associated with the discovery and use of patterns (such as periodicity, seasonality, or cycles), and prediction of future values (specifically termed forecasting in the time series context). One key difference between time series analysis and time series data mining is the large number of series involved in time series data mining. As shown in a vast volume of time series literature, traditional time series analysis and modeling tend to be based on non-automatic and trial-and-error approaches. It is very difficult to develop time series models using a non-automatic approach when a large number of time series are involved. To automatic model building, discovery of knowledge associated with events known or unknown a priori can provide valuable information toward the success of a business operation. Adding the time dimension to real-world databases produces Time Series Databases (TSDB) and introduces new aspects and difficulties to data mining and knowledge discovery.

Monica Chiş
SIEMENS Program and System Engineering (PSE) Romania

Soumya Banerjee
Department of Computer Science & Engineering, Birla Institute of Technology, International Center Mauritius

Aboul Ella Hassanien
Cairo University, Faculty of Computer and Information, Information Technology Department

A. Abraham et al. (Eds.): Foundations of Comput. Intel. Vol. 6, SCI 206, pp. 193–207.
springerlink.com © Springer-Verlag Berlin Heidelberg 2009

Time series arise in many important areas. Some of this areas are: the evolution of stock price indices, share prices or commodity prices, the sales figures for a particular good or service, macroeconomic and demographic indicators, image sequences, acceleration measurements by sensors on a bridge, ECG recordings (a test that records the electrical activity of the heart; with each heart beat, an electrical impulse travels through the heart) or EEG recordings (EEG recording shows rhythmical electrical activity, often called brain waves. The brain waves may be normal or show abnormalities in certain regions. In people with epilepsy, there may be "epileptic activity" on the EEG indicating their predisposition to seizures), gene expression measurements at consecutive time points in bioinformatics, are a part of the field in which time series play a very important role. Typically, the analysis of time series used on these areas is oriented to estimating a good model that can be used for monitoring, producing accurate forecasts, providing structural information and/or information about the influence of a particular set of inputs on the desired output. Time series data is perhaps the most frequently encountered type of data examined by the data mining community. Clustering is perhaps the most frequently used data mining algorithm, being useful in it's own right as an exploratory technique, and also as a subroutine in more complex data mining algorithms such as rule discovery, indexing, summarization, anomaly detection, and classification [18] [24] [25] [26].

Clustering of time series has attracted the interest of researchers from a wide range of fields, particularly from statistics [31], signal processing [12] and data mining [28]. A survey of current research in the field has been published in [39]. This growth in interest in time series clustering has resulted in the development of a wide variety of techniques designed to detect common underlying structural similarities in time dependent data.

The chapter is organized as follows. Section 2 presents the related works in time series clustering. In Section 3 the novel Evolutionary Time Series Clustering Algorithm (ETSC) is proposed. In Section 4 concludes the chapter and outlines some directions for future research in this direction.

2   Related Works: Time Series Clustering

The mining of time series data has attracted great attention in the data mining community in recent years ([1], [6], [7], [18], [19] [33] [40]). Clustering time series is an important topic, motivated by several research challenges including similarity search of bioinformatics sequences, as well as the challenge of developing methods to recognize dynamic change in time series ([15], [17], [34], [42]). In the field of marketing research and management problem the clustering time series is a very important topic. The mining of time series in general and the clustering of time series in particular has attracted the interest of researchers from a wide range of fields, particularly from statistics [31], signal processing [12] and data mining [30]. According to Keogh et al. [18] there are two main categories in time series clustering: *whole clustering* and *subsequence clustering*. *Whole clustering* is the clustering performed on many individual time series to group similar series into clusters.

Subsequence clustering is based on sliding window extractions of a single time series and aims to find similarity and differences among different time window of a single time series. Subsequence clustering is commonly used as a subroutine in many other algorithms, including rule discovery [11, 13, 16], indexing, classification, prediction [35, 38], and anomaly detection [43]. One of the most widely used clustering approaches is hierarchical clustering; due to the great visualization power it offers [19]. Hierarchical clustering produces a nested hierarchy of similar groups of objects, according to a pair wise distance matrix of the objects. One of the advantages of this method is its generality, since the user does not need to provide any parameters such as the number of clusters. However, its application is limited to only small datasets, due to its quadratic (or higher order) computational complexity. In [7], [32] was presented k-Means time series clustering which is considered to be the faster method to perform clustering. The basic intuition behind k-Means (and in general, iterative refinement algorithms) is the continuous reassignment of objects into different clusters, so that the within-cluster distance is minimized. For time-series, the objective is to find a representation at a lower dimensionality that preserves the original information and describes the original shape of the time-series data as closely as possible. Many approaches have been suggested in the literature, including the Adaptive Piecewise Constant Approximation [22], Piecewise Aggregate Approximation (PAA) [9, 46], Piecewise Linear Approximation [20] and the Discrete Wavelet Transform (DWT) [39]. All these approaches have shared the ability to produce a high quality reduced-dimensionality approximation of time series. Wavelets are unique in that their representation of data is intrinsically multi-resolution.

A novel version of partition clustering for time series is presented in [29]. The algorithm works by leveraging off the multi-resolution property of wavelets. Initializing the centers at each approximation level, using the final centers returned by the coarser representations, mitigated the dilemma of choosing the initial centers. In addition to casting the clustering algorithms as anytime algorithms, this approach has two other very desirable properties. By working at lower dimensionalities the author can efficiently avoid local minima. The quality of the clustering is usually better than the batch algorithm. In addition, even if the algorithm is run to completion, the proposed approach is much faster than its batch counterpart. Empirically demonstrate these surprising and desirable properties with comprehensive experiments on several publicly available real data sets. The algorithms presented in [29] have used a wavelet decomposition to perform clustering at increasingly finer levels of the decomposition, while displaying the gradually refined clustering results periodically to the user. Any wavelet basis (or any other multiresolution decomposition such as DFT) can be used. Haar Wavelet was used because of its simplicity. The Haar Wavelet decomposition is computed for all time-series data in the database. The complexity of the transformation is linear to the dimensionality of each object; therefore, the running time is reasonable even for large databases. The process of decomposition can be performed off-line, and needs to be done only once. The time series data can be stored in the Haar decomposition format, which takes the same amount of space as the original sequence. One important property of the decomposition is that it is a lossless transformation, since the original sequence can always be reconstructed from the decomposition. Once the Haar decomposition is computed, the k-Means clustering algorithm was performed, starting at the second level (each object at level i has $2(i-1)$ dimensions) and gradually progress to finer levels. Since the Haar decomposition is completely reversible, the approximation data can be reconstructed from the coefficients at any level and performs clustering on these data. Another categorization of clustering time series is the classification in model based or model free. Model based approaches (also called generative approaches) assume some form of the underlying generating process, estimate the model from each data then cluster based on similarity between model parameters. The most commonly assumed model forms are: polynomial mixture models [4]; [5] [41]; ARMA [18]; [45], Markov Chain and Hidden Markov Models (MC and HMM) [37]; [35]; [44]. Reviews of methods used to cluster time series can be found in [4].

Model free approaches involve the specification of a specific distance measure and/or a transformation of the data. Measures based on common subsequences (used in [10], shape parameters [27] and correlation [36] have been proposed. Transformations used in conjunction with time series data mining include fast Fourier transforms; dynamic time warping; wavelet transforms; and piecewise constant approximation. A large number of transformations are evaluated in [39]. In [3], Alcock and Manolopoulus instead of calculating the Euclidean distance directly between two sequences, the sequences are transformed into a feature vector and the Euclidean distance between the feature vectors is then calculated. Results show that this approach is superior for finding similar sequences. This is a different approach of time series similarity.

One common feature of time series is the likely presence of outliers. These uncharacteristic data can significantly affect the quality of clusters formed. In [4] a method of overcoming the detrimental effects of outliers is evaluated. Some of the alternative approaches to clustering time series are described and a particular class of model for experimentation with k-means clustering and a correlation based distance metric are presented. In Singhal & Seborg [43] a new methodology based on calculating the degree of similarity between multivariate time-series datasets using two similarity factors for clustering multivariate time-series data is proposed. One similarity factor is based on principal component analysis and the angles between the principal component subspaces. The other similarity factor is based on the Mahalanobis distance between the datasets. The standard K-means clustering algorithm is modified to cluster multivariate time-series datasets using similarity factors. The author showed with some comparisons with existing clustering methods several advantages of the proposed method. Another approach of time series clustering is the use of evolutionary computation techniques in order to cluster the data. In Chiş and Groşan [8] a new evolutionary algorithms for hierarchical time series clustering is proposed. A linear chromosome for solution representation is presented in the chapter.

3 Evolutionary Time Series Clustering Algorithm (ETSC)

Evolutionary algorithms are randomized search optimization techniques guided by the principles of natural biology evolution processes. They are effective and robust search processes provide a near optimal solution of a fitness function. Evolutionary algorithms play an important role in the various steps of data mining process. Evolutionary algorithms are used to improve the robustness and accuracy of the more traditional techniques used in feature extraction, feature selection, classification and clustering [32]. In this chapter a new evolutionary algorithms focused on whole clustering of time series, called *Evolutionary Time Series Clustering- ETSC* is presented.

A time series is the simplest form of temporal data. A time series is a sequence of real numbers collected regularly in time, where each number represents a value. A general representation of time series is listed below as an ordered set of m real-valued variables:

$$Y_t = x_1, ..., x_m \tag{1}$$

Time series can be described using a variety of qualitative terms such as seasonal, trending, noisy, nonlinear and chaotic.

Let $D = \{d_1, d_2,...,d_p\}$ be a time series data set with p the cardinality of the dataset. d_i is a time series with m real-valued. The dataset could be described in matrix as follows:

$$D = \begin{bmatrix} d_{11} & .... & d_{p1} \\ ... & ... & ... \\ d_{1m} & ... & d_{pm} \end{bmatrix} \tag{2}$$

The purpose of the presented algorithm is to classify all time series from dataset D, with respect to a similarity measure.

The individual length is equal to cardinality of classification data set, denoted by p.

An individual is represented as a vector:

$$c = (c_1,...,c_p)$$

Where c_j is an integer number from 1 to p.

Representation indicates how time series are assigned to classes. The value c_j of the gene j indicates the classes to which the object d_j is assigned. All the times series for dataset are assigned to a class (cluster). The number of classes is determining by the evolutionary algorithms [8].

The first step when apply an evolutionary algorithm for solving a problem is to find a suitable representation for candidate solutions by finding a good representation for clustering time series dataset. The fitness function will be considered a similarity measure of time series. Two time series are considered similar if they have enough non overlapping time ordered subsequences that are similar. The two subsequences are considered to be similar if one is enclosed within an envelope of a user defined width around another. Amplitude of one of the two sequences is scaled by a suitable amount to make the matching scale invariant.

Given a set of time series sequences, there are two types of similarity search. Subsequence matching finds all the data sequences that are similar to the given sequence, while whole matching finds those sequences that are similar to one other.

The following are some examples of the similarity queries:

- Identify companied with similar growth pattern.
- Determine products with similar selling patterns.
- Discover Stocks with similar movement in stock prices.
- Find if a musical score is similar to one of the copyrighted scores.
- Find portions of seismic waves that are not similar to spot geological irregularities.

Given two time series, a standard approach for similarity is to compute Euclidean distance.

In order to support efficient retrieval and matching of time series, indexing of time series data is needed. An indexing structure [29] was proposed for fast similarity searches over time-series, when the data as well as query sequences were of the same length. Discrete

Fourier Transformation (DFT) is used to map a time sequence to the frequency domain, the first few frequencies are retained and the retained frequencies are used to store the subsequences in the R* tree [47] structure, to create the index for efficient and fast search. This work was generalized [41] to allow subsequence matching. Data sequence could now be of different lengths and the query sequence could be smaller than any of the data sequence.

This work has the following limitations for employing it in practical applications:

- The problems of amplitude scaling and offset translation have not been addressed.
- The problem of ignoring unspecified portions of sequences while matching sequences is not addressed.

The above mentioned shortcomings are taken care in the technique proposed by Agrawal et.al [2]. According to the paper, two sequences are considered to be similar if they have enough non overlapping ordered pairs of subsequences that are similar. Amplitude of one of the two sequences is allowed to be scaled by any suitable amount and its offset adjusted.

Let us consider A cluster (class) of k time series is considered A = $\{d_1,..., d_k\}$. Each series of the cluster has m real valued terms.

Definition 1
The **mean time series of the cluster A** is denoted by M and is a **time series with** the real-valued M = $M_1, ..., M_m$ where of this series are given by:

$$M_j = \frac{\sum_{i=1}^{k} d_{ij}}{k} ; j = 1,..., m \qquad (3)$$

For each class detected by the evolutionary algorithms is calculated:

$$
f_{nc}(c) = \sum_{i=1}^{k} \left| \frac{m\sum\limits_{j=1}^{m} d_{ij} \cdot M_j - \sum\limits_{j=1}^{m} d_{ij} \cdot \sum\limits_{j=1}^{m} M_j}{\sqrt{\left[m\sum\limits_{j=1}^{m} d_{ij}^2 - \left(\sum\limits_{j=1}^{m} d_{ij}\right)^2 \right] \cdot \left[m\sum\limits_{j=1}^{m} M_j^2 - \left(\sum\limits_{j=1}^{m} M_j\right)^2 \right]}} \right| \tag{4}
$$

Where nc is the number of classes (the indices of classes) and k is the number of the time series in the class.

The first approach of fitness function proposed for this algorithm is

$$
F_1(c) = \sum_{nc=1}^{ntotc} f_{nc}(c) \tag{5}
$$

where $ntotc$ is the number of classes The fitness function will be maximized.

The mean of a cluster the mean of each term of the time series enter the cluster. We consider that if exist a similarity between a mean series of a time series dataset and all the series exist a similarity between series from dataset.

The second approach for fitness function use for this algorithm is the distance between time series in the classes.

For each class detected by the evolutionary algorithms is calculated:

$$
f_{nc}(c) = \sum_{i=1}^{k} \sum_{j=i+1}^{k} \| d_i - d_j \| \tag{6}
$$

where nc is the number of classes (the indices of classes) and k is the number of the time series in the class.

The second fitness function used is:

$$
F_2(c) = \sum_{nc=1}^{ntotc} f_{nc}(c) \tag{7}
$$

This second fitness function will be maximized. The fitness functions presented here are used in the algorithm. In case of series with the linear form the fitness F_1 was used. In other cases F_2 was used. The main steps of the evolutionary time series clustering algorithm are outlined below.

Evolutionary Time Series Clustering Algorithm (*ETSC*)

Step-1: **Input:** time series dataset;

Step-2: **Initialization:** Initialize the population $P(t)$. {Random initialization of the chromosome population}

Step-3: **For** each chromosome in the population **Do**

    Step-3.1: Calculate de mean of classes for the entire chromosome in the population;

Step-3.2: Calculate the fitness function;

Step-3.3: Apply selection for $P(t)$.

- {Let P^1 be the set of the selected solutions. The best 50 % individuals from current population are copied into $P(t+1)$. All individual from $P(t)$ compete for solution. Individuals from P^1 entering the mating pool based on tournament selection. Choose chromosomes from P^1 to enter the mating pool}.

Step 3.4: Apply the mutation operator to the solutions from the mating pool. Mutate solutions in P^1 offspring enter the next generation $P(t+1)$.

Step:4: Output: Classes of time series

4 Experimental Results

To show that our approach is useful for clustering time series and to see that the evolutionary algorithm are useful for time series clustering. A series of experiments on publicly available real datasets are performed.

4.1 Data Description

The data used are described below:

- **Dataset1:** We use first for tested some dataset generated randomly. Four time series are represented in figure 1. These form the pattern for data set. Using the 4 patterns time-series as seeds, a variation of the original patterns by adding small deviations (2-3% of the series term). The final dataset is form for 44 time series, each time series has 12 terms (the number of months).

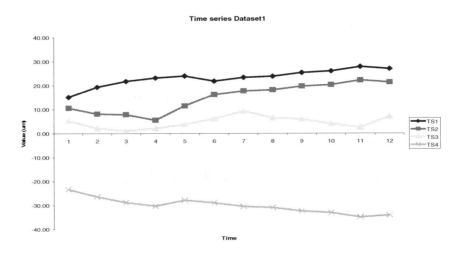

Fig. 1 Time series dataset1

- **Synthetic Control Chart Time Series (CC)** [23]: This data consists of synthetically generated control charts. This dataset contains 600 examples of control charts synthetically generated.. There are six different classes of control charts: Normal, Cyclic, Increasing trend, Decreasing trend, Upward shift, Downward shift. The dataset has 60 period is one of the most used dataset for time series clustering.

    The following image shows ten examples from each class:

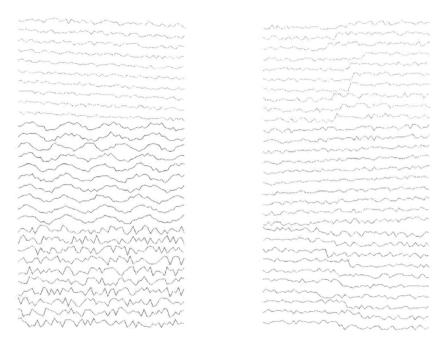

Fig. 2 Ten time series from each class of Synthetic Control Chart Time Series

- **Cylinder Bell Funnel Data (CBF):** This data set contains 300 series generated by implementing the Cylinder, Bell, and Funnel equations given in the UCR Time series Data Mining Archive [23].
- **Trace dataset (Trace):** this dataset contains the 4-class dataset with 200 instances, 50 for each class. The dimensionality of the data is 275 [23].

4.2 Algorithm Parameter

The proposed algorithm is evaluated in terms of clustering quality and clustering efficiency. The parameters of evolutionary algorithms for algorithm are given in Table 1:

Table 1 Evolutionary algorithms parameters

Parameter	**Value**							
Population Size	300	200	100	50	150	25	75	120
Chromosome length	The cardinality of dataset							
Crossover probability	0.9	0.8	0.7	0.6	0.5	0.5	0.8	0.8
Crossover type	Uniform crossover							
Mutation probability	0.9	0.8	0.7	0.6	0.5	0.4	0.7	0.7

The population size is better to be small if the data series are very large.

4.3 Results Evaluation

For data set 1 we obtained 4 classes. All the dataset generated from patterns time series are grouped together in the pattern class. The numbers of generation and the number of chromosome have increasing the possibility to find a better cluster structure. Fitness function F_1 is better to use when the time series have linear trend.

For Synthetic Control Chart Time Series after 15 runs we obtained 10, 12, and 8 classes. If the number of chromosome is very small the time series are not corrected clustered. The representation depends on the evolutionary algorithm parameters and the data from time series clustering. For Trace datasets the results are the best. After a number of 50 runs the dataset are clustered in 6-8 classes.

Evaluating clustering systems could be done because clustering is an unsupervised learning process in the absence of the information of the actual partitions. We used classified datasets and compared how good the clustered results fit with the data labels, which are the most popular clustering evaluation method [16]. Three objective clustering evaluation criteria were used in our experiments: Jaccard, Rand and FM [16].

Consider $G = G_1, G_2, ..., G_m$ as the clusters from a supervised dataset, and $A = A_1, A_2, ..., A_n$ as that obtained by a clustering algorithm under evaluations, in this case *Evolutionary Time Series Clustering*. Denote D as a dataset of original time series or features. For all the pairs of series (d_i, d_j) in D, we count the following quantities:

- *a* is the number of pairs, each belongs to one cluster in *G* and are clustered together in *A*.
- *b* is the number of pairs that are belong to one cluster in *G*, but are not clustered together in A.
- *c* is the number of pairs that are clustered together in *A*, but are not belong to one cluster in *G*.
- *d* is the number of pairs, each neither clustered together in *A*, nor belongs to the same cluster in *G*.

The used clustering evaluation criteria are defined as below:

1. *Jaccard* Score (Jaccard):

$$Jaccard = \frac{a}{a+b+c} \tag{8}$$

2. Rand statistic (Rand):

$$Rand = \frac{a+d}{a+b+c+d} \tag{9}$$

3. Folkes and Mallow index (FM):

$$FM = \sqrt{\frac{a}{a+b} \cdot \frac{a}{a+c}} \tag{10}$$

All the used clustering evaluation criteria have value ranging from 0 to 1, where 1 corresponds to the case when G and A are identical. A criterion value is the bigger, the more similar between A and G. Each of the above evaluation criterions has its own benefit and there is no consensus of which criterion is better than other criteria in data mining community. Table 2 describes the mean of the evaluation criteria values of 100 runs for K-means with original data.

In Table 2 , Table 3 and Table 4 we use the following notations:
-   CC is the Synthetic Control Chart Time Series data set;
-   CBF is the Cylinder Bell Funnel Data;
-   Trace is Trace dataset.

Table 2 The mean of the evaluation criteria values obtained from 100 runs of K-means algorithm with the original data

	CBF	CC	Trace
Jaccard	0.3490	0.4444	0.3592
Rand	0.6438	0.8529	0.7501
FM	0.5201	0.6213	0.5306

Table 3 gives the mean of the evaluation criteria values of 100 runs for K-means with extracted features.

Table 3 The mean of the evaluation criteria values obtained from 100 runs of K-means algorithm with the extracted features

	CBF	CC	Trace
Jaccard	0.3439	0.4428	0.3672
Rand	0.6447	0.8514	0.7498
FM	0.5138	0.6203	0.5400

Table 4 gives the mean of the evaluation criteria values of 100 runs for ETSC.

Table 4 The mean of the evaluation criteria values obtained from 100 runs of ETSC

	CBF	CC	Trace
Jaccard	0.3370	0.4425	0.3655
Rand	0.6500	0.8512	0.7152
FM	0.5622	0.6155	0.5510

5 Conclusions and Future Work

An evolutionary approach to perform clustering on time series data set was presented. Some different fitness functions are tested. The future work will be in the application of this method to data from different domains (financial, biological) and to find another fitness function that increase the clustering accuracy. Some new fitness functions according to the time series theory will be explored after analyzing dataset new representation of the chromosome, another special genetic operator will be proposed.

References

1. Aggarwal, C., Hinneburg, A., Keim, D.A.: On the Surprising Behavior of Distance Metrics in High Dimensional Space. In: Proceedings of the 8th Int'l Conference on Database Theory, London, UK, January 4-6, pp. 420–434 (2001)
2. Agrawal, R., Faloutsos, C., Swami, A.: Efficient Similarity Search in Sequence Databases. In: Proceedings of the 4th Int'l Conference on Foundations of Data Organization and Algorithms, Chicago, IL, October 13-15, pp. 69–84 (1993)
3. Alcock, R.J., Manolopoulos, Y.: Time-Series Similarity Queries Employing a Feature-Based Approach. In: 7th Hellenic Conference on Informatics, Ioannina, Greece, August 27-29 (1999)
4. Bagnall, A.J., Janacek, G., Iglesia, B.D., Zhang, M.: Clustering time series from mixture polynomial models with discretised data. In: Proceedings of the second Australasian Data Mining Workshop, pp. 105–120 (2003)
5. Bar-Joseph, G., Gerber, D., Gifford, T.: Jaakkola, & Simon I. A new approach to analyzing gene expression time series data. In: Proceedings of The Sixth Annual International Conference on Research in Computational Molecular Biology (RECOMB), pp. 39–48 (2002)
6. Bradley, P., Fayyad, U.: Refining initial points for k-means clustering. In: Proceedings of the 15th International Conference on Machine Learning, Madison, pp. 91–99 (1998)
7. Bradley, P., Fayyad, U., Reina, C.: Scaling Clustering Algorithms to Large Databases. In: proceedings of the 4th Int'l Conference on Knowledge Discovery and Data Mining, New York, NY, August 27-31, pp. 9–15
8. Chiş, M., Grosan, C.: Evolutionary Hierarchical Time Series Clustering. In: Sixth International Conference on Intelligent Systems Design and Applications (ISDA 2006), pp. 451–455 (2006)

9. Chu, S., Keogh, E., Hart, D., Pazzani, M.: Iterative Deepening Dynamic Time Warping for Time Series. In: Proceedings of the 2002 IEEE International Conference on Data Mining, Maebashi City, Japan, December 9-12 (2002)

10. Das, G., Gunopulos, D., Mannila, H.: Finding similar time series. In: Principles of Data Mining and Knowledge Discovery, pp. 88–100 (1997)

11. Das, G., Lin, K., Mannila, H., Renganathan, G., Smyth, P.: Rule discovery from time series. In: Proceedings of the 3rd International Conference of Knowledge Discovery and Data Mining, pp. 16–22 (1998)

12. Dermatas, E., Kokkinakis, G.: Algorithm for clustering continuous density HMM by recognition error. IEEE Tr. On Speech and Audio Processing 4(3), 231–234 (1996)

13. Fu, T.K., Chung, F., Ng, C.M.: Financial Time Series Segmentation based on Specialized Binary Tree Representation (2004)

14. Gafeney, S., Smyth, P.: Curve clustering with random effects regression mixtures. In: Bishop, C.M., Frey, B.J. (eds.) Proceedings of the Ninth International Workshop on Artificial Intelligence and Statistics (2003)

15. Guralnik, V., Srivastava, J.: Event detection from time series data. In: Proceedings of the Fifth ACM SIGKDD International Conference on Knowledge Discovery and Data Mining, pp. 33–42 (1999)

16. Halkidi, M., Batistakis, Y., Vazirgiannis, M.: On clustering validation techniques. Journal of Intelligent Information Systems 17(2-3), 107–145 (2001)

17. Kalpakis, K., Gada, D., Puttagunta, V.: Distance measures for effective clustering of ARIMA time-series. In: Proceedings of the 2001 IEEE International Conference on Data Mining (ICDM 2001), pp. 273–280 (2001)

18. Keogh, E., Lin, J., Truppel, W.: Clustering of Time Series Subsequences is Meaningless: Implications for Past and Future Research. In : The 3rd IEEE International Conference on Data Mining, Melbourne, FL, USA, pp. 19–22 (2003)

19. Keogh, E., Pazzani, M.: An Enhanced Representation of Time Series Which Allows Fast and Accurate Classification, Clustering and Relevance Feedback. In: Proceedings of the 4th Int'l Conference on Knowledge Discovery and Data Mining, August 27-31, pp. 239–241 (1998)

20. Keogh, E., Kasetty, S.: On the need for time series data mining benchmarks: A survey and empirical demonstration. Data Mining and Knowledge Discovery 7(4), 349–371 (2003)

21. Keogh, E., Chakrabarti, K., Pazzani, M., Mehrotra, S.: Locally Adaptive Dimensionality Reduction for Indexing Large Time Series Databases. In: Proceedings of ACM SIGMOD Conference on Management of Data, Santa Barbara, CA, pp. 151–162 (2001)

22. Keogh, E., Xi, X., Wei, L., Ratanamahatana, C.A.: The UCR Time Series Classification/Clustering Homepage: knowledge discovery, pp. 2–11. ACM Press, New York (2003),
    http://www.cs.ucr.edu/~eamonn/time_series_data/

23. Keogh, E., Smyth, P.: A probabilistic approach to fast pattern matching in time series databases. In: Proceedings of the 3rd International Conference of Knowledge Discovery and Data Mining, pp. 24–20 (1997)

24. Keogh, E., Chu, S., Hart, D., Pazzani, M.: Segmenting Time Series: A Survey and Novel Approach. In: an Edited Volume. Data Mining in Time Series Databases. Published by the World Scientific Publishing Company (2003)

25. Korn, F., Jagadish, H., Faloutsos, C.: Efficiently Supporting Ad Hoc Queries in Large Datasets of Time Sequences. In: Proceedings of the ACM SIGMOD Int'l Conference on Management of Data, Tucson, AZ, May 13-15, pp. 289–300 (1997)

26. Kosmelj, K., Batagelj, V.: Cross-sectional approach for clustering time varying data. Journal of Classification 7, 99–109 (1990)

27. Lin, J., Keogh, E., Lonardi, S., Chiu, B.: A symbolic representation of time series, with implications for streaming algorithms. In: Proceedings of the 8th ACM SIGMOD workshop on Research issues in data mining and knowledge discovery, pp. 2–11. ACM Press, New York (2003)

28. Lin, J., Vlachos, M., Keogh, E., Gunopulos, D.: Iterative Incremental Clustering of Time Series. In: Bertino, E., Christodoulakis, S., Plexousakis, D., Christophides, V., Koubarakis, M., Böhm, K., Ferrari, E. (eds.) EDBT 2004. LNCS, vol. 2992, pp. 106–122. Springer, Heidelberg (2004)

29. Lin, J., Keogh, E., Lonardi, S., Chiu, B.: A symbolic representation of time series, with implications for streaming algorithms. In: Proceedings of the 8th ACM SIGMOD workshop on Research issues in data mining and knowledge discovery, pp. 2–11. ACM Press, New York (2003)

30. Maharaj, E.A.: Clusters of time series. Journal of Classification 17, 297–314 (2000)

31. MacQueen, J.: Some Methods for Classification and Analysis of Multivariate Observation. In: Le Cam, L., Neyman, J. (eds.) Proceedings of the 5th Berkeley Symposium on Mathematical Statistics and Probability, Berkeley, CA, vol. 1, pp. 281–297 (1967)

32. Michalewicz, Z.: Genetic Algorithm + Data Structure = Evolutionary Programs. Springer, Berlin (1992)

33. Murthy, C.A., Chowdhury, N.: In search of optimal clusters using genetic algorithm. Pattern Recognition Letters 17(8), 825–832 (1996)

34. Oates, T., Firoiu, L., Cohen, P.R.: Using dynamic time warping to bootstrap HMM-based clustering of time series. In: Sun, R., Giles, C.L. (eds.) IJCAI-WS 1999. LNCS, vol. 1828, pp. 35–52. Springer, Heidelberg (2001)

35. Ormerod, P., Mounfield, C.: Localised structures in the temporal evolution of asset prices. In: New Approaches to Financial Economics. Santa Fe Conference (2000)

36. Park, S., Kim, S.W., Chu, W.W.: Segment-Based Approach for Subsequence Searches in Sequence Databases. In: Proceedings of the 16th ACM Symposium on Applied Computing (2001)

37. Piccolo, D.: A distance measure for classifying ARIMA models. Journal of Time Series Analysis 11(2), 153–164 (1990)

38. Popivanov, I., Miller, R.J.: Similarity Search over Time Series Data Using Wavelets. In: Proceedings of the 18th Int'l Conference on Data Engineering, San Jose, CA, Feburary 26-March 1, pp. 212–221 (2002)

39. Ramoni, M., Sebastiani, P., Cohen: Bayesian clustering by dynamics. Machine Learning 47(1), 91–121 (2002)

40. Ranganathan, M., Faloutosos, C., Manolopoulos, Y.: Fast subsequence matching in time-series databases. In: Proc. of the ACM SIGMOD Conference on Management of Data (May 1994)

41. Sarker, B.K., Mori, T., Uehara, K.: Parallel Algorithm for Mining Association in time Series Data. CS24-2002-1 Tech. report (2002)

42. Singhal, A., Seborg, D.E.: Clustering multivariate time-series data. Journal of Chemometrics 19, 427–438 (2005)

43. Steinback, M., Tan, P.N., Kumar, V., Klooster, S., Potter, C.: Temporal Data Mining for the Discovery and Analysis of Ocean Climate Indices. In: The 2nd Workshop on Temporal Data Mining, at the 8th ACM SIGKDD International Conference on Knowledge Discovery and Data Mining. Edmonton, Alberta, Canada, July 23 (2002)
44. Tong, P., Dabas, H.: Cluster of time series models: An example. Journal of Applied Statistics 17, 187–198 (1990)
45. Yi, B., Faloutsos, C.: Fast Time Sequence Indexing for Arbitrary Lp Norms. In: Proceedings of the 26th Int'l Conference on Very Large Databases, Cairo, Egypt, September 10-14, pp. 385–394 (2000); Database Management, Berlin, Germany, Jul 26-28, pp 55–68
46. Zhang, H., Ho, T.B., Lin, M.S.: An evolutionary K-means algorithm for clustering time series data. In: ICML 2004 (2004)

Support Vector Clustering: From Local Constraint to Global Stability

Bahman Yari Saeed Khanloo, Daryanaz Dargahi,
Nima Aghaeepour, and Ali Masoudi-Nejad

Summary. During recent years, numerous kernel based clustering algorithms have been proposed. Support Vector Machines, as standard tools for classification and clustering, have played an important role in this area. Information retrieval from the data in the absence of target labels, as an appealing aspect of unsupervised procedures, necessitates realization of a clear perception of the topological structure of the patterns. Moreover, the learner will require creating a balance between generalization and denoising abilities and regularization of the complexity as well. The Support Vector Clustering (SVC) approach, though globally informative, lacks mechanisms to take advantage of the inferences made from local statistics. Parameterizing the algorithm and embedding local methods is still an open problem in the SVC algorithm. In this chapter, the unsupervised support vector method for clustering is studied. Following the previous works, we will put forward recent efforts aimed at establishing a reliable framework for automating the clustering procedure and regularizing the complexity of the decision boundaries. The novel method takes advantage of the information obtained from a Mixture of Factor Analyzers (MFA) assuming that lower dimensional non-linear manifolds are locally linearly related and smoothly changing.

1 Introduction

Unsupervised learning algorithms involve discovering the structure of the data when the training patterns are partially or fully unlabeled. This often

Bahman Yari Saeed Khanloo, Daryanaz Dargahi, Nima Aghaeepour,
and Ali Masoudi-Nejad
Laboratory of Bioinformatics and Systems Biology (LBB), Department of
Bioinformatics, Institute of Biochemistry and Biophysics, University of Tehran
e-mail: amasoudin@ibb.ut.ac.ir, amasoudi@kais.kyoto-u.ac.jp

Ali Masoudi-Nejad
Center of Excellence in Biomathematics, Department of Computer Science,
College of Science, University of Tehran

A. Abraham et al. (Eds.): Foundations of Comput. Intel. Vol. 6, SCI 206, pp. 209–227.
springerlink.com   © Springer-Verlag Berlin Heidelberg 2009

requires the data to speak for itself which in turn calls for an appropriate representation of the patterns in the feature space. From statistics perspective, an unsupervised learning procedure is all about density estimation. Clustering is the act of grouping related data together often in an unsupervised manner using some similarity measurement criteria. These criteria could be inspired by a parametric or non-parametric model, a graph theoretic approach, distance measure, density estimation or a combination of them. In other words, clustering aims to reduce the number of objects to form regions or the so-called "clusters" that are interpretable to human perception.

The Support Vector approach, initially introduced for classification and regression purposes, has also been reformulated for unsupervised learning purposes. In an extreme point of view, the problem of clustering in the domain of support vector method is about partially estimating the underlying probability distribution function which has created the data [1]. In this principle, instead of trying to tackle the complex problem of density estimation, we will simply estimate a binary decision function which discriminates among two different subsets.

In the past decades, many clustering algorithms have been proposed to the domain of unsupervised learning. There, however, have been a few attempts to exploit kernel tricks to estimate functions that describe the underlying probability distributions in the feature space. Fortunately, a number of kernelized clustering techniques have been introduced recently which have shown to be successful especially in learning concave structures. These algorithms include Kernel K-Means [2], Spectral Graph Partitioning [3] and Spectral Clustering [4]. A framework for comparing these methods has been developed by Brand et al. [5].

Support Vector Clustering is a non-parametric clustering algorithm based on the original idea introduced by Schölkopf et al. [1, 6], Tax and Duin [7] where a support vector method was employed to delineate the support of an unknown high dimensional distribution. Ben-Hur et al. [8, 9, 10] then suggested enclosing the data with contours making no a priori assumption on the number and the shape of the clusters. They constructed a minimal hypersphere to form cluster boundaries in the data space. Although the algorithm proposed in [10] was capable of forming decision boundaries of arbitrary shape and parameterizing the clustering procedure, the choice of soft margin constant and Gaussian kernel width parameters was not addressed. The method, therefore, was deficient in controlling the number of clusters and dealing with noise whilst maintaining the generalization ability. In the following, Tax and Duin [11] suggested Support Vector Data Description (SVDD) where they presented a flexible boundary tightening mechanism for outlier or novelty detection and avoiding redundant space inside the contours. Lee et al. [12] further characterized the radius of the hypersphere as a function of Gaussian width and addressed the choice of kernel parameter by monotonically increasing it to obtain a greater or equal number of clusters than required. Their method, although noticeable in contribution, was blind as any other global density estimation method in the sense that they overlooked local properties of the data. Although the method incorporates an automatic procedure for clustering, it may result in some less interpretable contours in the case of sparse clusters or in the presence of noise. Lately, Yankov et al. [13] put one more step forward and automated the parameter selection procedure and satisfied obtaining a predefined number of clusters as a standard requirement for clustering algorithms. They employed a reduced dimensionality mixture of Gaussians to incorporate local statistical facts into the global estimation process. As a result, they moderated the tolerance to noise both locally and globally. Moreover, the regularized decision boundaries in their method, has contributed to superior generalization and hence, the ability of outlier detection.

The remainder of this chapter is organized as follows. In section 2, an overview of one-class SVM and its extension to clustering is given. Next, Factor Analysis using a mixture of factor analyzers is presented in 3 and its role in regularizing the complexity of cluster contours in the SVC algorithm is discussed in 4. In the following, we will experimentally compare the modified method with the original SVC algorithm in section 5. Finally, section 6 states the related work and the discussion while the conclusion is given in 7.

2   The SVC Algorithm

2.1   One-Class SVM

Support Vector Machines can be formulated for one-class classification. This unsupervised technique, also known as *Domain Description*, was originally introduced to estimate the support of a high dimensional distribution [1, 7] . In one-class SVM, regarding the whole patterns as positive ones, the input space is divided into two subspaces by estimating a binary decision function. This function takes a positive value inside the class region and negative elsewhere. As a result, the method is applicable for novelty detection [1], outlier detection [11], context change detection, etc.

In the following, we discuss the one-class support vector method. The formulation outlined here is based on the original one presented in [1].

Suppose we have a set of n independent and identically distributed observations $\mathbf{X} = \{\mathbf{x}_i\}_{i=1}^n$ from an unknown distribution function P. The one-class SVM algorithm seeks to find a minimal region R which surrounds almost all the data points. This approximated region is believed to enclose with high probability the test examples presuming that they are drawn from the same probability distribution P. Note also that in addition to being minimal, the region R is expected to demonstrate an acceptable generalization on unseen examples. As in support vector classification, using a non-linear transformation function Φ, the input space is mapped into a high dimensional feature space where the non-linear class boundaries in the input space are outlined by a hyperplane. This hyperplane, described by $w \cdot \Phi(\mathbf{x}) = b$, forms two subspaces one of which describes the region R and another which delineates the remaining feature space. The variables w and b mark the normal vector and the displacement from the origin respectively. Note that for all $\mathbf{x}_i$ the inequality $w \cdot \Phi(\mathbf{x}) \geq b$ must hold. In order to achieve the smoothest boundaries in the input space, we now look for the plane with minimum norm of the normal vector w. Hence the problem is to minimize

$$\min \quad \frac{1}{2}\|w\|^2 \tag{1}$$

$$\text{subject to} \quad w \cdot \Phi(\mathbf{x}_i) \geq b, \quad i = 1 \cdots n.$$

We employ the slack variable vector $\xi = (\xi_1, \cdots, \xi_n)$ to enable the decision boundary to exclude the noisy points residing in the neighborhood from regions with higher support for the density function. The constraint in (1) now becomes $w \cdot \Phi(\mathbf{x}_i) \geq b - \xi$. Hence we obtain:

$$\min_{w,b,\xi} \quad q(w,b,\xi) = \frac{1}{2}\|w\|^2 + \frac{1}{n\nu}\sum_{i=1}^{n}\xi_i - b \tag{2}$$

$$\text{subject to} \quad w \cdot \Phi(\mathbf{x}_i) \geq b - \xi_i, \quad \xi_i \geq 0, \quad i = 1 \cdots n$$

where $\nu \in [0,1]$ is the regularization term which controls the tradeoff between the outliers and the complexity of the region contours. The more the value of ν the more the number of outliers. If ν reaches 0, the slack variables will be severely penalized resulting in a hard margin problem [1]. Note that an appropriate value for the margin parameter ν satisfies maximization of margin and minimization of the classification error simultaneously. Introducing the nonnegative Lagrange multipliers we obtain:

$$L(w,b,\xi,\alpha,\beta) = \frac{1}{2}\|w\|^2 + \frac{1}{n\nu}\sum_{i=1}^{n}\xi_i - b - \sum_{i=1}^{n}\alpha_i((w \cdot \Phi(\mathbf{x}_i)) - b + \xi_i) - \sum_{i=1}^{n}\beta_i\xi_i \tag{3}$$

Setting the partial derivatives with respect to the primal variables w, b and ξ to zero, we have

$$w = \sum_i \alpha_i \Phi(\mathbf{x}_i), \tag{4}$$

$$\alpha_i = \frac{1}{n\nu}, \quad \beta_i \geq 0, \quad \sum_i \alpha_i = 1, \quad i = 1 \cdots n. \tag{5}$$

Replacing the optimum values in (2) reduces the quadratic optimization problem resulting in the following dual optimization problem:

$$\min \quad \frac{1}{2} \sum_{i,j=1}^{n} \alpha_i \alpha_j \Phi(\mathbf{x}_i) \cdot \Phi(\mathbf{x}_j) \tag{6}$$

$$\text{subject to} \quad \sum_{i=1}^{n} \alpha_i = 1, \quad 0 \le \alpha_i \le \frac{1}{n\nu}.$$

One can replace the dot product $\Phi(\mathbf{x}_i).\Phi(\mathbf{x}_j)$ with an appropriate Mercer's kernel. Unfortunately, there exist no rules of thumb to exactly determine the type of the kernel. The Gaussian kernel, however, has been successfully used in one-class SVM for its ability to form closed contours of arbitrary shape [10, 11]. Replacing the dot product with Gaussian kernel yields the following decision function:

$$f(\mathbf{x}) = sgn[\sum_{\mathbf{x}_i \in \{SV\}} \alpha_i e^{-\gamma \|\mathbf{x} - \mathbf{x}_i\|^2} - b] \tag{7}$$

where $\{SV\}$ denotes the set of support vectors, i.e. the data points $\mathbf{x}_i$ whose corresponding Lagrange multiplier α_i is positive. Figure 1 illustrates the decision boundaries for the Iris data set. The hills in the figure correspond either to a cluster or a sub-cluster and connected hills represent a cluster region.

The data points fall into three categories based on their corresponding Lagrange multiplier α_i. Support vectors (SVs) with $\alpha_i \in (0, \frac{1}{n\nu})$, bounded support vectors (BSVs) with $\alpha_i = \frac{1}{n\nu}$ and interior points with $\alpha_i = 0$. These three groups will lie on the surface, outside and inside the region (hypersphere) in the feature space, respectively.

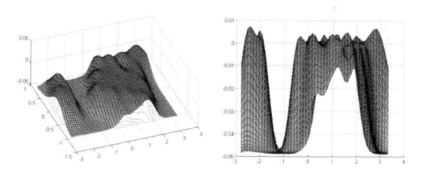

Fig. 1 The decision function of the Iris data set computed using the first two principal components (PCs) before applying the *sgn* operator. Left: $Z = f(\mathbf{x}')$ where $\mathbf{x}'$ is an example in the PCA space. Right: A profile representing the Z surface as a functional of PC_2. Notice that the region above Z=0 when mapped into $PC_1 PC_2$ space will form the cluster contours. In the case of concave clusters, starting from a point with a positive value of Z, if we move in a straight line along any direction in the original data space we will remain in the same cluster unless we reach negative boundaries of a valley

According to the formulation, support vectors lie on the surface of the separating hyperplane in the feature space. Therefore, to determine the value of the displacement b, one solves the equation $f(\mathbf{x}) = 0$. Hence, we have

$$b = \sum_{\mathbf{x}_i \in \{SV\}} \alpha_i e^{-\gamma \|\mathbf{x} - \mathbf{x}_i\|^2}. \tag{8}$$

In the test phase, in order to determine whether an example $\mathbf{x}_i$ belongs to region R, one substitutes it in equation (7). A positive value for $f(\mathbf{x})$ states that $\mathbf{x}_i$ falls within R while a negative f identifies the example $\mathbf{x}_i$ as an outlier, namely a bounded support vector.

Here, we investigate an alternative formulation for one-class SVM suggested in [14, 1, 10] which uses a ball to describe the data in the feature space. Using a non-linear transformation $\Phi : X \to F$ and aiming at enclosing most of the data with the smallest sphere, we are to solve

$$\min_{R, \boldsymbol{\xi}, C} \quad R^2 + \frac{1}{n\nu} \sum_i \xi_i \tag{9}$$

$$\text{subject to} \quad \|\Phi(x_i) - C\|^2 \leq R^2 + \xi_i, \quad \xi_i \geq 0, \quad i = 1 \cdots n$$

where $\boldsymbol{\xi} \in \mathbb{R}^n$ and $C \in F$ is the center of the sphere.
Introducing the nonnegative Lagrange multipliers we obtain:

$$L = R^2 + \frac{1}{n\nu} \sum_{i=1}^n \xi_i - \sum_{i=1}^n \alpha_i (R^2 + \xi_i - \|\Phi(x_i) - C\|^2) - \sum_{i=1}^n \beta_i \xi_i. \tag{10}$$

Setting the partial derivative of L with respect to R, C and ξ_i yields

$$\sum_i \alpha_i = 1, \tag{11}$$

$$C = \sum_i \alpha_i \Phi(\mathbf{x}_i) = 1, \tag{12}$$

$$\alpha_i = C - \beta_i. \tag{13}$$

Substituting the optimum results in (9) and replacing the dot product with an appropriate kernel yields to the following Wolfe dual optimization:

$$\min_\alpha \sum_{i,j} \alpha_i \alpha_j K(\mathbf{x}_i, \mathbf{x}_j) - \sum_i \alpha_i K(\mathbf{x}_i, \mathbf{x}_i) \tag{14}$$

$$\text{subject to} \quad 0 \leq \alpha_i \leq \frac{1}{n\nu}, \quad \sum_i \alpha_i = 1, \quad i = 1 \cdots n.$$

The center of the sphere in the feature space can be calculated by

$$C = \sum \alpha_i \Phi(\mathbf{x}_i).$$ (15)

The decision function is now in the form of

$$f(\mathbf{x}) = sgn[R^2 - \sum_{i,j} \alpha_i \alpha_j K(\mathbf{x}_i, \mathbf{x}_j) + 2 \sum_i \alpha_i K(\mathbf{x}_i, \mathbf{x}) - K(\mathbf{x}, \mathbf{x})].$$ (16)

Again, the condition $f(\mathbf{x}) = 0$ holds for every support vector. The term $K(\mathbf{x}, \mathbf{x})$ turns into a constant as we are utilizing the Gaussian kernel.

Exploiting the geometrical interpretation, the distance of the image of each point $\mathbf{x}$ from the center of the hypersphere in the feature space is described by

$$R^2(\mathbf{x}) = \|\Phi(\mathbf{x}) - C\|^2.$$ (17)

Replacing the dot products with kernels and substituting C using (15) yields

$$R^2(\mathbf{x}) = K(\mathbf{x}, \mathbf{x}) + \sum_{i,j} \alpha_i \alpha_j K(\mathbf{x}_i, \mathbf{x}_j) - 2 \sum_i \alpha_i K(\mathbf{x}, \mathbf{x}_i).$$ (18)

For each support vector $\mathbf{x}_S$, the distance of its image from the center will be $R = R(\mathbf{x}_S)$ i.e. the radius of the hypershere can be calculated by replacing any $\mathbf{x}_S$ into the above equation. As a result, the contours of the clusters will be determined using the set of support vectors.

One should note that the two formulations, though different in appearance, provide us with the same results. Schölkopf et al., however, have shown their equivalence in [2].

2.2 Support Vector Clustering

The binary decision function obtained from one-class SVM can be used to extend one-class support vector classification to a clustering method. This technique is based on a geometric interpretation in the feature space [10]. Consider a pair of examples belonging to different clusters. The line segment connecting these data points will exit the region R, i.e. the hypersphere in the feature space. Therefore, the line segment will contain a segment of points for which the decision function takes a negative value. Note, however, that an always positive value for the decision function assures that the two examples are of the same cluster whereas the opposite is not always true. For instance, there may exist some points on the line between two examples for which $f(\mathbf{x})$ is negative while they lie within the same component. This is the case either if the cluster regions in the input space are concave (as for p_2 and p_3) or the contours are too complex. The latter might result in a false negative value for the functional f when the line segment passes through the inner bottlenecks as in region R'. (see Figure 2) To overcome this problem, Yang et al. [15] have proposed a robust method for cluster assignment by introducing various proximity graph modeling strategies.

In order to extend one-class SVM to clustering, we first define an adjacency matrix A where

$$A_{ij} = \begin{cases} 1 \text{ if samples } \mathbf{x}_i \text{ and } \mathbf{x}_j \text{ are in the same component} \\ 0 \text{ otherwise.} \end{cases}$$

We can also express the first criterion using the ball interpretation. That is, we set $A_{ij} = 1$ if for all $\mathbf{x}_k$ connecting $\mathbf{x}_i$ to $\mathbf{x}_j$, $R(\mathbf{x}_k) \leq R$ holds. To determine whether the two points belong to the same cluster, one examines the decision function (7) for all points on the connecting line between them. In practice, regularly spaced points on the line segment between the two points are tested using the parametric formulation. Next, we find the connected components in the graph represented by A. Therefore, the number of clusters will be equal to the number of emerged components. In the following, the data points are labeled in a way that all the examples belonging to a component will be assigned the same label. Finally, bounded support vector are designated to their closest cluster regions to avoid unclassified regions.

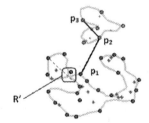

Fig. 2 The figure illustrates two cluster regions in the input space. The line segment connecting p_1 and p_2 will contain points for which $f(\mathbf{x})$ takes a negative value. Although f will take negative values on the line segment between p_2 and p_3, they belong to the same component

Aside from the original approach using the adjacency matrix, it might be useful to review the recent advances in the labeling process. To reduce the computational effort, Ben-Hur et al. [10] suggested inspecting the adjacencies only with support vectors. Yang et al. [15] improved the computational time while avoiding false negatives by introducing a new method for labeling using the Delaunay Diagram, the Minimum Spanning Tree and K-Nearest Neighbors approaches. Lee et al. [17] addressed the computational burden of the labeling process using a generalized gradient descent technique. They decomposed the data set into multiple disjoint groups and determined a stable equilibrium point for each group. Then, they assigned to each example the cluster label of its corresponding stable equilibrium point. Yılmaz et al. [16]

further developed a contour plotting algorithm to reduce the running time as well as decreasing the number of false negatives. They incorporated some heuristics into their method to determine the number of clusters and cluster labels as well.

3   Mixture of Factor Analyzers

Factor analysis (FA) is a statistical method for dimensionality reduction. Generally speaking, factor analysis seeks to form a set S' from a set of variables S that is more statistically "informative" where $|S'| \leq |S|$. This is done by investigating the linear relationships between the variables and determining a small number of unobservable variables called *latent variables*.

In practice, one incorporates simplifying presumptions into the projection scheme from S to S'. For instance, Ghahramani et al. [18] proposed an Expectation Minimization method for fitting a model to a mixture of factor analyzers. They used the following generative model and assumed the real valued observation $\mathbf{x}$ to be

$$\mathbf{x} = \Lambda\mathbf{z} + \mathbf{u}, \tag{19}$$

where $\mathbf{z}$ is the projecting dimensions with standard normal distribution i.e. $\mathbf{z} \sim N(0, I)$ and Λ is the factor loading matrix. In this model, $\mathbf{u} \sim N(0, \Psi)$ which plays the role of noise with independence of the observed variables as the key assumption. As a result, the covariance matrix Ψ is required to be diagonal. In the following, using a Maximum Likelihood Estimation (MLE) for each observation in the data set, Ψ and Λ matrices are iteratively updated [18].

They further extend the model to a Mixture of Factor Analyzers (MFA) allowing each individual factor analyzer have different mean μ_i and loading matrix Λ_i while maintaining the same assumption for the latent variable z for all the factor analyzers, i.e. $\mathbf{z_i} \sim N(\mu_i, \Psi)$. An ML estimation is now performed over every data point using $\mathbf{z_i}$ and the probability of it being projected by the i-th factor analyzer. In the mixture model, the estimated mean of the analyzers μ_i is updated in the same time as improved Ψ and Λ_i matrices take the place of their old values.

This method, iteratively performs dimensionality reduction while clustering the data set. One should note that the number of analyzers and the number of factors in each analyzer are to be selected carefully using some global method. Figure 3 demonstrates how the MFA algorithm delineates the lower dimensional manifolds in the input space. Local outliers i.e. those samples residing far from their corresponding manifold such as p_1 in our example, are perfectly detected using the mixture model. This helps us prevent the appearance of complex formations as well as eliminating the bridging examples [13].

Fig. 3 A mixture of 16
factor analyzers estimate
the structure of the data
in the input space. The
ellipses enclose two standard deviations from the
center of the analyzers.
The data points p_1 and
p_2 are believed to be de-
viating from the main
trajectory of the data.
Examples such as p_2 are
called bridging elements
since they will cause the
manifolds in the neigh-
borhood to be merged
and form a unifying clus-
ter which is not always
desirable

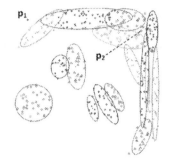

4 From Local Constraint to Global Stability

In section 3 we explained how the mixture model can help determine the
local outliers. One may argue that ruling out such examples from the data
set might considerably increase the performance of the SVC method [13]. The
local estimation models alone, however, might be unstable subject to noise.

To highlight the effect of parameter selection in the SVC algorithm, it
might be useful to analyze the effects of parameterization for a while. One
should note that large values of γ or small values of ν will result in appearance of several tight boundaries. On the contrary, small γ or a large value
of ν yields the emergence of multiple non-descriptive clusters. As previously
discussed, the slack variable vector ξ in (2) provides us with an outlier detection mechanism. As a result, one may allow for some misclassified examples
while bringing about superior generalization ability and vise versa. Every ξ_i
variable, however, penalizes its corresponding example proportional to its distance from the hyperplane with the same weighting factor $\frac{1}{n\nu}$. A modification
to this method may incorporate a priori knowledge about each example into
the method by determining the probability that the corresponding example
is an outlier.

Having the trained mixture of analyzers at hand, we can estimate the
significance of each data point by estimating its distance from the mean of
the factor analyzer to which it belongs. Introducing the Mahalanobis distance
and denoting the projections of the examples belonging to the j-th mixture
by $\mathbf{z}_j = (z_1^j, z_2^j, \cdots, z_{r_j}^j)$, the distance of each example projection z_i^j form its
corresponding factor analyzer can be defined by

$$\mathbf{d}_j = [(\mathbf{z}_j - \mu_j)'(\Lambda_j \Lambda_j' + \Psi)(\mathbf{z}_j - \mu_j)]^{\frac{1}{2}}. \tag{20}$$

In the following, we shall explain how the local constraint by the MFA when combined with the SVC algorithm will help discover the structure of the data effectively and achieve an excellent stability for this global method[1].

In order to regularize the complexity of decision boundaries in the one-class SVM, one can put forward a weighting scheme. A reasonable weighting suggests that the more the probability that a certain example is an outlier, the less it should be penalized so that its impact on the decision function will be minimized. If such a weighting is forces, sparse regions residing in the neighborhood of dense manifolds are less likely to be identified as a cluster by the SVC algorithm. Including the above weighting procedure in (2) the problem will be in the form of:

$$\min_{w,b,\xi} \quad q(w,b,\xi) = \frac{1}{2}\|w\|^2 + \frac{1}{n\nu} \sum_{i=1}^{n} \frac{1}{d_i} \xi_i - b \tag{21}$$

$$\text{subject to} \quad w \cdot \Phi(\mathbf{x}_i) \geq b - \xi_i, \quad \xi_i \geq 0, \quad i = 1 \cdots n$$

where d_i marks the distance from the sample $\mathbf{x}_i$ to the mixture component that its projection belongs. The Lagrangian is now rewritten as

$$L = \frac{1}{2}\|w\|^2 + \frac{1}{n\nu} \sum_{i=1}^{n} \frac{1}{d_i} \xi_i - b - \sum_{i=1}^{n} \alpha_i((w \cdot \Phi(\mathbf{x}_i)) - b + \xi_i) - \sum_{i=1}^{n} \beta_i \xi_i. \tag{22}$$

Taking the partial derivatives with respect to w, b and ξ_i and substituting the results in equation (21) yields

$$\min \quad \frac{1}{2} \sum_{i,j} \alpha_i \alpha_j \Phi(\mathbf{z}_i) \cdot \Phi(\mathbf{z}_j) \tag{23}$$

$$\text{subject to} \quad \sum_{i=1}^{n} \alpha_i = 1, \quad 0 \leq \alpha_i \leq \frac{1}{d_i n\nu}.$$

Note that formulations (6) and (23) are different in the sense that in the latter, the upper-bounds are set independently and based on the probability that the corresponding sample is an outlier and that instead of the samples, their projections are used.

To be able to reconstruct the underlying structure of the data, Yankov et al. [13] recommended taking advantage of greater or equal number of mixture components than the predetermined number of clusters. They also argued that a feasible outline of the cluster's topology might be guaranteed if several analyzers for each cluster is used. This is justifiable considering the fact that more than one analyzer is required to realize the structure of a concave or a

---

[1] This method has been termed Locally Constrained Support Vector Clustering (LSVC) [13].

complex formation. The performance of the LSVC algorithm, however, seems not to be deteriorated by the choice of large number of mixture components. They also suggested the number of analyzers to be one tenth the size of the data set.

Schölkopf et al. [1] set a constraint on the fraction of the outliers to be no greater than the constant ν in formulation (2). If we estimate the outliers using the information obtained from the factor analyzers, we can express an approximation for the tradeoff parameter by $\nu = \frac{1}{n} \sum_{j=1}^{n} s_j$ where s_j denotes the number of examples located farther than two empirical standard deviations of the Mahalanobis distance d_{ij} from the mean of the j-th analyzer [13].

5 Experiments

In the following, we will evaluate the performance of the SVC and LSVC algorithms. In our experiments [2], we set a fixed value 10 for the number of factor analyzers and 2 for the number of factors in each analyzer. The variable ν for both SVC and LSVC is set equal to each other and is found using the MFA method [3] and the number of clusters k is determined.

To estimate the value of γ, we perform an iterative procedure similar to [12] for both methods. In our experiments, we iteratively modified the value of $\log_2 \gamma$ in the range of -8 and 8. As a result, we examined different values for γ in this interval doubling the value in each step. The step size, however, was further reduced if a more accurate search was required. We observed the appearance of a monotonically increasing number of clusters. The procedure is terminated as soon as $k' \geq k$ number of clusters emerge. If $k' > k$ holds, we identify the k largest clusters (major clusters) and append the remaining clusters to their nearest major cluster. Finally, we assign each bounded support vector to its closest cluster.

Table 1 Specifications of the Data Sets

Data Set	Class Specifications
Synthetic Data Set I	size:300 class1:39 class2:51 class3:45 class4:165
Synthetic Data Set II	size:450 class1:150 class2:300
Iris	size:150 class1:50 class2:50 class3:50
Swiss Role	size:900 class1:700 class2:200

Four data sets are used in this study with two of them being standard benchmarks and the remaining are synthetic databases each of which representing different structural patterns. Refer to table 1 for details about specifications of data sets.

---

[2] Some experiments and procedures are adopted from [10] and [13].
[3] The code was obtained from ftp://ftp.cs.toronto.edu/pub/zoubin/mfa.tar.gz.

5.1 Synthetic Data Sets

Our first experiment studies the performance of the SVC and LSVC methods in the case of well separated cluster regions. The data set is composed of 300 two dimensional samples in 4 classes. Three of the clusters are in the form of circles with the fourth dense cluster enclosing the 3 circular clusters from one side (See Figure 4). The clusters contain 39, 51, 45 and 165 examples respectively.

The variable k is set to 4 and ν is estimated to be 0.063. With $\gamma = 2$ for SVC and $\gamma = 1$ for LSVC, $k' > k$ is satisfied. Both of the algorithms succeeded to perform the clustering perfectly. The complexity of the cluster boundaries i.e. the number of support vectors is roughly equal to the total number of examples in the case of SVC.

The second experiment is on another synthetic data set with 450 two dimensional examples as two concentric circles. The data set used here is the same as [10] and [13]. The inner circle is composed of 150 examples from a Gaussian distribution whereas the outer contains 300 data points from a radial Gaussian distribution and a uniform angular distribution.

The number of required clusters k is set to 2 and $\nu = 0.08$ is calculated. Both SVC and LSVC detect $k' > k$ clusters for $\gamma = 8$. LSVC gained an accuracy of 99% against 66% for SVC. We could achieve similar results for the SVC as well. This, of course, was obtained by an exhaustive search procedure. Figure 5 illustrates the results of the automatic procedure.

5.2 Real-World Data Sets

The third experiment is on the Fisher Iris data. This data set contains iris flower feature vectors of size four. There are three clusters in this data set

Fig. 4 The SVC (top) and LSVC (bottom) algorithms are applied to the 4-class synthetic data set (Synthetic Data Set I). The left shows cluster contours and the right shows the cluster regions after the merging procedure. Both methods obtained an accuracy of 100%. LSVC identified 77 SVs and 7 BSVs versus 300 SVs and 24 BSVs for the SVC. Both of the algorithms detected 5 clusters

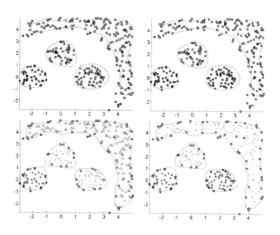

Fig. 5 The SVC (top) and LSVC (bottom) applied to the 2-class synthetic data set (Synthetic Data Set II). The left shows cluster contours and the right shows the clusters obtained after merging. The SVC algorithm is unable to expose the structure of the data well. LSVC identified 211 SVs and 43 BSVs versus 196 SVs and 2 BSVs for the SVC. LSVC detected 4 clusters against 5 clusters in the case of SVC

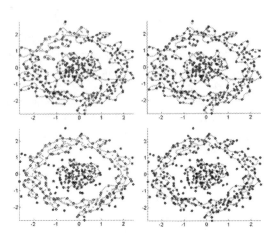

each of which contains 50 examples. One of the clusters is linearly separable from the others and the other two are significantly overlapping. (See figure 6)

We performed a preprocessing on the data set using PCA before running the algorithms. Using the first two principal components with $k = 3$ and ν estimated to be 0.07 and $\gamma = 8$ for both algorithms, SVC and LSVC correctly labeled 75% of the examples.

Our last experiment is based on the Swiss Role data set. It is composed of 900 examples represented by three features which create two strongly overlapping wrapped clusters. The larger cluster is composed of 700 examples

Fig. 6 Top left: Iris data set represented using the first two principal components. Top right: clusters determined by SVC before (and after) merging. The algorithm tries to accommodate all the data points. Bottom: The left shows cluster contours obtained using LSVC and the right depicts the clusters after merging. LSVC identified 62 SVs and 11 BSVs versus 60 SVs and no BSVs for the SVC. LSVC detected 4 clusters versus 3 clusters by the SVC

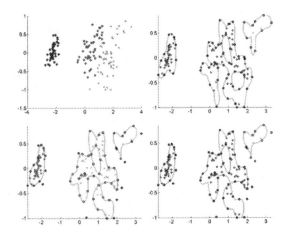

Fig. 7 The SVC (top) and LSVC (bottom) applied to the Swiss Role data set. The left shows cluster contours and the right shows the clusters obtained after merging. The SVC algorithm fails to discover the smaller cluster as well as the noticeable noise in the data. Trying to include all the data points, SVC detects complex and rather uninformative cluster boundaries which leads to false identification of a closed ring. LSVC identified 289 SVs and 211 BSVs versus 162 SVs and 105 BSV for the SVC. SVC detected $k' = 3$ clusters while LSVC identified $k' = 4$ clusters

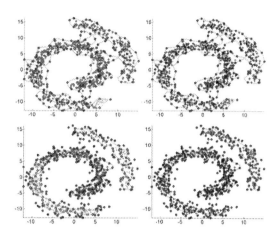

while the outer cluster is made up 200. A few elements in the overlapping area are removed and some Gaussian noise is added [13] as shown in figure 7.

We applied the PCA preprocessor to the data set. Then the two first principal components was used for clustering. k was set to 3 and ν was estimated to be 0.07. $\gamma = 0.25$ for SVC and $\gamma = 0.5$ for the LSVC was automatically determined. LSVC achieved superior accuracy of 99% against 75% for SVC.

Table 5.2 summarizes the experimental results.

Table 2 Experimental Results

Data Set	Method	%RR	#SV	#BSV	#Classes
Synthetic Data Set I	SVC	100	300	24	5
	LSVC	100	77	7	5
Synthetic Data Set II	SVC	66	196	2	5
	LSVC	99	211	43	4
Iris	SVC	75	60	0	3
	LSVC	75	62	11	4
Swiss Role	SVC	75	162	105	3
	LSVC	99	289	211	4

6   Discussion and Related Work

In this chapter, an overview of support vector clustering and recent modifications to different stages of this algorithm was presented.

Support vector clustering is a powerful method which offers a flexible clustering scheme capable of forming arbitrary shaped contours. The uniqueness of the solution and the ability to deal with outliers have made the SVC algorithm preferable to most of the previous clustering techniques.

Parameterization is definitely the most important concern in the support vector method for clustering. Many attempts have been made to directly determine the Gaussian width parameter γ and the tradeoff parameter ν for the SVC algorithm. Ben-Hur et al. [10] considered the SVC approach as a *divisive* clustering algorithm. Although they did not aim at solving the parameterization problem, they suggested an initial value for the the kernel parameter to be

$$\gamma = \frac{1}{max_{ij}\|\mathbf{x}_i - \mathbf{x}_j\|^2}.$$

They argued that although increasing the value of γ causes more clusters to emerge since it forces the clusters to split, using the bounded support vectors, the procedure will not be necessarily similar to a hierarchical clustering scheme. Instead, to regularize the cluster boundaries, they suggested to increase γ and ν in a way that a minimal number of support vectors is achieved.

Lee et al. [12] proposed a sequence generator algorithm to determine the optimal value of γ assuming that the value of tradeoff constant is kept unchanged. Consequently, the choice of a proper ν was not addressed in their model.

It can be argued that the procedure for determining the value of γ proposed in [12] and [13] does not necessarily yield a better accuracy. Allowing more number of trivial clusters (i.e. choosing larger γ) might result in better clustering precision after the merging procedure is performed. This can be particularly useful when the structure of the data is very complex or a small number of examples is available.

Competitive co-evolutionary support vector clustering was another attempt to establish an efficient parameterization for the SVC algorithm [19]. In this approach, a genetic algorithm based on competition was proposed. The optimal values for the SVC parameters γ and ν were obtained implicitly using a fitness function which was also used for the evaluation of the clustering results. The optimal number of clusters was also estimated in this approach. The same similarity measurements as the K-means algorithm was exploited to describe the distance between centers of the clusters as well as the distance between a cluster center and a data point.

Yılmaz et al. [16] tackled the parameterization problem using a simulated annealing framework. They developed a performance measurement criterion to indirectly tune the SVC parameters. In their method, an index was optimized to achieve the best values for γ and ν. Also, they examined the performance of their proposed framework using four different indexes including YAS index, Silhouette coefficient, Davis-Bouldin index and the Dunn's index. The requirement for an extra tuning of the margin parameter ν can be considered as a disadvantage of this model [16].

A further study by Yankov et al. [13] emphasized that local and global estimation methods offer complementary information about the underlying structure of the data. Taking advantage of a mixture of factor analyzers, they introduced a model for estimating the confidence level of a particular example to be identified as an outlier. Hence, an estimation for ν was obtained. As a result, monotonically increasing the kernel width, they tackled the problem of parameterization of SVC and addressed the choice of ν and γ in the same time.

7 Conclusion

Global optimization methods such as genetic algorithms and simulated annealing as well as local methods such as MFA have been used to fully parameterize the SVC algorithm [13, 16, 19]. It might be useful to develop a framework for comparing these methods.

In this chapter, the goal was to focus on utilizing local methods in the SVC algorithm. Hence, the LSVC algorithm with the mixture of factor analyzers as its local method was studied. The performance of LSVC algorithm might be improved by determining the number of factor analyzers and the number of analyzers in each factor. Using local methods other than the MFA model may also increase the performance of the support vector clustering technique. Determining a mechanism for controlling the number of clusters before the merging procedure while preserving a promising computational complexity seems inevitable. This can be done by setting a stopping criterion for bifurcation in the SVC algorithm. Moreover, determining the number of clusters in the mixture model can contribute to a fully automated clustering procedure.

Finally, one should keep in mind that many questions arisen in the field of clustering, originates with the fact that the concept of clustering is not so clear-cut. The definition of cluster and the similarity measurement criteria used in clustering, if forced to comply with the human perception, might prevent us from devising methods that are able to discover complex intrinsic patterns in the data. For instance, a human might not able to correctly recognize three clusters in the Iris data set even if the number of cluster is known formerly and the data is visualized properly. Nevertheless, many automatic clustering methods can distinguish among the two overlapping components and perform an excellent clustering.

Acknowledgements. This work was supported partially by Institute of Biophysics and Biochemistry (IBB) and Center of Excellence in Biomathematics, Department of Computer Science , College of Science, University of Tehran. This publication only reflects the authors' views. Also, we are very grateful to D. Yankov and his co-workers at University of California Riverside for the code snippets and their valuable comments.

References

1. Schölkopf, B., Platt, J.C., Shawe-Taylor, J., Smola, A.J., Williamson, R.C.: Estimating the support of a high-dimensional distribution. Neural Computation 13, 1443–1471 (2001)
2. Schölkopf, B., Smola, A.: Learning with Kernels. MIT Press, Cambridge (2002)
3. Fiedler, M.: Algebraic connectivity of graphs. Czechoslovak Mathematical Journal 23(98), 298–305 (1973)
4. Ng, A., Jordan, M., Weiss, Y.: On spectral clustering: Analysis and an algorithm. In: Advances in Neural Information Processing Systems (NIPS), vol. 14 (2001)
5. Brand, M., Huang, K.: A unifying theorem for spectral embedding and clustering. In: Proc. of the 9-th International Workshop on Artificial Intelligence and Statistics (2003)
6. Schölkopf, B., Williamson, R.C., Smola, A.J., Shawe-Taylor, J., Platt, J.: Support vector method for novelty detection. In: Solla, S.A., Leen, T.K., Muller, K.R. (eds.) Advances in Neural Information Processing Systems 12: Proceedings of the 1999 Conference (2000)
7. Tax, D.M.J., Duin, R.P.W.: Support vector domain description. Pattern Recognition Letters 20, 1991–1999 (1999)
8. Ben-Hur, A., Horn, D., Siegelmann, H.T., Vapnik, V.: A support vector clustering method. In: International Conference on Pattern Recognition (2000)
9. Ben-Hur, A., Horn, D., Siegelmann, H.T., Vapnik, V.: A support vector clustering method. In: Leen, T.K., Dietterich, T.G., Tresp, V. (eds.) Advances in Neural Information Processing Systems 13: Proceedings of the 2000 Conference (2001)
10. Ben-Hur, A., Horn, D., Siegelmann, H.T., Vapnik, V.: Support vector clustering. Journal of Machine Learning Research 2, 125–137 (2001)
11. Tax, D., Duin, R.: Support vector data description. Mach. Learn. 54(1), 45–66 (2004)
12. Lee, S., Daniels, K.: Gaussian kernel width generator for support vector clustering. In: Proc. of the International Conference on Bioinformatics and its Applications. Series in Mathematical Biology and Medicine, vol. 8, pp. 151–162 (2005)
13. Yankov, D., Keogh, E., Kan, K.F.: Locally Constrained Support Vector Clustering. In: Seventh IEEE International Conference on Data Mining, pp. 715–720 (2007)
14. Tax, D.M.J., Duin, R.P.W.: Data domain description by support vectors. In: Verleysen, M., Brussels, D.F. (eds.) Proceedings ESANN, Brussels, D Facto, pp. 251–256 (1999)

15. Yang, J., Estvill-Castro, V., Chalup, S.K.: Support vector clustering through proximity graph modelling. In: Neural Information Processing 2002, ICONIP 2002, pp. 898–903 (2002)
16. Yilmaz, Ö., Achenie, L.E.K., Srivastava, R.: Systematic tuning of parameters in support vector clustering. Mathematical Biosciences Journal 205(2), 252–270 (2007)
17. Lee, J., Lee, D.: An improved cluster labeling method for support vector clustering. IEEE Transactions on Pattern Analysis and Machine Intelligence 27(3), 461–464 (2005)
18. Ghahramani, Z., Hinton, G.: The EM algorithm for mixtures of factor analyzers. Technical Report CRG-TR-96-1 (1996)
19. Jun, S.H., Oh, K.W.: A Competitive Co-evolving Support Vector Clustering. In: ICONIP (1), pp. 864–873 (2006)
20. Du, S., Chen, S.: Weighted support vector machine for classification. In: Proc. International Conference on Systems, Man and Cybernetics, vol. 4, pp. 3866–3871 (2005)
21. Roweis, S., Saul, L.: Nonlinear dimensionality reduction by locally linear embedding. Science 290(5500), 2323–2326 (2000)
22. Roweis, S., Saul, L., Hinton, G.: Global coordination of local linear models. In: Advances in Neural Information Processing Systems (NIPS), vol. 14 (2002)
23. Tenenbaum, J., de Silva, V., Langford, J.: A global geometric framework for nonlinear dimensionality reduction. Science 290, 2319–2323 (2000)
24. Girolami, M.: Mercer kernel-based clustering in feature space. IEEE Transactions on Neural Networks 13(3), 780–784 (2002)
25. Camastra, F., Verri, A.: A Novel Kernel Method for Clustering. IEEE Transactions on Pattern Analysis and Machine Intelligence 27(5), 801–805 (2005)
26. Abe, S.: Support Vector Machines for Pattern Classification. Springer, London (2005)
27. Cristianini, N., Shawe-Taylor, J.: Introduction to Support Vector Machines: And Other Kernel-Based Learning Methods. Cambridge University Press, Cambridge (2000)
28. Vapnik, V.: The Nature of Statistical Learning Theory. Springer, New York (1995)
29. Loehlin, J.C.: Latent Variable Models: An Introduction to Factor, Path, and Structural Equation Analysis, 4th edn. LEA, London (2004)
30. van der Heijden, F., Duin, R.P.W., de Ridder, D., Tax, D.M.J.: Classification, Parameter Estimation and State Estimation: An Engineering Approach Using MATLAB. Wiley, U.K. (2004)

New Algorithms for Generation Decision Trees—Ant-Miner and Its Modifications

Urszula Boryczka and Jan Kozak

Abstract. In our approach we want to ensure the good performance of Ant-Miner by applying the well-known (from the ACO algorithm) two pheromone updating rules: local and global, and the main pseudo-random proportional rule, which provides appropriate mechanisms for search space: exploitation and exploration. Now we can utilize an improved expression of this classification rule discovery system as an Ant-Colony-Miner. Further modifications are connected with the simplicity of the heuristic function used in the standard Ant-Miner. We propose to employing a new heuristic function based on quantitative, not qualitative parameters used during the classification process. The main transition rule will be changed dynamically as a result of the simple frequency analysis of the number of cases from the point of view characteristic partitions. This simplified heuristic function will be compensated by the pheromone update in different degrees, which helps ants to collaborate and is a good stimulant on ants' behavior during the rule construction. The comparative study will be conducted using 5 data sets from the UCI Machine Learning repository.

1 Introduction

Data mining is the process of extracting useful knowledge from real-world data. Among several data mining tasks — such as clustering and classification — this paper focuses on classification. The aim of the classification

Urszula Boryczka
Institute of Computer Science, University of Silesia, Będzińska 39, 41–200
Sosnowiec, Poland, Phone/Fax: (+48 32) 291 82 83
e-mail: urszula.boryczka@us.edu.pl

Jan Kozak
Institute of Computer Science, University of Silesia, Będzińska 39, 41–200
Sosnowiec, Poland, Phone/Fax: (+48 32) 291 82 83
e-mail: j.m.kozak@gmail.com

A. Abraham et al. (Eds.): Foundations of Comput. Intel. Vol. 6, SCI 206, pp. 229–262.
springerlink.com   © Springer-Verlag Berlin Heidelberg 2009

algorithm is to discover a set of classification rules. One algorithm for solving this task is Ant-Miner, proposed by Parpinelli et al. [66], which employs ant colony optimization techniques [1, 33] to discover classification rules. Ant Colony Optimization is a branch of a newly developed form of artificial intelligence called swarm intelligence. Swarm intelligence is a form of emergent collective intelligence of groups of simple individuals: ants, termits or bees, in which we can observe a form of indirect communication via pheromone. Pheromone values encourage following ants to build good solutions of the analyzed problem and the learning process occurring in this situation is called positive feedback or autocatalysis.

The application of ant colony algorithms to rule induction and classification is a research area that still is not explored and tested very well. The appeal of this approach, similarly to the evolutionary techniques, is that it provides an effective mechanism for conducting a more global search. These approaches are based on a collection of attribute-value terms. Consequently, it can be expected that these approaches will also cope with attribute interactions a little bit better than greedy induction algorithms [35]. What is more, these applications require minimum understanding of the problem domain; the main components are: a heuristic function and an evaluation function, both of which may be employed in the ACO approach in the same form as in the existing literature on deterministic rule induction algorithms.

Ant-Miner is an ant-based system, which it is more flexible and robust than traditional approaches. This method incorporates a simple ant system, in which a heuristic value based on an entropy measure is calculated. Ant-Miner has produced good results when compared with more conventional data mining algorithms, such as C4.5 [69], ID3 and CN2 [14, 15]. Moreover, it is still a relatively recent algorithm, which motivates us to try to amend it. This work proposes some modifications to Ant-Miner to improve it. In the original Ant-Miner, the goal of the algorithm was to produce an ordered list of rules, which was then applied to test data in order in which they were discovered.

The original Ant-Miner was compared to CN2 [14, 15], a classification rule discovery algorithm that uses the strategy for generating rule sets similar to that of the heuristic function used in the main rule of ants' strategy in Ant-Miner. The comparison was made using 6 data sets from the UCI Machine Learning repository that is available at www.ics.uci.edu/ mlearn/MLRepository.html. The results were analyzed according to the predictive accuracy of the rule sets and the simplicity of the discovered rule set, which is measured by the number of terms per rule. While Ant-Miner had better predictive accuracy than CN2 on 4 of the data sets and worse on only one of the data sets, the most interesting result is that Ant-Miner returned much simpler rules than CN2. Similar conclusions could also be drawn from the comparison of Ant-Miner to C4.5, a well-known decision tree algorithm [69].

This chapter is organized as follows. Section 1 comprises an introduction to the subject of this work. Section 2 presents the fundamentals of the knowledge based systems. In section 3, the Swarm Intelligence and Ant Colony Optimization is introduced. In section 4, Ant Colony Optimization in Rule Induction is presented. Section 5 describes the modifications and extensions of original Ant-Miner. In section 6 our further improvements are shown. Then the computational results from five tests are reported. Finally, we conclude with general remarks on this work and outline further research directions.

2 Knowledge-Based Systems

One of the most successful areas of Artificial Intelligence applications are expert systems or, more generally, knowledge-based systems. A popular approach to building knowledge-based systems is using the production system model. There are three main components in the production system model:

- the production rule (long term memory),
- working memory (short term memory),
- the recognize-act cycle.

Knowledge-based systems [13] are usually not able to acquire new knowledge or improve their behavior. It is an important fact, because intelligent agents should be capable of learning. Machine learning is a field of Artificial Intelligence. To deal with such kind of problems, machine learning techniques can be incorporated into knowledge-based systems. Learning, in our considerations, refers to positive changes toward improved performance. When symbol-based machine learning is used, a learner must search the concept space to find the desired concept. These specific improvements must be introduced in programs to direct and order of search, as well as to use of available training data and heuristics to search efficiently. Because of that we will analyze the pheromone trail as a learning possibility.

The main idea of machine learning can be demonstrated through inductive reasoning, which refers to the process of deriving conclusions from given facts. A well-known tree induction algorithm adapted from machine learning is ID3 [67], which employs a process of constructing a decision tree in a top-down fashion. A decision tree is a hierarchical representation that can be used to determine the classification of an object by testing its properties for certain values. ID3 has been proven a very useful method, yet there are many restrictions that make this algorithm not applicable in many real world situations. C4.5 was developed to deal with these problems, and can be considered a good solution when using bad or missing data, continuous variables, as well as data of large size.

Many useful forms of this algorithm have been developed, ranging from simple data structures to more complicated ones . We can distinguish four representations of decision trees:

- a binary decision tree,
- linear decision trees,

- classification and regression trees (CART) — a binary decision trees, which split a single variable at each node. CART performs a recursively search on all variables to find an optimal splitting rule for each node.
- a Chi-squared automatic interaction detector (CHAID) — where multiple branches can be produced.

Several methods have been proposed for the rule induction process, such as: ID3 [67], C4.5 [69], CN2 [14, 15], CART [6], and AQ15 [60]. There are two categories of these approaches: sequential covering algorithms and simultaneous covering algorithms. The latter group is represented by: ID3 and C4.5.

Algorithm 1. Algorithm ID3

```
1  ID3_tree (examples, properties)
2  if all entries in examples are in the same category of the decision variable then
3      return the leaf node labeled with that category;
4  end
5  else
6      Calculate information gain;
7      Select the property P with highest information gain;
8      Assign the root of a current tree = P;
9      Assign properties = properties - P;
10     for each value V of P do
11         Create a branch of the tree labeled with V;
12         Assign examples_V = the subset of examples with values V for property P;
13         Applied ID3_tree (examples_V, properties) to branch V;
14     end
15 end
```

C4.5 is an improved version of ID3. C4.5 follows a, divide and conquer" strategy to build a decision tree through recursive partitioning of a training set. The process of building the decision tree by C4.5 begins with choosing an attribute (a corresponding variablesof this attribute) to split the data set into subsets. Selecting the best splitting attribute is based on the heuristic criteria. This methodology includes: Information Gain [69], Information Gain Ratio [69], Chi-square test [68], and Gini-index [6]. The Information Gain of an attribute A relative to a set of examples S is defined as follows:

$$Gain(S, A) = Entropy(S) - \sum_{v \in values(A)} \frac{S_v}{S} Entropy(S_v),$$

where:

- $values(A)$ is a set of all possible values for A,
- S_v is a subset of S for which attribute A has value v.

$Entropy(S)$ is the entropy of S, which characterizes the purity of a set of examples. If a class attribute has k different values, the $Entropy(S)$ is defined as:

$$Entropy(S) = -\sum_{c=1}^{k} \frac{n_c}{N} log_2 \frac{n_c}{N}$$

where:

- n_c is the number of examples for the cth class,
- $N = \sum_{c=1}^{k} n_c$ is the total number of examples in the data set.

The higher the entropy is, more incidentally the k values are distributed all over the data set. Unfortunately, the Information Gain favors attributes with many values over those with a small number of values. So, an alternative criterion — Information Gain Ratio is employed in the C4.5 approach.

The Information Gain Ratio is calculated as follows:

$$GainRatio(S, A) = \frac{Gain(S, A)}{SplittingInfo(S, A)}$$

where $SplittingInformation$ is used to penalize attributes with too many values; we can calculate it as follows:

$$SplittingInfo(S, A) = \sum_{i=1}^{k} \frac{S_i}{S} log_2 \frac{S_i}{S},$$

where S_i is a subset of training set S, which is represented by attribute A.

Our approach is not associated with CN2, which uses accuracy as a heuristic criterion to determine the best rule, but analyzing only the rule accuracy and thereforecan result in less general rules. Consequently, the Laplace error estimate [14] is used to establish generic rules, which cover a large number of examples:

$$LaplaceAccuracy = \frac{n_c + 1}{n_{tot} + k}$$

where:

- k is the number of classes in the domain,
- n_c is the number of examples in the predicted class c covered by the rule,
- n_{tot} is the total number of examples covered by the rule.

3 Swarm Intelligence and Ant Colony Optimization

Ant colonies exhibit very interesting collective behaviors: even if a single ant has only simple capabilities, the behavior of the whole ant colony is highly structured. This is the result of co-ordinated interactions and co-operation between ants (agents) and represents Swarm Intelligence. this term is applied to any work involving the design of algorithms or distributed problem-solving devices inspired by the collective behavior of social insects [4].

3.1 Swarm Intelligence

Swarm Intelligence (SI) is an innovative distributed intelligent paradigm for solving optimization problems that originally took its inspiration from the

biological examples by swarming, self-organizing foraging phenomena in social insects. There are many examples of swarm optimization techniques, such
as: Particle Swarm Optimization, Artificial Bee Colony Optimization and Ant
Colony Optimization (ACO). The last approach deals with artificial systems,
inspired by the natural behaviors of real ants, especially foraging behaviors
based on the pheromone substances laid on the ground.

The fundamental concept underlying the behavior of social insects is self-
organization. SI systems are complex systems — collections of simple agents
that operate in parallel and interact locally with each other and their environment to produce emergent behavior.

The basic characteristic of metaheuristics from nature could be summarized as follows [1]:

- they model a phenomenon in nature,
- they are stochastic,
- in the case of multiple agents, they often have parallel structure,
- they use feedback information for modifying their own parameters — they
  are adaptive.

Developing algorithms that utilize some analogies with nature and social
insects to derive non-deterministic metaheuristics capable of obtaining good
results in hard combinatorial optimization problems could be a promising
field of research.

The optimization algorithm we propose in this paper was inspired by the
previous works on Ant Systems and, in general, by the term – stigmergy.
This phenomenon was first introduced by P.P. Grasse [45, 46]. Stigmergy is
easily overlooked, as it does not explain the detailed mechanism by which
individuals co-ordinate their activities. However, it does provide a general
mechanism that relates individual and colony-level behavior: individual behavior modifies the environment, which in turn modifies the behavior of other
individuals. The synergetic effect is understood as a result of natural social
behavior among individuals connected by the main goal.

Many features of the collective activities of social insects are self-organized.
The theories of self-organization (SO) [20], originally developed in the context
of physics and chemistry to describe the emergence of macroscopic patterns
out of processes and interactions defined at the macroscopic level. They can
be extended to social insects to show that complex collective behavior may
emerged from interactions among individuals that exhibit simple behavior:
in these cases, there is no need to refer to individual complexity to explain
complex collective behavior. Recent researches show that SO is indeed a
major component of a wide range of collective phenomena in social insects [4].

Self-organization in social insects often requires interaction among insects:
such interactions can be either direct and indirect. Direct interactions are the
,,obvious" interactions: antennation, trophallaxis (food or liquid exchange),
mandibular contact, visual contact, chemical contact (the odour of nearby
nestmates) etc. Indirect interactions are more subtle: two individuals interact

indirectly when one of them modifies the environment and the other responds to the new environment, at a later time. Such an interaction is an example of stigmergy (from Greek stigma: sting, and ergon: work) employed to explain task co-ordination and regulation in the context of nest reconstruction in termites of the genus Macrotermes.

The ant systems – the first version of the ant colony optimization systems mimic the nature and take advantage of various observations made by people, who studied ant colonies. Especially the ACO algorithms were inspired by the experiment run by Goss et al. [44] using a colony of real ants. Deneubourg et al. [4] using a special experimental setup showed that the selection of a path to a food-source in the Argentine ant is based on self-organization. Individual ants deposit a chemical substance called pheromone as they move from a food source to their nest and foragers follow such pheromone trails. The process in which an ant is influenced toward a food source by another ant or by a chemical trail is called recruitment, and recruitment based solely on chemical trails is called mass recruitment.

3.2 Ant Colony Optimization as a New Metaheuristics

In this paper we defined an ant algorithm to be a multi–agent system inspired by the observation of real ant colony behavior exploiting the stigmergic communication paradigm. The optimization algorithm we propose in this paper was inspired by the previous works on Ant Systems and, in general, by the term — stigmergy. This phenomenon was first introduced by P.P. Grasse [45, 46].

The last two decades have been highlighted by the development and the improvement of approximative resolution methods, usually called heuristics and metaheuristics. In the context of combinatorial optimization, the term heuristic is used as a contrast to methods that guarantee to find a global optimum, such as branch and bound or dynamic programming. A heuristic is defined by [70] as a technique which seeks good (i.e. near-optimal) solutions at a reasonable computational cost without being able to guarantee either feasibility or optimality, or even in many cases to state how close to optimality a particular feasible solution is. Often heuristics are problem-specific, so that a method which works for one problem cannot be used to solve a different one. In contrast, metaheuristics are powerful techniques applicable generally to a large number of problems. A metaheuristic refers to an iterative master strategy that guides and modifies the operations of subordinate heuristics by combining intelligently different concepts for exploring and exploiting the search space [43, 64]. A metaheuristic may manipulate a complete (or incomplete) single solution or a collection of solutions at each iteration. The family of metaheuristics includes, but is not limited to, Constraint Logic Programming, Genetic Algorithms, Evolutionary Methods, Neural Networks,

Simulated Annealing, Tabu Search, Non–monotonic Search Strategies, Scatter Search, and their hybrids. The success of these methods is due to the capacity of such techniques to solve in practice some hard combinatorial problems.

Also some traditional and well-established heuristic optimization techniques, such as Random Search, Local Search [3], or the class of Greedy Heuristics (GH) [3] may be considered as metaheuristics.

Considering that Greedy Heuristics are available for most practical optimization problems and often produce good results, it seems, that in these cases, it is less expensive with regard to the development costs to further improve their solution quality by extending them to repetitive procedures rather than to replace them by iterative heuristics which follow completely different optimization strategies. So it seems desirable to have a constructive and repetitive metaheuristics including GH as a special (boundary) case.

An essential step in this direction was the development of Ant System (AS) by Dorigo et al. [25, 31, 28], a new type of heuristic inspired by analogies to the foraging behavior of real ant colonies, which has proven to work successfully in a series of experimental studies. Diverse modifications of AS have been applied to many different types of discrete optimization problems and have produced very satisfactory results [27]. Recently, the approach has been extended by Dorigo and Di Caro [26] to a full discrete optimization metaheuristics, called the Ant Colony Optimization (ACO) metaheuristics.

AS, which was the first ACO algorithm [16, 29, 31], was designated as a set of three ant algorithms differing in the way the pheromone trail was updated by ants. Their names were: ant-density, ant-quantity, and ant-cycle. A number of algorithms, including the metaheuristics, were inspired by ant-cycle, the best performing of the ant algorithms.

The Ant Colony System (ACS) algorithm has been introduced by Dorigo and Gambardella [28, 30] to improve the performance of Ant System [30, 39], which allowed to find good solutions within a reasonable time for small size problems only. The ACS is based on 3 modifications of Ant System:

- a different node transition rule,
- a different pheromone trail updating rule,
- the use of local and global pheromone updating rules (to favor exploration).

The node transition rule is modified to allow explicitly for exploration. An ant k in city i chooses the city j to move to following the rule:

$$j = \begin{cases} \arg \max_{u \in J_i^k} \{[\tau_{iu}(t)] \cdot [\eta_{iu}]^\beta\} & \text{if } q \leq q_0 \\ J & \text{if } q > q_0 \end{cases}$$

where q is a random variable uniformly distributed over $[0, 1]$, q_0 is a tunable parameter $(0 \leq q_0 \leq 1)$, and $J \in J_i^k$ is a city that is chosen randomly according to a probability:

$$p_{iJ}^k(t) = \begin{cases} \dfrac{\tau_{iJ}(t) \cdot [\eta_{iJ}]^\beta}{\sum\limits_{l \in J_i^k} [\tau_{il}(t)] \cdot [\eta_{il}]^\beta} \end{cases}$$

which is similar to the transition probability used by Ant System. We see therefore that the ACS transition rule is identical to Ant System's one, when $q > q_0$, and is different when $q \leq q_0$. More precisely, $q \leq q_0$ corresponds to the exploitation of the knowledge available about the problem, that is, the heuristic knowledge about distances between cities and the learned knowledge memorized in the form of pheromone trails, whereas $q > q_0$ favors more exploration.

In Ant System all ants are allowed to deposit pheromone after completing their tours. By contrast, in the ACS only the ant that generated the best tour since the beginning of the trail is allowed to globally update the concentrations of pheromone on the branches. The updating rule is:

$$\Delta_{ij}(t+n) = (1-\alpha) \cdot \tau_{ij}(t) + \alpha \cdot \Delta\tau_{ij}(t, t+n)$$

where (i, j) is the edge belonging to T^+, the best tour since the beginning of the trail, α is a parameter governing pheromone decay, and:

$$\Delta\tau_{ij}(t, t+n) = \frac{1}{L^+}$$

where L^+ is the length of the T^+.

The local update is performed as follows: when, while performing a tour, ant k is in city i and selects city $j \in J_i^k$ to move to, the pheromone concentration on edge (i, j) is updated by the following formula:

$$\tau_{ij}(t+1) = (1-\rho) \cdot \tau_{ij}(t) + \rho \cdot \tau_0$$

The value of τ_0 is the same as the initial value of pheromone trails and it was experimentally found that setting $\tau_0 = (n \cdot L_{nn})^{-1}$, where n is the number of cities and L_{nn} is the length of a tour produced by the nearest neighbor heuristic, produces good results [28, 39].

3.3 Hybrid Approaches in ACO

A single algorithm cannot be the best approach to solve quickly every atomization problem. One way to deal with this issue is to hybridize an algorithm with more standard procedures, such as greedy methods or local search procedures. Individual solutions obtained so far can be improved using another searching procedures, and then placed back in competition with other solutions not improved yet. There are so many ways to hybridize that there is a common tendency to overload hybrid systems with too many components.

Some modern hybrid systems contain fuzzy-neural-evolutionary components, together with some other problem-specific heuristics. Ant Colony Optimization algorithms are extremely flexible and can be extended by incorporating diverse concepts and alternative approaches. For example Stüzle and Hoos [75] have implemented version of AS-QAP based on their MAX-MIN Ant System (MMAS). They discussed the possibility of incorporating a local search (2-opt algorithm and short Tabu Search runs) into the Ant Algorithm in the case of the QAP. Ant Colony Optimization technique can also be enhanced by including new points in a standard version of this algorithm. Gambardella et al. [4] proposed a new Ant Colony Optimization algorithm, called the hybrid Ant System (HAS-QAP), which differs significantly from the ACO algorithm. There are two novelties:

- ants modify solutions, as opposed to building them,
- pheromone trails are used to guide the modifications of solutions, and not as an aid to direct their construction.

In the initialization phase, each ant $k, k = 1, \ldots, m$, is given a permutation Pi^k which consists of a randomly generated permutation to its local optimum by the local search procedure. This procedure examines all possible ways in a random order and accepts any swap that improves the current solution. Gambardella and Dorigo [39] have developed the Hybrid Ant System (called HAS-SOP), which is a modified version of ACS plus a local optimization procedure specifically designed to solve the sequential ordering problem (SOP). Results obtained with HAS-SOP were very useful (effective). HAS-SOP, which was tested on an extensive number of problems, outperforms existing heuristics in terms of solution quality and computational time. Also, HAS-SOP has improved many of the best known results on problems maintained in TSPLIB.

The ACO metaheuristics has been successfully applied to many discrete optimization problems, as listed in tab. 1 [4]. ACO algorithms results turned out to be competitive with the best available heuristic approaches. In particular, the results obtained by the application of ACO algorithms to the TSP are very encouraging. They are often better than those obtained using other general purpose heuristics like Evolutionary Computation or Simulated Annealing. Also, when adding to ACO algorithms Local Search procedures based on 3-opt, the quality of the results is close to that obtainable by other state-of-the-art methods. ACO algorithms are currently one of the best performing heuristics available for the particularly important class of quadratic assignment problems which model real world problems. AntNet, an ACO algorithm for routing in packet switched networks, outperformed a number of state-of-the-art routing algorithms for a set of benchmark problems. AntNet-FA, an extension of AntNet for connection oriented network routing problems, also shows competitive performance. HAS-SOP, an ACO algorithm coupled to a local search routine, has improved many of the best known results on a wide set of benchmark instances of the sequential ordering problem (SOP), i.e. the

Table 1 Current applications of ACO algorithms

Problem name	Algorithm	Main references
Traveling salesman	AS	Dorigo [31]
		Dorigo [25]
		Dorigo [32]
	Ant-Q	Gambardella [37]
	ACS and	
	ACS-3-opt	Dorigo [30]
		Dorigo [29]
		Gambardella [38]
	AS	Stützle [75]
		Stützle [74]
	ASrank	Bullheimer [8]
Quadratic assignment	AS-QAP	Maniezzo [58]
	HAS-QAP	Gambardella [41]
		Gambardella [42]
	MM AS-QAP	Stützle [76]
	ANTS-QAP	Maniezzo [54]
	AS-QAP	Maniezzo [56]
Scheduling problems	AS-JSP	Colorni [17]
	AS-FSP	Stützle [73]
	ACS-SMTTP	Bauer [2]
	ACS-SMTWTP	Denbesten [19]
Vehicle routing	AS-VRP	Bullheimer [10]
		Bullheimer [7]
		Bullheimer [9]
	HAS-VRP	Gambardella [40]
Connection-oriented network routing	ABC	Schoonderwoerd [72]
		Schoonderwoerd [71]
	ASGA	White [81]
	AntNet-FS	Di Caro [23]
	ABC-smart ants	Bonabeau [5]
Connection-less network routing	AntNet and	
	AntNet-FA	Di Caro [21]
		Di Caro [22]
		Di Caro [24]
	Regular ants	Subramanian [77]
	CAF	Heusse [47]
	ABC-backward	Vanderput [78]
Sequential ordering	HAS-SOP	Gambardella [39]
Graph coloring	ANTCOL	Costa [18]
Shortest common supersequence	AS-SCS	Michel [61]
		Michel [62]

Table 1 (*continued*)

Problem name	Algorithm	Main references
Frequency assignment	ANTS-FAP	Maniezzo [55] Maniezzo [57]
Generalized assignment	MMAS-GAP	Ramalhinho [53]
Multiple knapsack	AS-MKP	Leguizamon [50]
Optical networks routing	ACO-VWP	Navarro [79]
Redundancy allocation	ACO-RAP	Liang [51]

problem of finding the shortest Hamiltonian circle on a graph which satisfies
a set of precedence constraints on the order in which cities are visited. ACO
algorithms have also been applied to a number of other discrete optimization problems like the shortest common supersequence problem, the multiple
knapsack, single machine total tardiness, and others, with very promising
results.

Multiple Ant Colony System for Vehicle Routing Problem with Time Windows (MACS-VRPTW) is a new ACO based approach to solve vehicle routing
problems with time windows. The general idea of adapting ACS for multiple
objectives in VRPTW is to define two colonies, each dedicated to the optimization of a different objective function. In the MACS-VRPTW algorithm
the minimization of the number of tours (or vehicles) and the minimization of
the total travel time are optimized simultaneously by two ACS based colonies
— ACS-VEI and ACS-TIME. The goal of the first colony is to reduce the
number of vehicles used, whereas the second colony optimises the feasible
solutions found by ACS-VEI. Each colony uses its own pheromone trails but
collaborates with another one by sharing information about the best results
obtained so far. MACS-VRPTW is shown to be competitive with the best
existing methods both in terms of solution quality and computation time.
Moreover, MACS-VRPTW improves good solutions for a number of problem
instances known from the literature.

4   Ant Colony Optimization in Rule Induction

The adaptation of ant colony optimization to rule induction and classification is a research area still not well explored and examined. Ant-Miner is a
sequential covering algorithm that merged concepts and principles of ACO
and rule induction. Starting from a training set, Ant-Miner generates a set

of ordered rules through iteratively finding an appropriate rule that covers a subset of the training data, adds the formulated rule to the induced rule list and then removes the examples covered by this rule until the stopping criteria are reached.

ACO has a number of features that are important for computational problem solving [34]:

- it is relatively simple and easy to understand and then to implement
- it offers emergent complexity to deal with other optimization techniques
- it is compatible with the current trend towards greater decentralization in computing
- it is adaptive and robust and it is enables coping with noisy data.

There are numerous characteristics of ACO which are crucial in data mining applications. ACO, in contrary to deterministic decision trees or rule induction algorithms, during rule induction tries to alleviate the problem of premature convergence to local optima because of a stochastic element which prefers a global search in problem's search space. Secondly, ACO metaheuristic is a population–based one. It permits the system to search in many independently determined points in the search space concurrently and to use the positive feedback between ants as a search mechanism [65].

Ant-Miner was invented by Parpinelli et al. [66, 65]. It was the first Ant algorithm for rule induction and it has been shown to be robust and comparable with the CN2 [14] and C4.5 [69] algorithms for classification. Ant-Miner generates solutions in the form of classification rules. The original Ant–Miner has a limitation that it can only process discrete values of attributes.

We present a short review of the main aspects of the rule discovery process by Ant-Miner together with the description of the Ant-Miner algorithm. Ant-Miner is a sequential covering approach to discover a list of classification rules, by discovering one rule at a time until all or almost all the examples in the training set are covered by the discovered rules. When the algorithm begins to work, the training set holds all the examples and the discovered rule list is empty. Each ant builds one rule. At the end of the **While** loop, the best rule is added to the discovered rule list. Examples having the class predicted by the rule, are removed from the training set before the next iteration of the **While** loop. The process of rule discovering is repeated as long as the number of uncovered examples in the training set is less than a user-specified threshold.

Three main phases may be distinguish in every iteration of the **Repeat** — **Until** loop: rule construction, rule pruning and pheromone updating. In the first step, Ant_t starts with an empty rule without term in the antecedent, and adds one term at time until one of the two criteria is satisfied:

- there is a term added to the current rule R_t, which makes the rule cover a number of examples less than a user specified parameter — *MinExamplesPerRule*,

- there are no more terms which can be added to the rule antecedent by the current Ant_t. Please notice that no rule can contain any attribute twice (with different values).

During the construction of the partial rule, Ant_t builds a path, and every term added to the partial rule represents the direction of how the path is being extended. The next term will be added to the partial rule according to the probability of a term being selected. This value depends on the value of the heuristic function and the amount of the pheromone associated with the term. After this phase, pruning will be performed. The aim of this process is to remove all irrelevant terms and improve the rule R_t. The importance of this process is due to the probabilistic mechanism performed during the construction of the rule. We want to eliminate the drawback of ignoring interactions between attributes. Rule pruning iteratively remove one term at a time from the rule as long as this improves the quality of the rule. This process will be performed till there is only one term in the rule, or no term improves the quality of the rule.

The next phase — pheromone updating is performed according to the quality of the rule (predictive accuracy) — each term in the antecedent of the just-pruned rule is affecting this value. The process of reducing the value of pheromone will be performed for other terms, which are not present in the rule antecedent.

The main loop is repeated until one of the following termination criteria is achieved:

- the number of constructed rules is equal or greater than the number of ants,
- the rule constructed by Ant_t is equal to $No\_rules\_converg-1$ rules (defined by the user).

All cells in the pheromone table are initialized equally to the following value:

$$\tau_{ij}(t=0) = \frac{1}{\sum_{i=1}^{a} b_i}$$

where:

- a — the total number of attributes,
- b_i — the number of values in the domain of attribute i.

The probability is calculated for all of the attribute–value pairs, and the one with the highest probability is added to the rule. The transition rule in Ant-Miner is given by the following equation:

$$p_{ij} = \frac{\tau_{ij}(t) \cdot \eta_{ij}}{\sum_i^a \sum_j^{b_i} \tau_{ij}(t) \cdot \eta_{ij}}, \forall i \in I$$

where:

Algorithm 2. Algorithm Ant-Miner

```
 1  TrainingSet = {all training examples};
 2  DiscoveredRuleList = [ ]; /* rule list is initialized with an empty list */
 3  while TrainingSet > MaxUncoveredExamples do
 4      t = 1;/* ant index */
 5      j = 1;/* convergence test index */
 6      Initialize all trails with the same amount of pheromone;
 7      repeat
 8          Ant_t starts with an empty rule and incrementally constructs a classification rule
            R_t by adding one term at a time to the current rule;
 9          Prune rule R − t;
10          Update the pheromone amount of all trails by increasing pheromone in the trail
            followed by Ant_t (proportional to the quality of R_t) and decreasing pheromone
            amount in the other trails (simulating pheromone evaporation);
11          /* update convergence test */
12          if R_t is equal to R_t − 1 then
13              j = j + 1;
14          end
15          else
16              j = 1;
17          end
18          t = t + 1;
19      until (t ≥ No_of_ants) OR (j ≥ No_rules_converg) ;
20      Choose the best rule R_{best} among all rules R_t constructed by all the ants;
21      Add rule R_{best} to DiscoveredRuleList;
22      TrainingSet = TrainingSet - (set of examples correctly covered by R_{best});
23  end
```

- η_{ij} is a problem-dependent heuristic value for each term,
- τ_{ij} is the amount of pheromone currently available at time t on the connection between attribute i and value j,
- I is the set of attributes that are not yet used by the ant,
- Parameter β is equal to 1.

In Ant-Miner, the heuristic value is supposed to be an information theoretic measure for the quality of the term to be added to the rule. For prefering the quality is measured in terms of entropy term to the others, and the measure is given as follows:

$$\eta_{ij} = \frac{log_2(k) - InfoT_{ij}}{\sum_i^a \sum_j^{b_j} (log_2(k) - InfoT_{ij})}$$

where the function $Info$ is similar to another function employed in C4.5 approach:

$$InfoT_{ij} = - \sum_{w=1}^{k} \left[\frac{fregT_{ij}^w}{T_{ij}} \right] log_2 \left[\frac{fregT_{ij}^w}{|T_{ij}|} \right]$$

where: k is the number of classes, $|T_{ij}|$ is the total number of cases in partition T_{ij} (the partition containing the cases, where attribute A_i has the value V_{ij}), $freqT_{ij}^w$ is the number of cases in partition T_{ij} with class w, b_i is a number of values in the domain of attribute A_i (a is the total number of attributes).

The higher the value of $InfoT_{ij}$, is the less likely is that the ant will choose $term_{ij}$ to add to its partial rule.

Please note that this heuristic function is a local method and it is sensitive to attribute interaction. The pheromone values assigned to the term have a more global nature. The pheromone updates depend on the evaluation of a rule as a whole, i.e. we must take into account interaction among attributes appearing in the rule.

The heuristic function employed here comes from the decision tree world and it is similar to the method used in algorithm C4.5. There are many other heuristic functions that can be adapted and used in Ant-Miner. We can derive them from information theory, distance measures or dependence measures.

The rule pruning procedure iteratively removes the term whose removal will cause the maximum increase in the quality of the rule. The quality of a rule is measured using the following formula:

$$Q = \left(\frac{TruePos}{TruePos + FalseNeg} \right) \cdot \left(\frac{TrueNeg}{FalsePos + TrueNeg} \right)$$

where:

- TruePros – the number of cases covered by the rule and having the same class as the one predicted by the rule,
- FalsePros – the number of cases covered by the rule and having a different class from the one predicted by the rule,
- FalseNeg – the number of cases that are not covered by the rule while having a class predicted by the rule,
- TrueNeg – the number of cases that are not covered by the rule which have a different class from the class predicted by the rule.

The quality measure of a rule is determined by:

$$Q = sensitivity \cdot specificity.$$

We can say that accuracy among positive instances determines sensitivity, and the accuracy among negative instances determines specificity. Now we take into account only the rule accuracy, but it can be changed to analyze the rule length and interestingness.

Once each ant completes the construction of the rule, pheromone updating is carried out as follows:

$$\tau_{ij}(t+1) = \tau_{ij}(t) + \tau_{ij}(t) \cdot Q, \forall term_{ij} \in \text{the rule}$$

The amounts of pheromones of terms belonging to the constructed rule R are increased in proportion to the quality of Q. To simulate pheromone evaporation τ_{ij}, the amount of pheromone associated with each $term_{ij}$ which does not occur in the constructed rule must be decreased. The reduction of pheromone of an unused term is performed by dividing the value of each

τ_{ij} by the summation of all τ_{ij}. The pheromone levels of all terms are then normalized.

5 Modifications

The authors of Ant-Miner [66, 65] suggested two directions for future research:

1. Extension of Ant-Miner to cope with continuous attributes;
2. The investigation of the effects of changes in the main transition rule:

   a. the local heuristic function,
   b. the pheromone updating strategies.

Recently, Galea [36] proposed a few modifications in Ant-Miner. Firstly, the population size of ants was changed (from 1 to 50 ants). Secondly, new stopping criteria were examined. Compared to the classical Ant-Miner, the implementation of Ant-Miner proposed by Galea [36] produces slightly lower predictive accuracy then Parpinelli et al.'s. Galea also examined the effect of employing the pseudo random proportional rule. Galea suggests to study the relationship between the characteristics of the dataset, such as:

- the distribution of the data set and correlation between attributes,
- the selection of algorithms: stochastic or deterministic algorithms.

The Ant-Miner modifications and extensions have been presented in many articles, as listed in Tab.2.

This subsection gives a brief overview of Ant-Miner extensions. Some of the propositions are relatively simple and, as a result, give the same type of classification rules discovered by this algorithm. Another modifications cope with the problem of attributes having ordered categorical values, some of them improve the flexibility of the rule representation language. Finally, more sophisticated modifications have been proposed to discover multi-label classification rules and to investigate fuzzy classification rules. Certainly there are still many problems and open questions for future research.

6 Experiments and Remarks

An Ant Colony Optimization technique is in essence, a system based on agents which simulate the natural behavior of ants, incorporating a mechanism of cooperation and adaptation, especially via pheromone updates. When solving different problems with the ACO algorithm we have to analyze three major functions. Choosing these functions appropriately helps to create better results and prevents stucking in local optima of the search space.

The first function is a problem-dependent heuristic function (η) which measures the quality of terms that can be added to the current partial rule. The

Table 2 Modifications of the Ant-Miner algorithm

1	Modifications or extractions:
	The class is known during the rule construction. All ants construct rules predicting the same class [12, 36, 59, 48]

Authors	Year	Open questions:
Chen, Chen & He,	2006	This approach leads to: a new heuristic function
Galea & Shen,	2006	and new pheromone update strategies.
Martens et al.,	2006	
Smaldon & Freitas,	2006	

2	Modifications or extensions:
	A new heuristic function. The replacement of the entropy reduction heuristic function by a simpler heuristic function [12, 52, 59, 63, 48, 80]

Authors	Year	Open questions:
Chen, Chen & He,	2006	There is no guarantee that the use of pheromones
Liu, Abbas & Mc Kay,	[12]	would completely compensate the use of a less ef-
Martens et al.,	2004	fective function.
Oakes,	2006	The heuristic function can be computed with re-
Smaldon & Freitas,	2004	spect to the number of cases and attributes (only
Wang & Feng,	2006	once in the initialization of the algorithm).
	2004	

3	Modifications or extensions:
	Using a pseudo random proportional transition rule. Using parameter q_0 [12, 52, 80]

Authors	Year	Open questions:
Chen, Chen & He,	2006	This transition rule has an advantage that it al-
Liu, Abbas & Mc Kay,	2004	lows the user to have an explicit control over the
Wang & Feng,	2004	exploitation versus exploration trade-of.
		It requires to choose a good value for the param-
		eter q_0 in empirical way.

4	Modifications or extensions:
	A new rule Quality measure. Ant-Miner's rule quality is based on the product of sensitivity and specificity. The replacement of the rule quality function by measures that are essentially based on the confidence and coverage of a rule [12, 59]

Authors	Year	Open questions:
Chen, Chen & He,	2006	It would be interesting to perform extensive ex-
Martens et al.,	2006	periments comparing the effectiveness of these
		kinds of rule quality measures.

5	Modifications or extensions:
	New pheromone updating rules. The use of an explicit pheromone evap- oration rate. A self adaptive parameter. Different procedures to update pheromone as a function of the rule quality [52, 59, 48, 80]

Table 2 (*continued*)

Authors	Year	Open questions:
Liu, Abbas & Mc Kay,	2004	It is important to change the original Ant-Miner's formula to update pheromone in order to cope better with law-quality rules. We treat evaporation trails as a form of a learning factor.
Martens et al.,	2006	
Smalton & Freitas,	2006	
Wang & Feng,	2004	

6   Modifications or extensions:
Ant-Miner can cope with ordered values of categorical attributes (not continuous) [63, 59]

Authors	Year	Open questions:
Oakes	2004	It is important to point out that the connection with the Rough Sets theory in the contrast to Fuzzy Sets is a new augmentation.
Martens et al.,	2006	

7   Modifications or extensions:
Dropping the rule pruning procedure. The removal of pruning makes the Ant-Miner+ significantly faster [59]

Authors	Year	Open questions:
Martens et al.,	2006	It would be interesting to evaluate, whether for the discovered by Ant-Miner+ rules the predictive accuracy could be increased by the use of a deterministic rule pruning procedure driven by the rule quality.

8   Modifications or extensions:
Discovering Fuzzy Classification Rules. FRANTIC-SRL maintains multiple populations of ants, each of them discovers rules predicting a different class (fixed for all ants belongs to one colony) [36]

Authors	Year	Open questions:
Galea & Shen	2006	The evaluation of the rule sets could be reduced by considering only the combinations of the best rules (the delaying reinforcement (perhaps to a late) — the update of the pheromones).

9   Modifications or extensions:
Discovering rules for multi-label classification. The simultaneous prediction of the value of two or more class attributes rather than just 1 class attribute [11]

Authors	Year	Open questions:
Chan & Freitas	2006	MuLAM — each ant constructs a set of rules. MuLAM uses a pheromone matrix for each of the class attributes. Authors update the pheromone matrix of the class attributes (depending on the terms in its antecedent) predicted by the rule. The model of communication via pheromone on different levels is worth to analyze.

heuristic function stays unchanged during the algorithm run in the classical approach. We want to investigate whether the heuristic function depends on the previous well-known approaches in the data-mining area (C4.5, CN2) and can influence the behavior of the whole colony, or not. A rule for pheromone updating, which specifies how to modify the pheromone trail (τ_{ij}) and guarantee the communication between ants, is an important aspect of learning via pheromone values. Consequently the changing scheme of this value may play a significant role in establishing a suitable form of cooperation. Finally, the probabilistic transition rule based on the value of the heuristic function and on the contents of the pheromone trail matrix (that is used to iteratively construct rules) can also be analyzed as a form of specific two-way switch between deterministic and stochastic rule induction in Ant-Miner. These problems will be analyzed carefully in the following experimental studies.

6.1 Data Sets Used in Our Experiments

The evaluation of the performance behavior the behavior of Ant-Miner was performed using 5 public-domain data sets from the UCI (University of California at Irvine) data set repository available from: `http://www.ics.uci.edu.~mlearn/MLRepository.html`. Table 3 shows the main characteristics of the data sets, which were the data sets similar to ones used to evaluate the original Ant-Miner. Please note that Ant-Miner cannot cope directly with continuous attributes (i.e. continuous attributes have to be discretized in a preprocessing step, using the RSES program (`http://logic.mimuw.edu.pl/~rses/`). For the data sets marked with an asterisk in table 3 we used the discretized version of the data. In the original Ant-Miner and Galea implementation [35], the discretization was carried out using a method called C4.5-Disc [49]. C4.5-Disc is an entropy-based method that applies the decision-tree algorithm C4.5 to obtain discretization of the continuous attributes.

Both the original Ant-Miner and our proposal have some parameters. The first one — the number of ants will be examined during the experiments. The rest of parameters are presented in tab. 4. There are the following values of the parameters:

- the minimum number of cases per rule,
- the maximum number of uncovered cases in the training set,
- the number of rules used to test convergence of the ants.

In order to obtain reliable performance estimates ten-fold cross-validations were carried out to produce each of the statistics in the tables below. As some of the tests required changes to the parameters for each of the data sets, for example the number of ants or q_0-value was changed during appropriate tests. The results obtained are summarized after 300 ten-fold cross-validations, summarized and discussed below. Possible follow-up investigations suggested by the results are also mentioned.

The authors of the original version of Ant-Miner used 2 sets of evaluation criteria for the comparison with other rule induction algorithms: C4.5 and CN2. We usually used the first criterion: a predictive accuracy. The second criterion determines the rule set size:

- the number of rules in a rule set,
- an individual rule size (the number of terms per rule).

Table 3 The properties of data sets

Dataset	Number of Instances	Number of Attributes	Number of Values of Attributes	Number of Decision Classes
Breast cancer	296	9+1	51	2
Wisconsin breast cancer	699	9+1	90	2
Dermatology	366	34+1	*137	6
Hepatitis	155	19+1	*71	2
Tic-tac-toe	958	9+1	27	2

Table 4 Original parameters in data sets

Test Set	50%
Training Set	50%
Min. Cases per Rule	5
Max. Uncovered Cases	10
Rules for Convergence	10
Number of Iterations	100

6.2   TEST 1: Changing the Population Size

This test was carried out to see whether a variable number of ants as compared to a single ant in Ant-Miner has an influence on the predictive accuracy. The number of ants was changed from 1 to 50, with the step equal to 5. The aim of this experiment was to observe the obvious effect of a changing of the number of ants for an iteration and how does it affct the predictive accuracy.

Two tables: 5 and 6 show the results obtained for all of the data sets and compare them with the result obtained with a population size equal to 1. Similarly to Galea results [35], in two of the data sets — Wisconsin Breast Cancer

Table 5 Comparing the predictive accuracy for the different number of ants' population. The numbers next to the ,,±" symbol are the standard deviations of the corresponding average predictive accuracies

| Dataset | Best Achieved result | | Predictive |
	Population size (%)	Predictive Accuracy (%)	Accuracy with a Population Size of one Ant (%)
Breast cancer	5	74.69 (± 1.76)	74.36 (± 1.69)
Wisconsin breast cancer	1	92.13 (± 1.06)	92.13 (± 1.06)
Dermatology	40	94.02 (± 1.77)	87.84 (± 3.20)
Hepatitis	50	79.17 (± 2.64)	78.79 (± 3.16)
Tic-tac-toe	1	73.80 (± 1.43)	73.80 (± 1.43)
	10	73.80 (± 1.61)	

Table 6 The simplicity of the rule sets generated by Ant-Miner, C4.5 and CN2. The numbers next to the ,,±" symbol are the standard deviations of the corresponding average predictive accuracies

| Dataset | Average No. of Rules | | |
	Ant Miner	C4.5	CN2
Breast cancer	5.48 (± 0.36)	6.2 (± 3.21)	55.4 (± 2.07)
Wisconsin breast cancer	9.92 (± 0.62)	11.1 (± 1.45)	18.6 (± 1.45)
Dermatology	7.11 (± 0.43)	23.2 (± 1.99)	18.5 (± 0.47)
Hepatitis	4.42 (± 0.31)	4.4 (± 0.93)	7.2 (± 0.25)
Tic-tac-toe	7.90 (± 0.86)	83.0 (± 14.1)	39.7 (± 2.52)

and Tic-Tac-Toe the algorithm achieved the best average predictive accuracy for one ant only. In the rest of the analyzed data sets the improvements can be seen when the population size was increased to the greater values, close to 50. This number of ants depends on the particular characteristics of the data sets.

During these investigations, one thing to note is that choosing the best rule from all of the created within one iteration comes a superficial and purposeless task. It can be helpful only for later ants (using pheromone updates), which could create different rules that might also could have an improved quality.

It is worth to make further experiments to check the pheromone updating rules force the ants to explore different regions in the problem search space and create different rules. This may be accomplished by testing different strategies of pheromone reinforcements or evaporations according to local and global rules. The pheromone evaporation create a possibility, or situation, when other ants in the same iteration select different terms and hence create different, more valuable rules. More detailed investigation were carried out in the next tests.

6.3 TEST 2 — Changing the Pheromone Values

There are various ways in which a pheromone updating can be performed — the initialization of the pheromone and setting it at minimum/maximum bounds of the levels reached. In the original Ant-Miner the rule induction is performed with a population of one ant. In the ant colony algorithm, where the population size is greater than one, choices need to be made as to how to update the pheromone values. All the ants may have to be reinforced or only the k best ants are used, in short, the elitist strategy should be used for pheromone updating, ensuring that exploration is more channelized. As it was mentioned before in the original Ant-Miner algorithm the pheromone values of terms were updated after each ant create the solution while in our implementation the pheromone valueis changed at the end of an iteration based on the best ant from the current iteration (a global updating rule). Meanwhile, the gathered pheromone values evaporate according to the local pheromone updating rule. This makes the pheromone updating of terms more prejudicing and therefore allows more control over the search of ants in successive iterations.

$$\Delta\tau_{ij}(t) = (1 - \alpha)\tau_{ij}(t - 1) + \tau_{ij}(t - 1) \cdot Q, \forall i, j \in R,$$

where the parameters are the same as mentioned in the second section of this article.

Our approach also imitate the same pheromone updating rule as in the original Ant-Miner:

$$\tau_{ij}(t) = \tau_{ij}(t - 1) + \tau_{ij}(t - 1) \cdot Q = (1 + Q)\tau_{ij}(t - 1)$$

Meanwhile, the effect of evaporation for unused terms is achieved by dividing the value of each $\tau_{ij}(t)$ by the summation of all τ_{ij} (a specific normalization process).

In the case, when pheromone values were generally too low, and only few good rules were being generated and reinforced by new pheromone values, we propose another method of changing the values. This process causes that the pheromone is increased, but the rest of the terms have decreasing pheromone values. Because the results indicate that the pheromone values are not in the correct boundaries, we propose to employed the another method of normalization in changeable the scheme of pheromone values. This idea was presented by V. Maniezzo as Approximate Nondeterministic Tree Search (ANTS) [33], where the pheromone rule is presented as follows:

$$\Delta\tau_{ij}(t) = \tau_{ij}(t-1) \cdot \left(1 - \frac{Q - LB}{avgQ - LB}\right), \forall i, j \in R,$$

where LB is the value of the lower bound on the optimal solution value computed at the start of the algorithm and we have $LB = 0.05$ and $avgQ$ is the moving average of the last $l = 10$ evaluated solutions according to the Q function; l is a parameter of the algorithm.

Another point that needs to be clarified is whether the normalization is correctly performed. We examine this procedure by changing the upper bound of the normalization (in these circumstances we establish the boundary equal to 3). We also reinforced the addition value of the Q function by multiplying this value by $(1 - \alpha)$, which leads to $(1 + Q) \cdot 0.9$ updating the previous values of pheromone:

$$\tau_{ij}(t) = \tau_{ij}(t-1) \cdot 0.9 + \tau_{ij}(t-1) \cdot 0.9 \cdot Q,$$

where α is the evaporation factor in the global updating rule (equal to 0.1).

The best achieved predictive accuracies are presented in the tab. 7 in bold. Unfortunately, the interesting ANTS approach is not as promising as we expected. To sum up all the algorithm changes and developments we can

Table 7 The influence of the pheromone updates on the predictive accuracy (%). The numbers below, next tor the ,,±" symbol are the standard deviations of the corresponding average predictive accuracies

Dataset	Standard update	Normalize 0-3	Increase c. 0.9	Minus c. 0.01	ANTS
Breast cancer	74.69 ($\pm$ 1.76)	72.99 ($\pm$ 2.06)	73.93 ($\pm$ 2.46)	**74.14** ($\pm$ 1.86)	71.97 ($\pm$ 2.88)
Wisconsin breast cancer	92.13 ($\pm$ 1.06)	**91.98** ($\pm$ 1.10)	91.58 ($\pm$ 1.39)	91.93 ($\pm$ 1.59)	91.57 ($\pm$ 1.35)
Dermatology	94.02 ($\pm$ 1.77)	93.27 ($\pm$ 1.64)	**93.96** ($\pm$ 1.41)	92.71 ($\pm$ 2.18)	88.32 ($\pm$ 2.42)
Hepatitis	79.17 ($\pm$ 2.64)	79.58 ($\pm$ 2.80)	79.22 ($\pm$ 3.11)	**79.82** ($\pm$ 1.92)	76.98 ($\pm$ 3.26)
Tic-tac-toe (1)	73.80 ($\pm$ 1.43)	**74.26** ($\pm$ 1.87)	73.31 ($\pm$ 1.02)	74.23 ($\pm$ 2.24)	71.80 ($\pm$ 1.97)
Tic-tac-toe	73.80 ($\pm$ 1.61)	72.46 ($\pm$ 1.96)	72.71 ($\pm$ 1.11)	**73.17** ($\pm$ 1.41)	71.29 ($\pm$ 1.69)

observed that the pheromone values for some term in some of the data sets were getting extraordinarily low. Several solutions were possible to include in our approach:

- setting a minimum value for the pheromone values of terms, similarly to the Min-Max approach proposed by Stützle and Hoos [76],
- ignore the terms with small, uninteresting values, i.e. do not consider them for inclusion in the currently created rule.

6.4 TEST 3 — Changing the Main Transition Rule

In this experimental study we want to see whether altering the main transition rule (selection) from the random proportional selection to the pseudo random proportional selection has an effect on the predictive accuracy. A pseudo random proportional selection requires a setting of the q parameter value that enables the exploitation or the exploration of the problem search space.

Different values of q were tested, ranging from 0.1 to 1.0. A setting of $q = 0.0$ means no exploitation of owned knowledge and is the same as using a random proportional rule. A setting $q = 1.0$ turns the algorithm into a deterministic approach and a term with a higher probability is always chosen.

The following tab. 8 shows the results for the different q_0-values settings.

Table 8 Comparing predictive accuracy for different q_0 values

Dataset	Expl./Explor. Rate (q_0)	Predictive Accuracy (%)	Exploration Accuracy (%)
Breast cancer	0.9	**74.71** ($\pm$ 1.79)	74.69 ($\pm$ 1.76)
Wisconsin breast cancer	0.6	**92.69** ($\pm$ 1.16)	92.13 ($\pm$ 1.06)
Dermatology	0.0 (s.)	94.02 ($\pm$ 1.77)	94.02 ($\pm$ 1.77)
Hepatitis	1.0 (d.)	**80.31** ($\pm$ 2.79)	79.17 ($\pm$ 2.64)
Tic-tac-toe (1 ant)	0.9	**74.87** ($\pm$ 1.82)	73.80 ($\pm$ 1.43)
Tic-tac-toe	0.9	**74.33** ($\pm$ 1.94)	73.80 ($\pm$ 1.61)

It is especially interesting to note that for the Dermatology data set the stochastic algorithm achieved the best predictive accuracy. On contrary, in the Hepatis data set we observe the best performance for the fully deterministic approach.

It can be an intriguing aspect of future research to adjust specific features of data sets to the nature of this stochastic/deterministic approach. In order to achieve better predictive accuracy than in the standard C4.5 and CN2 we can combine this study with the population size examination. Another question is the dynamic or adaptive scheme of changing the q_0 value. This possibility enable the employment of coarse-grained or fine-grained methodology during the searching process.

6.5   TEST 4 — Changing the Q Values (Modifications of Pruning)

The rule pruning procedure may be slightly different from Ant-Miner's implementation. We have introduced three extensions to the classical approach.

Accuracy Convergence.   It is a compromise between accuracy and convergence:

$$Q = \frac{TP}{TP+FP} \cdot \frac{TP+FP}{TP+FP+TN+FN}$$

$$Q = \frac{TP}{TP+FP+TN+FN}$$

Laplace Accuracy.   This evaluation function is based on the Laplace error estimate (we explain it using the standard notation, first used in Q function):

$$Q = \frac{TP+1}{TP+FP+k}$$

where k — the number of decision cases.

Elite 0.2.   This formula is derived from the elitist strategy commonly applied in many other combinatorial optimization problems. We have analyzed 20% of the population size (when this size is greater or equal to 10).

In the case of Ant-Miner and rule induction we search a fitness function that assesses how well a rule constructed by an ant fulfills our expectations.

From the study summarized in the tab. 9, we see that the Laplace accuracy is used successfully in almost every type of data sets (with an exception of the Breast cancer). It should be emphasized that in this type of modification the number of rules increases regularly in all the analyzed data sets (see tab. 10). We also observe a smaller value of accuracy in the first column for the Dermatology data set. As in classification, the question arises as to whether the loss of accuracy is due to the wrong method of pruning and consequently the rule constructed or to the increased difficulty in classifying cases in this data set. We achieved the best results for the Tic-Tac-Toe data set with a pruning procedure utilizing the Laplace error estimate.

6.6   TEST 5 — Changing the Heuristic Function

According to the proposition concerning the heuristic function [52], we also analyze the simplicity of this part of the main transition rule in Ant-Miner.

Table 9 Modifications of pruning

Dataset	Accuracy Convergence (%)	Laplace Accuracy (%)	Elite 0.2 (%)	Standard Q (%)
Breast cancer	70.73 (± 2.14)	65.15 (± 2.60)	—	**74.69** (± 1.76)
Wisconsin breast cancer	79.27 (± 1.26)	**93.95** (± 1.02)	—	92.13 (± 1.06)
Dermatology	47.49 (± 2.50)	91.97 (± 2.02)	—	**94.02** (± 1.77)
Hepatitis	**79.82** (± 2.93)	76.46 (± 2.17)	79.03 (± 3.70)	79.17 (± 2.64)
Tic-tac-toe (1)	69.71 (± 1.29)	**97.77** (± 1.31)	—	73.80 (± 1.43)
Tic-tac-toe	70.10 (± 0.95)	**99.80** (± 0.19)	73.42 (± 1.84)	73.80 (± 1.61)

Table 10 Simplicity of rule sets generated in the context of Q

| Dataset | Average No. of Rules | | | |
	Accuracy Convergence	Laplace Accuracy	Elite 0.2	Standard Q
Breast cancer	3.24 (± 0.24)	18.64 (± 1.00)	—	5.40 (± 0.46)
Wisconsin breast cancer	5.98 (± 0.38)	14.06 (± 1.06)	—	9.82 (± 0.50)
Dermatology	2.96 (± 0.08)	9.32 (± 0.88)	—	7.02 (± 0.38)
Hepatitis	2.54 (± 0.34)	6.66 (± 0.74)	4.44 (± 0.28)	4.42 (± 0.34)
Tic-tac-toe (1)	4.10 (± 0.10)	19.64 (± 2.64)	—	6.94 (± 0.78)
Tic-tac-toe	4.08 (± 0.08)	14.24 (± 0.56)	7.94 (± 0.98)	8.18 (± 0.70)

The motivation is as follows: in ACO approaches we do not need sophisticated information in the heuristic function, because of the pheromone value, which compensate some mistakes in term selections. Our intention is to explore the effect of using a simpler heuristic function instead of a complex one, originally proposed by Parpinelli [66].

Only pheromone.   We propose to investigate a simple ant algorithm without the heuristic function. The pheromone information should govern the behavior of agents.

$$P_{ij}(t) = \frac{\tau_{ij}(t)}{\sum\limits_{i}^{a} x_i \sum\limits_{j}^{b_i} \tau_{ij}(t)}.$$

Density η_{ij} I.   This proposition is derived from the density measure:

$$\eta_{ij} = \frac{max(d_{ij})}{\sum\limits_{k}^{d_{ij}} d_{ijk}},$$

where $max(d_{ij})$ is the biggest number of objects belonging to the decision class for a specific term T_{ij}, and $\sum\limits_{k}^{d_{ij}} d_{ijk}$, is the sum of all objects from all decision classes concerning this term.

Density η_{ij} II.   This version of the density measure is another interpretation of the proposition of using a simpler version of the heuristic function:

$$\eta_{ij} = \frac{max(d_{ij})}{\sum\limits_{l}^{a} x_l \sum\limits_{m}^{b_m} \sum\limits_{k}^{d_{lm}} d_{lmk}},$$

where $max(d_{ij})$ is similar to the majority class of T_{ij}, and a denominator $\sum\limits_{l}^{a} x_l \sum\limits_{m}^{b_m} \sum\limits_{k}^{d_{lm}} d_{lmk}$ describes all the objects belonging to every decision class for all terms used in the examined rule. We consider it the same rule as proposed in [52], i.e.:

$$\eta_{ij} = \frac{max(d_{ij})}{|T_{ij}|}.$$

This new notation should help to better understand the differences in our two density modifications.

Table 11 shows the accuracy rates for rule sets produced in different approaches. It can be seen that in general these modifications are similar to the original Ant-Miner in the context of effectiveness. There are two reasons of disparity in the comparison between the presented results and the proposal

Table 11 Modifications of the heuristic function — predictive accuracy

Dataset	Only pheromone (%)	Density η_{ij} I (%)	Density η_{ij} II (%)	Standard function (%)
Breast cancer	71.23 (± 2.95)	72.20 (± 2.97)	72.85 (± 2.28)	74.69 (± 1.76)
Wisconsin breast cancer	91.39 (± 0.96)	91.56 (± 0.96)	92.05 (± 1.08)	92.13 (± 1.06)
Dermatology	92.72 (± 1.90)	93.63 (± 1.83)	93.98 (± 1.48)	94.02 (± 1.77)
Hepatitis	75.89 (± 2.81)	77.16 (± 2.75)	75.95 (± 3.07)	79.17 (± 2.64)
Tic-tac-toe (1)	71.21 (± 1.67)	72.23 (± 1.68)	71.93 (± 1.54)	73.80 (± 1.43)
Tic-tac-toe	72.36 (± 2.00)	72.22 (± 1.57)	71.81 (± 1.99)	73.80 (± 1.61)

in [52]. Firstly, more ants should be employed in this version of the density oriented heuristic function. Secondly, this version requires more running time for a reliable comparison.

We can conclude that looking for the appropriate scheme of changing the analyzed functions and rules independently was the wrong way. A more satisfactory procedure was found that consisted of two key elements:

- tunning the procedure of effective pruning with simultaneous changing the pheromone matrix,
- simplifying the main transition rule by incorporating a new heuristic function based on density measures with a learning procedure via pheromone values precisely performed.

7 Conclusions

Simple ants, following very simple rules, interact with each other and represent an example of swarm intelligence behavior. As we know: ,, the whole is more than the sum of the parts", so this kind of intelligence is an example of an emergent phenomenon. The employment of the ant colony metaphor for discovering rules, classification is a very underestimated research area. In other research areas the ACO algorithms have been presented as an effective approach, which produce good solutions for different combinatorial optimization problems.

The important advantage in the context of data mining and rule induction is that Ant-Miner produces rule sets much smaller than the rule sets created by the classical approaches: C4.5 and CN2. In this chapter new modifications based on the ant colony metaphor were incorporated in the original Ant-Miner rule producer. Compared to the previous implementations and settings of Ant-Miner, these different experimental studies show how these extensions improve, or sometimes deteriorate, the performance of Ant-Miner. We presented the efficiency of our Ant-Miner was presented via this comparative study.

Examining the previous modifications and extensions (first applied in the combinatorial optimization field) to find appropriate and satisfying decision rules in analyzed data sets was an interesting directions of research.

This approach is the result of a more detailed and deepened review of the previous approaches. We observed that these modifications analyzed separately tend not to be sufficiently motivated. There is still room for improvement in two directions. Firstly, it is still not clear wheather this approach consistently improves its efficiency in maintaining multiple colonies of ants. Secondly, this algorithm is not fully examined in the version without pruning procedure. We also plan to examine this method when used for rule induction in the bigger data sets.

References

1. Corne, D., et al.: New Ideas in Optimization. Mc Graw-Hill, Cambridge (1999)
2. Bauer, A., Bullnheimer, B., Hartl, R.F., Strauss, C.: An Ant Colony Optimization approach for the single machine total tardiness problem. In: Proceedings of the 1999 Congress on Evolutionary Computation, pp. 1445–1450. IEEE Press, Piscataway (1999)
3. Boffey, B.: Multiobjective routing problems. Top 3(2), 167–220 (1995)
4. Bonabeau, E., Dorigo, M., Theraulaz, G.: Swarm Intelligence. In: From Natural to Artificial Systems. Oxford University Press, Oxford (1999)
5. Bonabeau, E., Henaux, F., Guérin, S., Snyers, D., Kuntz, P., Théraulaz, G.: Routing in telecommunication networks with "Smart" ant–like agents telecommunication applications. Springer, Heidelberg (1998)
6. Breiman, L., Friedman, J.H., Olshen, R.A., Stone, C.J.: Classification and Regression Trees, Belmont C.A., Wadsworth (1984)
7. Bullnheimer, B., Hartl, R.F., Strauss, C.: An improved Ant System algorithm for the Vehicle Routing Problem. Technical Report POM–10/97, Institute of Management Science, University of Vienna (1997)
8. Bullnheimer, B., Hartl, R.F., Strauss, C., Bullnheimer, B., Hartl, R.F., Strauss, C.: A new rankbased version of the Ant System: A computational study. Technical Report POM–03/97, Institute of Management Science, University of Vienna (1997)
9. Bullnheimer, B., Hartl, R.F., Strauss, C.: Applying the Ant System to the Vehicle Routing Problem. In: Martello, S., Osman, I.H., Voß, S., Martello, S., Roucairoll, C. (eds.) MetaHeuristics: Advances and Trends in Local Search Paradigms for Optimization, pp. 109–120. Kluwer Academics, Dordrecht (1998)
10. Bullnheimer, B., Strauss, C., Bullnheimer, B., Hartl, R.F., Strauss, C.: Instituts für Betriebwirtschaftslehre, Universität Wien (1996)

11. Chan, A., Freitas, A.A.: A new ant colony algorithm for multi-label alssification with applications in bioinformatics. In: Proceedings of Genetic and Evolutionary Computation Conf (GECCO 2006), San Francisco, pp. 27–34 (2006)
12. Chen, C., Chen, Y., He, J.: Neural network ensemble based ant colony classification rule mining. In: Proceedings of First Int. Conf. Innovative Computing, Information and Control (ICICIC 2006), pp. 427–430 (2006)
13. Chen, Z.: Data Mining and uncertain reasoning. An integrated approach. John Wiley and Sons, Chichester (2001)
14. Clark, P., Boswell, R.: Rule induction with CN2: some recent improvements. In: Kodratoff, Y. (ed.) EWSL 1991. LNCS (LNAI), vol. 482, pp. 151–163. Springer, Heidelberg (1991)
15. Clark, P., Niblett, T.: The CN2 rule Induction algorithm. Machine Learning 3(4), 261–283 (1989)
16. Colorni, A., Dorigo, M., Maniezzo, V.: Distributed optimization by ant colonies. In: Vavala, F., Bourgine, P. (eds.) Proceedings First Europ. Conference on Artificial Life, pp. 134–142. MIT Press, Cambridge (1991)
17. Colorni, A., Dorigo, M., Maniezzo, V., Trubian, M.: Ant system for job–shop scheduling. Belgian Journal of Operations Research, Statistics and Computer Science (JORBEL) 34, 39–53 (1994)
18. Costa, D., Hertz, A.: Ants can colour graphs. Journal of the Operational Research Society 48, 295–305 (1997)
19. Den Besten, M., Stützle, T., Dorigo, M.: Scheduling single machines by ants. Technical Report 99–16, IRIDIA, Université Libre de Bruxelles, Belgium (1999)
20. Deneubourg, J.–. L., Goss, S., Franks, N.R., Pasteels, J.M.: The Blind Leading the Blind: Modelling Chemically Mediated Army Ant Raid Patterns. Insect Behaviour 2, 719–725 (1989)
21. DiCaro, G., Dorigo, M.: AntNet: A mobile agents approach to adaptive routing. Technical report, IRIDIA, Université Libre de Bruxelles (1998)
22. DiCaro, G., Dorigo, M.: AntNet: Distributed stigmergetic control for communications networks. Journal of Artificial Intelligence Research (JAIR) 9, 317–365 (1998)
23. DiCaro, G., Dorigo, M.: Extending AntNet for best–effort Quality–of–Service routing. In: ANTS 1998 – From Ant Colonies to Artificial Ants: First International Workshop on Ant Colony Optimization, October 15–16 (1998) (Unpublished presentation)
24. DiCaro, G., Dorigo, M.: Two ant colony algorithms for best–effort routing in datagram networks. In: Proceedings of the Tenth IASTED International Conference on Parallel and Distributed Computing and Systems (PDCS 1998), pp. 541–546. IASTED/ACTA Press (1998)
25. Dorigo, M.: Optimization, Learning and Natural Algorithms (in Italian). PhD thesis, Dipartimento di Elettronica, Politecnico di Milano, IT (1992)
26. Dorigo, M., DiCaro, G.: The ant colony optimization meta–heuristic. In: Corne, D., Dorigo, M., Glover, F. (eds.) New Ideas in Optimization. McGraw–Hill, London (1999)
27. Dorigo, M., DiCaro, G., Gambardella, L.: Ant algorithms for distributed discrete optimization. Artif. Life 5(2), 137–172 (1999)
28. Dorigo, M., Gambardella, L.: A Study of Some Properties of Ant–Q. In: Proceedings of Fourth International Conference on Parallel Problem Solving from Nature, PPSNIV, pp. 656–665. Springer, Berlin (1996)
29. Dorigo, M., Gambardella, L.: Ant Colonies for the Traveling Salesman Problem. Biosystems 43, 73–81 (1997)

30. Dorigo, M., Gambardella, L.: Ant Colony System: A Cooperative Learning Approach to the Traveling Salesman Problem. IEEE Trans. Evol. Comp. 1, 53–66 (1997)
31. Dorigo, M., Maniezzo, V., Colorni, A.: Positive feedback as a search strategy. Technical Report 91–016, Politechnico di Milano, Italy (1991)
32. Dorigo, M., Maniezzo, V., Colorni, A.: The Ant System: Optimization by a Colony of Cooperating Agents. IEEE Trans. Syst. Man. Cybern. B26, 29–41 (1996)
33. Dorigo, M., Stützle, T.: Ant Colony Optimization. MIT Press, Cambridge (2004)
34. Freitas, A.A., Johnson, C.G.: Research cluster in swarm intelligence. Technical Report EPSRC Research Proposal GR/S63274/01 — Case for Support, Computing Laboratory, Computing Laboratory, Laboratory of Kent, Kent (2003)
35. Galea, M.: Applying swarm intelligence to rule induction. MS thesis, University of Edingbourgh (2002)
36. Galea, M., Shen, Q.: Simultaneous ant colony optimization algorithms for learning linguistic fuzzy rules. In: Agraham, A., Grosan, C., Ramos, V. (eds.) Swarm Intelligence in Data Mining. Springer, Berlin (2006)
37. Gambardella, L.M., Dorigo, M.: AntQ.Ant–Q. A Reinforcement Learning Approach to the Traveling Salesman Problem. In: Proceedings of Twelfth International Conference on Machine Learning, pp. 252–260. Morgan Kaufman, Palo Alto (1995)
38. Gambardella, L.M., Dorigo, M.: Solving symmetric and asymmetric TSPs by ant colonies. In: Proceedings of the IEEE Conference on Evolutionary Computation, ICEC 1996, pp. 622–627. IEEE Press, Los Alamitos (1996)
39. Gambardella, L.M., Dorigo, M.: HAS–SOP: Hybrid Ant System for the Sequential Ordering Problem. Technical Report 11, IDSIA Lugano (1997)
40. Gambardella, L.M., Taillard, E., Agazzi, G.: MACS–VRPTW: A Multiple Ant Colony System for Vehicle Routing Problems with Time Windows. Technical Report 06–99, IDSIA, Lugano, Switzerland (1999)
41. Gambardella, L.M., Taillard, E.D., Dorigo, M.: Ant colonies for the QAP. Technical Report 4–97, IDSIA, Lugano, Switzerland (1997)
42. Gambardella, L.M., Taillard, E.D., Dorigo, M.: Ant colonies for the QAP. Journal of the Operational Research Society (JORS) 50(2), 167–176 (1999)
43. Glover, F., Laguna, M.: Tabu Search. Kluwer Academic Publishers, Dordrecht (1997)
44. Goss, S., Beckers, R., Denebourg, J.L., Aron, S., et al.: How Trail Laying and Trail Following Can Solve Foraging Problems for Ant Colonies. In: Hughes, R.N. (ed.) Behavioural Mechanisms for Food Selection, vol. G20. Springer, Berlin (1990)
45. Grasse, P.-P.: La Reconstruction du Nid et les Coordinations Inter–Individuelles chez Bellicositermes Natalensis et Cubitermes sp. La Theorie de La Stigmerie. Insects Soc. 6, 41–80 (1959)
46. Grasse, P.-P.: Termitologia, vol. II, Paris, Masson (1984)
47. Heusse, M., Guérin, S., Snyers, D., Kuntz, P.: Adaptive agent–driven routing and load balancing in communication networks. Technical Report RR–98001–IASC, Départment Intelligence Artificielle et Sciences Cognitives, ENST Bretagne, ENST Bretagne (1998)
48. Smaldon, J., Freitas, A.A.: A new version of the Ant-Miner algorithm discovering unordered rule sets. In: Proceedings of Genetic and Evolutionary Computation Conf (GECCO 2006), San Francisco, pp. 43–50 (2006)

49. Kohavi, R., Sahami, M.: Error-based and entropy-based discretization of continuous features. In: Proc. 2nd Intern. Conference Knowledge Discovery and Data Mining, pp. 114–119 (1996)
50. Leguizamón, G., Michalewicz, Z.: A new version of Ant System for subset problems. In: Proceedings of the 1999 Congress on Evolutionary Computation, pp. 1459–1464. IEEE Press, Piscataway (1999)
51. Liang, Y.-C., Smith, A.E.: An Ant System approach to redundancy allocation. In: Proceedings of the 1999 Congress on Evolutionary Computation, pp. 1478–1484. IEEE Press, Piscataway (1999)
52. Liu, B., Abbas, H.A., Mc Kay, B.: Classification rule discovery with ant colony optimization. IEEE Computational Intelligence Bulletin 1(3), 31–35 (2004)
53. Ramalhinho Lourenço, H., Serra, D.: Adaptive approach heuristics for the generalized assignment problem. Technical Report EWP Series No. 304, Department of Economics and Management, Universitat Pompeu Fabra, Barcelona (1998)
54. Maniezzo, V.: Exact and approximate nondeterministic tree–search procedures for the quadratic assignment problem. Technical Report CSR 98–1, C. L. In: Scienze dellInformazione, Universita di Bologna, sede di Cesena, Italy (1998)
55. Maniezzo, V., Carbonaro, A.: An ANTS heuristic for the frequency assignment problem. Technical Report CSR 98–4, Scienze dell Informazione, Universita di Bologna, Sede di Cesena, Italy (1998)
56. Maniezzo, V., Colorni, A.: The Ant System applied to the Quadratic Assignment Problem. IEEE Trans. Knowledge and Data Engineering (1999)
57. Maniezzo, V., Colorni, A.: An ANTS heuristic for the frequency assignment problem. Future Generation Computer Systems 16, 927–935 (2000)
58. Maniezzo, V., Colorni, A., Dorigo, M.: The Ant System applied to the Quadratic Assignment Problem. Technical Report 94–28, IRIDIA, Université Libre de Bruxelles, Belgium (1994)
59. Martens, D., De Backer, M., Haesen, R., Baesens, B., Holvoet, T.: Ants constructing rule-based classifiers. In: Agraham, A., Grosan, C., Ramos, V. (eds.) Swarm Intelligence in Data Mining. Springer, Berlin
60. Michalski, R., Mozetic, J., Hong, J., Lavrac, N.: The multi-purpose incremental learning system AQ15 and its testing application to three medical domains. In: AAAI 1986, vol. 2, pp. 1041–1045 (1987)
61. Michel, R., Middendorf, M.: An island model based Ant System with lookahead for the Shortest Supersequence Problem. In: Eiben, A.E., Back, T., Schoenauer, M., Schwefel, H.-P. (eds.) Proceedings of PPSN–V, Fifth International Conference on Parallel Problem Solving from Nature, pp. 692–701. Springer, Heidelberg (1998)
62. Michel, R., Middendorf, M.: An ACO algorithm for the Shortest Common Supersequence Problem. In: Corne, D., Dorigo, M., Glover, F. (eds.) New Methods in Optimisation. McGraw-Hill, New York (1999)
63. Oakes, M.P.: Ant colony optimization for stylometry: the federalist papers. In: Proceedings of Recent Advances in Soft Computing (RASC 2004), pp. 86–91 (2004)
64. Osman, I., Laporte, G.: Metaheuristics: A bibliography. Annals of Operations Research 63, 513–623
65. Parpinelli, R.S., Lopes, H.S., Freitas, A.A.: An ant colony algorithm for classification rule discovery. In: Abbas, H., Sarker, R., Newton, C. (eds.) Data Mining: a Heuristic Approach. Idea Group Publishing, London (2002)

66. Parpinelli, R.S., Lopes, H.S., Freitas, A.A.: Data mining with an ant colony optimization algorithm. IEEE Transactions on Evolutionary Computation, Special issue on Ant Colony Algorithms 6(4), 321–332 (2004)
67. Quinlan, J.R.: Introduction of decision trees. Machine Learning 1, 81–106 (1986)
68. Quinlan, J.R.: Generating production rules from decision trees. In: Proc. of the Tenth International Joint Conference on Artificial Intelligence, pp. 304–307. Morgan Kaufmann, San Francisco (1987)
69. Quinlan, J.R.: C4.5: Programs for Machine Learning. Morgan Kaufmann, San Francisco (1993)
70. Reeves, C.: Modern Heuristic Techniques for Combinatorial Problems. In: Advanced Topics in Computer Science. McGrawHill, London (1995)
71. Schoonderwoerd, R., Holland, O., Bruten, J.: Ant–like agents for load balancing in telecommunications networks. In: Proceedings of the First International Conference on Autonomous Agents, pp. 209–216. ACM Press, New York (1997)
72. Schoonderwoerd, R., Holland, O., Bruten, J., Rothkrantz, L.: Ant–based load balancing in telecommunications networks. Adaptive Behavior 5(2), 169–207 (1996)
73. Stützle, T.: An ant approach to the Flow Shop Problem. Technical Report AIDA–97–07, FG Intellektik, FB Informatik, TH Darmstadt (September 1997)
74. Stützle, T., Hoos: Improvements on the Ant System: Introducing MAX–MIN Ant System. In: Improvements on the Ant System: Introducing MAX–MIN Ant System Algorithms, pp. 245–249. Springer, Heidelberg (1997)
75. Stützle, T., Hoos: The MAX–MIN Ant System and Local Search for the Traveling Salesman Problem. In: Baeck, T., Michalewicz, Z., Yao, X. (eds.) Proceedings of IEEE–ICEC–EPS 1997, IEEE International Conference on Evolutionary Computation and Evolutionary Programming Conference, pp. 309–314. IEEE Press, Los Alamitos (1997)
76. Stützle, T., Hoos: MAX–MIN Ant System and Local Search for Combinatorial Optimisation Problems. In: Proceedings of the Second International conference on Metaheuristics MIC 1997, Kluwer Academic, Dordrecht (1998)
77. Subramanian, D., Druschel, P., Chen, J.: Ants and Reinforcement Learning: A case study in routing in dynamic networks. In: Proceedings of IJCAI 1997, International Joint Conference on Artificial Intelligence. Morgan Kaufmann, San Francisco (1997)
78. van der Put, R.: Routing in the faxfactory using mobile agents. Technical Report R&D–SV–98–276, KPN Research (1998)
79. Navarro Varela, G., Sinclair, M.C.: Ant Colony Optimisation for virtual–wavelength–path routing and wavelength allocation. In: Proceedings of the 1999 Congress on Evolutionary Computation, pp. 1809–1816. IEEE Press, Piscataway (1999)
80. Wang, Z., Feng, B.: Classification rule mining with an improved ant colony algorithm. In: Webb, G.I., Yu, X. (eds.) AI 2004. LNCS (LNAI), vol. 3339, pp. 357–367. Springer, Heidelberg (2004)
81. White, T., Pagurek, B., Oppacher, F.: Connection management using adaptive mobile agents. In: Arabnia, H.R. (ed.) Proceedings of the International Conference on Parallel and Distributed Processing Techniques and Applications (PDPTA 1998), pp. 802–809. CSREA Press,

Part III
Data Mining Applications

Automated Incremental Building of Weighted Semantic Web Repository*

Martin Řimnáč and Roman Špánek

Summary. The chapter introduces an incremental algorithm creating a self-organizing repository and it describes the processes needed for updates and inserts into the repository, especially the processes updating estimated structure driving data storage in the repository. The process of building repository is foremost aimed at allowing the well-known Semantic web tools to query data presented by the current web sources. In order to respect features of current web documents, the relationships should be at least weighted by an additional indirect criteria, which allow the query result to be sorted accordingly to an estimated quality of data provided by web sources. The relationship weights can be based on relationship soundness or on the reputation of the source providing them. The extension of the relationships by the weights leads to the repository able to return a query result as complete as possible, where (possibly) inconsistent parts are sorted by the relationships weights.

1 Introduction

With the amount of data available on the Web rapidly growing during the past years, the ability to find relevant data on the Web by current methods mostly based on approaches known from information retrieval becomes even more hard but also crucial task.

Martin Řimnáč and Roman Špánek
Institute of Computer Science, AS CR, Prague, Czech Republic
e-mail: {rimnacm,spanek}@cs.cas.cz

Roman Špánek
Technical University of Liberec, Hálkova 6, Czech Republic

* The work was supported by the project 1M0554 "Advanced Remedial Processes and Technologies" and partly by the Institutional Research Plan AV0Z10300504 "Computer Science for the Information Society: Models, Algorithms, Applications".

The *information retrieval engines* [1] build large indices storing a word occurrence in (Web) documents, and data searching algorithms use these indices to find the documents corresponding to an end-user query given as a list of keywords.

The first information retrieval engines used only a simple algorithm for evaluating queries returning just a set of links to (relevant) documents. Link relevance was evaluated by a cosine measure, which compared a vector representing words in a document with a vector representing keywords in a query. The amount of returned links and their various quality were the most severe drawbacks.

A solution has been brought by *Google* that has proposed a novel approach sorting the links according to source quality estimated indirectly by so called the *Page-Rank* [2]. The Page-rank is based on the assumption that sound (interesting and high quality) documents are referenced more by (sound) documents. In this way, the Page-Rank tries to extend the classical cosine measure by a quality consideration in order to put the most relevant and high quality documents at top positions in the query result.

Today information retrieval engines used on the Web mostly return thousands of links, but only few of them is analyzed by a human user. Therefore seeking a complete information on the web is currently quite a difficult task.

The *Semantic Web vision* [3] is the inspiration for many currently studied approaches for searching relevant data on the Web. The Semantic Web documents provide data as (well defined) relationships between well-defined entities, which are called resources. While resource meaning is often defined in the external ontologies, data relationships are provided by independent local sources. This leads to a need for preprocessing queries by software agents analyzing all relevant documents and trying to provide a complete answer to the user query. Searching data on the Semantic Web makes engines to manage inverse indices of resources and their related relationships (instead of words) provided by a given document. The limited amount of Semantic Web documents is one of the drawbacks of the current Semantic Web.

Most of current web documents are presented in human user friendly form. This fact leads to a strong motivation for developing new machine learning methods handling current web documents and analyzing their content in order to estimate meaning of the documents. These methods [4, 5] try to propagate estimated relationships from Web documents into *extensional definitions* of Semantic Web resources. All the estimated relationships are stored in the proposed *repository*. As the web documents presenting the relationships are often updated, all proposed methods handling processes in the repository are designed incrementally.

Since the repository includes relationships from various web documents, basic *data integration* should be supported in the repository. Using semiautomatic approaches to estimate *integration rules*, a special kind of relationships connecting sources, may cause a proposal including conflicting rules. Therefore all the rules should be weighted by measures based on analysis of structural relationships in data. These measures can be used to define a weight (i.e. sureness) of the proposed relationship. The weight usability will be demonstrated on an examples including inconsistent data. At more general level, the weights can be also affected by a *source reputation*, handling a source behavior in the past. This higher level weights provide a similar benefit as Page-Rank in the case of classical information retrieval engines.

2   Building a Repository

Since the Semantic Web allows to define a relationship type by a resource, a structure of the presented data can be very complex and very difficult to be automatically estimated. This fact leads to a need for a much more simple formalism for data description. A used formalism is inspired by a relational database theory [6], and it distinguishes the following kinds of resources [7]:

- domain $\mathscr{D}$ - denoting a set of values, for example 'Prague', 'Czech Republic';
- attributes $\mathscr{A}$ - denoting a set of names, defining (at least) a context of presented value in a tuple, for example *Capital, State*;
- elements $\mathscr{E}$ defined as attribute-value pairs $\mathscr{E} \subseteq \mathscr{A} \times \mathscr{D}$ standing for elementary modeling units

Moreover *attribute domain* $\mathscr{D}(A) \subseteq \mathscr{D}$ is defined for each attribute $A \in \mathscr{A}$.

While tuples $t_k \in \mathscr{T}$ are composed of a set of elements $t_k \subseteq \mathscr{E}$, data in the repository are internally represented by relationships between the elements. These relationships, further called *instances*, can be seen as associative rules or implications $\mathscr{I}$, for example,

$$(Capital, Prague) \rightarrow (State, Czech\,Republic)$$

Generally, all the binary relationships can be expressed by a binary matrix [7]. Concretely, a set of instances can be expressed by a binary square *repository matrix*

$$\Phi = [\phi_{ij}] : \phi_{ij} = \begin{cases} 1 & \text{if } e_i \rightarrow e_j \in \mathscr{I}, \ e_i, e_j \in \mathscr{E} \\ 0 & \text{otherwise} \end{cases} \tag{1}$$

Similarly, the *domain matrix* Δ assigning elements $e_i \in \mathscr{E}$ to attributes $A_J \in \mathscr{A}$ is defined as

$$\Delta = [\delta_{iJ}] : \delta_{iJ} = \begin{cases} 1 & \text{if } \exists e_i = (A_J, \star) \in \mathscr{E} \\ 0 & \text{otherwise} \end{cases} \tag{2}$$

Since the set of the instances at general does not satisfy any structure, its usage and result interpretation by humans can be obviously difficult. Therefore the repository stores only instances, which correspond to a relationship at a higher, more general level. This level should identify commonly valid relationships, for instance warranting ability to uniquely derive a value of the attribute on the right side of the instance by a given value of the attribute on the left side. These relationships between the attributes can be described by a binary matrix as

$$\Omega = [\omega_{IJ}] : \omega_{IJ} = \begin{cases} 1 \text{ if } (A_I \to A_J) \\ 0 \text{ otherwise} \end{cases} \tag{3}$$

Because these relationships correspond to functional dependencies, the matrix will be called *functional dependency matrix*.

In order to keep the repository consistent, only instances related to functional dependencies are allowed. Schematically, a consistent repository matrix Φ can be obtained from any repository matrix Φ' using[1]

$$\Phi = \Delta \Omega \Delta^T \odot \Phi' \tag{4}$$

The repository defined by matrices above can be easily transformed into a Semantic Web document $\mathcal{W}$, which composes from triples corresponding to all activated positions in the matrices Φ, Δ, Ω by a rule

$$\forall M = [m_{ij}] \in \{\Phi, \Delta, \Omega\} \atop \forall i, j : m_{ij} = 1 \rightsquigarrow (\text{name}(i, M), \text{name}(M), \text{name}(j, M)) \in \mathcal{W} \tag{5}$$

where *name* is a function assigning a human readable identifier to a resource. For instance, an example triple can be expressed by

(element-1-Prague, in-domain, attribute-2-Capitol)

When keeping an original structure of the repository is not necessary, only relationships between elements can be exported using the rule

$$\forall \phi_{ij} = 1 : {e_i = (A_I, \star) \atop e_j = (A_J, \star)} : (\text{name}(i), \text{name}(A_I \to A_J), \text{name}(j)) \in \mathcal{W} \tag{6}$$

where $A_I \to A_J$ identifies the kind of relationship between the elements e_i, e_j by a functional dependency, for instance

(Capitol-Prague, hasAState, State-CzechRepublic)

Data can be queried in an exported way by standard Semantic Web tools and languages (SPARQL [8]) or directly in the matrix formalism using

- *generalization* operator by
$$\mathbf{y}^G = \Phi \cdot \mathbf{x} \tag{7}$$

- *specialization* operator by
$$\mathbf{y}^S = \Phi^T \cdot \mathbf{x} \tag{8}$$

---

[1] The operator $\odot$ represent item per item multiplication, i.e. $A = B \odot C \Leftrightarrow a_{ij} = b_{ij} \cdot c_{ij}$.

where $\mathbf{x}, \mathbf{y}$ are *query* and *answer vectors* respectively. The vector $\mathbf{x}$ sets only positions $x_i = 1$ corresponding to activated elements $e_i \in \mathscr{E}$, all other positions are zero.

For optimizing the operator evaluation, if the repository matrix describes a monotonic reasonable data (i.e. $\phi_{ii} = 1$), the querying mechanisms can be rewritten as follows (for generalization)

$$
\begin{aligned}
&\mathbf{y} = \mathbf{x}; \ \mathbf{x}' = \mathbf{0}; \\
&\text{for } \forall i : x_i > 0 \land x_i \neq x_i' &&(9)\\
&\quad \text{for } \forall \phi_{ij} > 0 \\
&\qquad y_j = \phi_{ij} \cdot x_i; \ x_i' = x_i &&(10)
\end{aligned}
$$

The result is initialized by a query vector $\mathbf{x}$ and the list of already evaluated elements, implemented by vector $\mathbf{x}'$, is empty. For each activated and yet to processed element e_i (cycle 9), activations y_j are evaluated for all instances $\phi_{ij} \geq 0$ having element e_i on the left side. In the consistent repository, the maximal amount of activations is bounded by $\mathcal{O}(|\mathscr{A}|)$.

2.1 Data Structure Estimation

The relational database theory defines a functional dependency, denoted as $f = A_L \rightarrow A_R \in \mathscr{F}$, as a logical (intensional) relationship between (possible complex) attributes $A_L \subseteq \mathscr{A}$ and $A_R \subseteq \mathscr{A}$. The consequence of such a relationship is existence of an injection between domains of corresponding attributes, i.e.

$$
A_L \rightarrow A_R \in \mathscr{F} \Rightarrow \exists \mathscr{I}_{LR} : \mathscr{D}(A_L) \rightarrow \mathscr{D}(A_R) \tag{11}
$$

The approach for functional dependency discovery [9, 10, 11, 5] deals with a task of estimating a functional dependency system directly from data. Data often cover only a part of attribute domains - such a domain part is called *attribute active domain* $\mathscr{D}_\alpha^{\mathscr{R}}(A) \subseteq \mathscr{D}(A)$ over am input data relation $\mathscr{R}$. The estimation process is driven by an assumption that the implication (11) is valid for both directions and only data from active domains of the attributes can be covered:

$$
A_L \rightarrow A_R \in \mathscr{F}^{\mathscr{R}} \Leftrightarrow \exists \mathscr{I}_{LR} : \mathscr{D}_\alpha^{\mathscr{R}}(A_L) \rightarrow \mathscr{D}_\alpha^{\mathscr{R}}(A_R) \tag{12}
$$

In other words, the existence of the injection $\mathscr{I}_{LR}$ over data in $\mathscr{R}$ implies existence of the extensional (in respect to $\mathscr{R}$) functional dependency. It can be shown for data respecting constraints given by (intensional) functional dependencies $\mathscr{F}$, that the estimated functional dependency set $\mathscr{F}^{\mathscr{R}}$ always includes all intensional functional dependencies. On the other hand, for unrepresentative input data it may also cover some functional dependencies by fault.

$$\forall \mathscr{R} : \mathscr{F}^{\mathscr{R}} \supseteq \mathscr{F} \tag{13}$$

The relation $\mathscr{R}$, which leads to $\mathscr{F}_{\mathscr{R}} = \mathscr{F}$, is called *Armstrong Relation* [9]. In inverse way, when a learning relation $\mathscr{R}$ can be mapped onto Armstrong relation corresponding to a given reality, the estimated functional dependency system equals the intensional.

Naïve Algorithm and Its Improvements

The naïve algorithm [10, 9] for functional dependency discovery takes into account all possible functional dependencies generated by a given set of attributes. The functional dependencies with a complex attribute on the right side are not considered, because they are always trivial; i.e. they can be simplified into ones having only (single) attribute on the right side.

$$\mathscr{F}_{\mathscr{R}} = \emptyset$$
$$\text{for } \forall A_R \in \mathscr{A}_{\mathscr{R}}$$
$$\quad \text{for } \forall A_{\mathbb{L}} \in \mathscr{P}(\mathscr{A}_{\mathscr{R}} - A_R) \tag{14}$$
$$\quad \text{if } \exists \mathscr{I} : \mathscr{D}_\alpha^{\mathscr{R}}(A_{\mathbb{L}}) \to \mathscr{D}_\alpha^{\mathscr{R}}(A_R) \text{ then} \tag{15}$$
$$\quad \mathscr{F}_{\mathscr{R}} := \mathscr{F}_{\mathscr{R}} \cup \{A_{\mathbb{L}} \to A_R\}$$

The naïve algorithm tests functional dependencies between all single attributes on the right side and all complex attributes (from a powerset generated by $\mathscr{A}$) on the left side. Respecting extensional definition (12), only relationships passing test (15) for the existence of the injection between attribute active domains are marked as functional dependencies. Due to a whole powerset (14) is tested, the naïve algorithm belongs to NP-complete tasks.

Advanced methods [5, 11] or methods using a greedy search [12] use basics of the naïve algorithm, but restrict the amount of tested functional dependencies by analyzing only nontrivial ones. In other words, if any functional dependency $\mathscr{A}_{\mathbb{L}} \to \mathscr{A}_R$ is valid for input data, no functional dependency having left side attribute $A_{\mathbb{L}'}$ extended from $A_{\mathbb{L}} \subset A_{\mathbb{L}'}$ will be tested. Such a functional dependency $\mathscr{A}_{\mathbb{L}'} \to \mathscr{A}_R$ is trivial for the relationship it presents is already covered by a simpler $\mathscr{A}_{\mathbb{L}} \to \mathscr{A}_R$. These improved methods return all the necessary functional dependencies describing the data model, which can be used for an automatic design of a relational database.

Incremental Algorithm

The main drawback of the methods described above is inability to update the functional dependency system with new tuples inserted. These methods Principally require to repeat the test for all functional dependencies. This is sufficient for onetime design of the model, but it is inappropriate for incremental building of the repository covering dynamic changes on the web.

In order to simplify the description of the method handling changes in the repository, let assume that all the elements, attributes and domains are defined before calling the method, i.e. active domain matrix and sizes of matrices are constant[2]. Further, let consider only functional dependencies between simple attributes. The functional dependencies with the complex attributes can be naturally expressed by a set of simple attribute functional dependencies under each tuple can be identified by an unique key value [7].

Finally, tuples are restricted to ones covering exactly one value per attribute, i.e. meeting conditions:

$$\forall t_k \in \mathcal{T} : \quad \forall e_i = (A_I, \star) \, \forall e_j = (A_J, \star') : i \neq j \wedge I = J \quad (16)$$

$$\forall t_k \in \mathcal{T} : \quad |t_k| = |\mathcal{A}| \quad (17)$$

The condition (16) warrants that the estimated model will follow the third Codd normal form. Considering only non-empty values (condition extension (17)) leads to a transitivity property to be satisfied in the repository.

The repository is built in the following way:

Firstly, the repository is initialized by tuple t_0, which is represented by a vector $\mathbf{t}_0$. If the repository includes only one tuple, the repository matrix will be called *tuple t repository matrix* and denoted Φ_t. The tuple repository matrix always contains all instances generated by a cartersian product of all activated elements. Schematically, *tuple t_0 repository matrix* is generated by

$$\Phi_{t_0} = \mathbf{t}_0 \cdot \mathbf{t}_0^T \quad (18)$$

With respect to condition (16) - tuple t_0 covers one element per one attribute, all functional dependencies are marked as valid; all elements are uniquely implied from any element in the tuple. Because of condition (17) holds,

$$\Omega_{t_0} = \mathbf{1} \quad (19)$$

Because a corrupted functional dependency remains corrupted, it is recommended to use a *corrupted functional dependency matrix* $\mho$ instead of a list of (currently) valid functional dependencies stored in Ω. If condition (17) is satisfied, then

$$\mho = [\omega_{IJ}^{\ominus}] = \mathbf{1} - \Omega \quad (20)$$

The insertion next tuple t_k is quite difficult. Let be a state of the repository after insertion of $k - 1$ tuple described by pair of matrices Φ_{k-1} and $\mho_{k-1}$. The tuple t_k repository matrix is obtained using formula (18). Once this tuple repository matrix (18) is merged with the original repository matrix Φ_{k-1}, some functional dependencies might be corrupted. The functional dependency ω_{IJ} is corrupted, when any previous tuple $t_{k'}$ includes elements $e_i = (A_I, v)$ and $e_j = (A_J, v')$ and tuple t_k includes the same element $e_i \in t_k$ as $t_{k'}$ and

---

[2] Not providing this assumption results in update of matrix sizes and active domains first.

element $e_{j'} = (A_J, v'') \in t_k$. Element $e_{j'}$ has assigned the same attribute A_J as element e_j, but has assigned a different value $v'' \neq v'$. Therefore from definition (12), functional dependency ω_{IJ} is corrupted: it is not possible to uniquely derive an element of attribute A_J from the element of attribute A_I. Formally:

$$\exists k' < k : \begin{array}{l} e_i = (A_I, \star) \in t_{k'}, \; e_j = (A_J, \star') \in t_{k'} \\ e_i \in t_k, \; e_j = (A_J, \star'') \in t_k \end{array} \rightsquigarrow \omega_{IJ}^{\ominus} = 1$$

The tuples $\{t_{k'}, t_k\}$ causes that the value of attribute A_J can not be uniquely derived from a value of attribute A_I (e_i implies e_j as well as $e_{j'}$), which requires to mark the functional dependency ω_{IJ} corrupted (by setting $\omega_{IJ}^{\ominus} = 1$). Schematically, a list of corrupted functional dependencies has to be updated by[3]

$$\mho_k = \mho_{k-1} + \Delta((\Phi_{k-1} + \Phi_{t_k})^T \Delta^T) > 1) \tag{21}$$

In order to keep the repository in a consistent state, instance of any corrupted functional dependency should not be stored. Formula (4) can be rewritten as

$$\Phi_k = (\Phi_{k-1} + \Phi_{t_k}) \odot \Delta \cdot (1 - \mho_k) \cdot \Delta^T \tag{22}$$

2.2 Optimizing Repository

Because steps (18, 21, 22) can be implemented by matrix multiplication, the whole algorithm has polynomial complexity, mostly given by the number of elements $|\mathscr{E}|$. The final repository contains $\mathscr{O}(|\mathscr{E}|^2)$ instances of $\mathscr{O}(|\mathscr{A}|^2)$ functional dependencies. The repository organized in this way represents all the obtained relationships between the elements.

Because the transitivity property is warranted if condition (17) holds,

$$\forall i, j, k : \phi_{ik} = \phi_{ij} \cdot \phi_{jk} \tag{23}$$

the repository incrementally built by (22) represents a transitive closure generated by ground instances. The ground instances are subset of instances, which can not be derived using the transitivity rule (23). The repository including only ground instances is called *reduced form repository* and its reduced repository matrix will be denoted $\Phi_k^\flat$. Functional dependency matrix $\Omega_k^\flat$ is reduced in the same way into *model skeleton*, and only the ground instances of (ground - model skeleton) functional dependencies $\Omega_k^\flat$ will be stored in the repository - i.e. the repository satisfies (4). The original, *full repository form* can be schematically obtained by

---

[3] The operator $>$ represents a matrix items comparison to a number, i.e. $A = B > n \Leftrightarrow a_{ij} = \begin{cases} 1 & \text{if } b_{ij} > n \\ 0 & \text{otherwise} \end{cases}$

$$\Phi_k = (\Phi_k^\flat)^\kappa$$
$$\Omega_k = (\Omega_k^\flat)^\kappa \quad \kappa < |\mathscr{A}| \tag{24}$$

Selection of the ground functional dependencies can be implemented by methods *finding a transitive reduction* [13, 14] in a graph. These methods are based on step by step insertion of vertices from original graph $\mathscr{G}$ into its transitive reduction $\mathscr{G}^\flat$. If a given vertex $v \in \mathscr{G}$ is already covered by the transitive closure of the reduced graph $\mathscr{G}^\flat$, such a vertex is skipped, otherwise the vertex is added into $\mathscr{G}^\flat$.

As given in the graph theory, finding a transitive reduction of an input graph is a task with ambiguous solution. This ambiguity can be used to define an additional criterion preferring one kind of vertices. In order to obtain a repository organized in the most efficient way, the criterion should lead to the storage of minimal number of instances[4].

$$\Phi_k^\flat = \arg \min_{\Phi_k^\flat : \, (\Phi_k^\flat)^\kappa = \Phi_k} \{||\Phi_k^\flat||\} \tag{25}$$

If condition (17) holds, the number of instances of functional dependency ω_{IJ} is given by the active domain size of the left side attribute . This can be used to modify a classical reduction algorithm applied on the full form functional dependency matrix Ω:

$$\Omega^\flat = \mathbf{0}; \, \Omega^C = \mathbf{0};$$
$$\text{for } \forall I, J : \omega_{IJ} > 0 \text{ sorted by } |\mathscr{D}_\alpha^{\mathscr{I}}(A_I)| \tag{26}$$
$$\text{if } \omega_{IJ}^C = 0 \text{ then} \tag{27}$$
$$\omega_{IJ}^\flat = \omega_{IJ}; \, \Omega^C = \Omega^C \Omega^\flat; \tag{28}$$
$$\Phi^\flat = \Delta \Omega^\flat \Delta^T \odot \Phi \tag{29}$$

Firstly, at step (26), the functional dependencies ω_{IJ} are sorted accordingly to the criterion preferring functional dependencies with smaller active domains of the left side attributes. Afterwards, functional dependency ω_{IJ} is tested (27) if it is derivable via others (e.g. whether it is already in the closure Ω^C). If it is not, a given functional dependency is inserted into reduction $\Omega^\flat$ as given by (28) and the closure Ω^C is updated with respect to newly inserted dependency (i.e. all dependencies derivable via ω_{IJ} are inserted into Ω^C). Finally, only instances of selected functional dependencies will be stored in the reduced form repository matrix during step (29). A set of functional dependencies selected in the reduced functional dependency matrix $\Omega^\flat$ will be called a *skeleton model*.

Using the model skeleton for storing the instances causes

$$1 - \mho_k = \Omega_k \geq \Omega_k^\flat = (\Delta^T \Phi^\flat \Delta) > 0 \tag{30}$$

---

[4] The norm of the matrix $||A||$ is defined as the number of non-zero items, i.e. $||A|| = |\{\forall i, j : a_{ij} \neq 0\}|$.

Because the repository is described by a pair of matrices $\mho_k$ and $\Phi^\flat$ and the first unevenness in (30) holds, the reduced functional dependency matrix $\Omega_k^\flat$ should be reconstructed from the repository matrix $\Phi_k^\flat$. The reconstruction rule is based on the fact, that all instances in the repository matrix can be assigned to any valid functional dependency.

Incremental Building of Reduced Form Repository

Insertion of a new tuple can change the sizes of attribute active domains, which requires to re-evaluate a transitive reduction of the functional dependency system. Because evaluating a whole reduction is not effective, this section introduces an incremental algorithm handling the process: the model skeleton is initialized by first tuple repository matrix and then it is only updated according to corruption of functional dependencies.

The skeleton model can be initialized in any of following variants:

- symmetric skeleton

  - *chain model* maximizing amount of attributes having two input and two output functional dependencies (Figure 1);
  - *star model* maximizing amount of attributes having one input and one output functional dependency (Figure 2); and

- asymmetric *cycle model* minimizing amount of total functional dependencies (Figure 3).

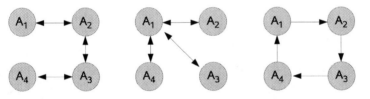

Fig. 1 Chain Model   **Fig. 2** Star Model   **Fig. 3** Cycle Model

Using reduced forms requires only $\mathscr{O}(|\mathscr{E}|)$ instances to be stored in the repository, and only $\mathscr{O}(|\mathscr{A}|)$ functional dependencies have to be verified in injection test (21). It can be easily proved that corruption of any functional dependency excluded from the skeleton model leads to corruption of some functional dependency in the skeleton model.

The steps of the tuple inserting algorithm (18, 21, and 22) can be rewritten into the reduced form as follows:

- Unchanged remains creation of a tuple repository matrix Φ_{t_k}.
- Detection of corrupted functional dependencies has to be repeated while any corrupted functional dependency in the skeleton exists:

$$\mho_k = \mho_{k-1}; \; \Omega^\flat = (\Delta^T \Phi^\flat \Delta) > 0$$
$$\text{do } \{$$
$$\mho = \Delta((\Phi_{k-1} + \Phi_{t_k})^T \Delta^T) > 1); \tag{31}$$
$$\text{if } \mho \neq \mathbf{0} \text{ then}$$
$$\Omega^\flat = \text{update}^\Omega(\Omega^\flat, \mho); \tag{32}$$
$$\Phi_{k-1} = (\Phi_{k-1})^2 \odot \Delta\Omega^\flat\Delta^T \tag{33}$$
$$\mho_k = \mho_k + \mho;$$
$$\} \text{ until } \mho \neq \mathbf{0}$$

If any functional dependency is corrupted, the skeleton is firstly updated (32) and corresponding changes are propagated in the repository matrix (33); all the instances originally derivable via the corrupted functional dependency have to be reconstructed.

The functional dependencies inserted into the model skeleton in (32) can be also corrupted, so test (31) has to be recalled since no new corruption is detected.

- Merging the instances and removing ones corresponding to corrupted dependencies is similar to (22), but only instances of dependencies in the model skeleton are stored:

$$\Phi_k = (\Phi_{k-1} + \Phi_{t_k}) \odot \Delta \cdot ((1 - \mho_k) \odot \Omega^\flat) \cdot \Delta^T \tag{34}$$

Updating Skeleton and the Repository

The functional dependency ω_{IJ} in the model skeleton principally represents a whole set of functional dependencies, which can be derived from it by the transitivity rule. When the functional dependency is corrupted, the possibility of derivation of these functional dependencies has to be preserved. In other words, if any skeleton dependency corrupts, it has to be "*bridged*" by ones, which have been originally derivable via the corrupted functional one. Such a situation is illustrated in Figure 4, where corruption of functional dependency $A_2 \rightarrow A_3$ was detected.

For symmetric *chain* or *star* models, let a corruption of the functional dependency ω_{IJ} has been detected. Before detecting the corruption, the functional dependency ω_{IJ} has been a part of the model skeleton. It has provided,

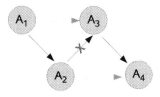

Fig. 4 Detected Corruption

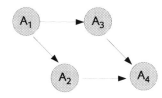

Fig. 5 Bridging the Corruption

together with connected functional dependencies $\omega_{I'I}$, respective $\omega_{JJ'}$ ability to derive functional dependencies $\omega_{I'J}$, respective $\omega_{IJ'}$. Since the functional dependency ω_{IJ} is corrupted and it can not be further used, functional dependencies $\omega_{I'J}$, respective $\omega_{IJ'}$ have to be inserted into the model skeleton in order to keep all the originally derivable dependencies. For example, functional dependencies $A_1 \rightarrow A_2$, $A_2 \rightarrow A_3$ and $A_3 \rightarrow A_4$ are in the model skeleton, and functional dependencies $A_1 \rightarrow A_3$ and $A_2 \rightarrow A_4$ can be derived using these ground dependencies. When functional dependency $A_1 \rightarrow A_3$ is detected as corrupted (Figure 4), this functional dependency has to be removed from the model skeleton. Functional dependency $A_1 \rightarrow A_3$ will be inserted into the model skeleton in order to keep a "path" $A_1 \rightarrow A_2 \nrightarrow A_3$, and functional dependency $A_2 \rightarrow A_4$ will be inserted into the model skeleton in order to keep a "path" $A_2 \nrightarrow A_3 \rightarrow A_4$ as given on Figure 5.

Formally, the skeleton update function can be implemented by

$$\Omega^b = \text{function update}^{\Omega}(\Omega^b, \mho)$$
$$\{$$
$$\quad \text{for } \forall I, J : \omega_{IJ}^{\ominus} = \mathbb{I}$$
$$\quad\quad \text{for } \forall I' : \omega_{I'I}^b > 0 \tag{35}$$
$$\quad\quad \omega_{I'J}^b = \omega_{I'I}^b \cdot \omega_{IJ}^b;$$
$$\quad\quad \text{for } \forall J' : \omega_{JJ'}^b > 0 \tag{36}$$
$$\quad\quad \omega_{JJ'}^b = \omega_{JJ'}^b \cdot \omega_{IJ}^b;$$
$$\quad \omega_{IJ}^b = 0;$$
$$\}$$

The complexity of the skeleton update function for one corrupted functional dependency is given by the amount of functional dependencies originally derivable through it. This is bounded by $\mathcal{O}(|\mathscr{A}|)$. Note that the functional dependency inserted into the skeleton can be corrupted as well. Therefore the test at step (31) has to be repeated until no new corruption is detected.

For example let be two tuples from Table 1 inserted into the repository, where attribute A_1 stand for City, A_2 for Region, A_3 for State, and A_4 for Location:

Let be a chain model in use, then the tuple t_0 initializes the repository by the skeleton as given in Figure 1. The insertion of new tuple t_1 the functional dependency $A_3 \rightarrow A_2$ is not valid and has to be removed (Figure 6); the

Table 1 Example 1 Input Data Set

tuple	A_1	A_2	A_3	A_4
t_0	Prague	Prague	Czech Republic	Central Europe
t_1	Brno	South-Moravian	Czech Republic	Central Europe

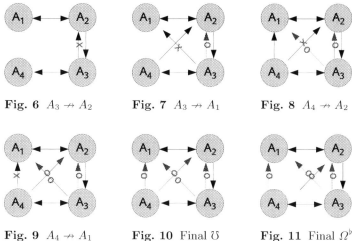

Fig. 6 $A_3 \nrightarrow A_2$ **Fig. 7** $A_3 \nrightarrow A_1$ **Fig. 8** $A_4 \nrightarrow A_2$

Fig. 9 $A_4 \nrightarrow A_1$ **Fig. 10** Final $\tilde{U}$ **Fig. 11** Final $\Omega^\flat$

Czech Republic leads to the activation of Prague as well as South-Moravian. The functional dependencies originally derivable through $A_3 \rightarrow A_2$ have to be kept; in this case $A_3 \rightarrow A_1$ using (35) and $A_2 \rightarrow A_4$ using (36) as illustrated in Figure 7. The test (31) is called again and functional dependency $A_3 \rightarrow A_1$, inserted into the skeleton in the previous step is detected as corrupted. Similarly, the overlapping dependency $A_4 \rightarrow A_1$ is inserted but it is also detected as invalid (Figure 9). Figure 10 shows all corrupted functional dependencies and (Figure 11) shows the final skeleton model.

A completely different situation is in the case of a *cycle model* for the asymmetric initial functional dependencies. Usage of the same mechanism as in the previous case leads to swapping the attributes as illustrated in Figure 13. The attribute swapping principally leads to attribute separation into two groups $\mathbb{A}_1 = \{A_{1\star}\}$, $\mathbb{A}_2 = \{A_{2\star}\}$ connected into one cycle $A_{11} \rightarrow A_{1n} \rightarrow A_{21} \rightarrow A_{2m} \rightarrow A_{11}$, where functional dependencies $A_{2\star} \rightarrow A_{1\star}$ are detected as corrupted. Because dependency $A_{2m} \rightarrow A_{11}$ closing the cycle will be also marked as corrupted, it will interrupt the cycle. Because the cycle has to be kept (to keep the possibility to derive all valid functional dependencies in the groups), functional dependency $A_{2m} \rightarrow A_{11}$ will be replaced by the nearest valid one closing the cycle, i.e. $A_{1m} \rightarrow A_{11}$.

Finding a new functional dependency to be inserted into the skeleton instead of corrupted one (35/36) has to be generalized to finding the nearest functional dependency closing a cycle by

$$\text{for } \forall I' : \omega^\flat_{I'I} > 0 \tag{37}$$
$$\text{for } \forall I'' \in \prec (I', J)$$
$$\omega^\flat_{I''J} = \omega^\flat_{I''I} \cdot \omega^\flat_{IJ};$$

where

$$\prec (I', J) = \begin{cases} \{I'\} & \text{if } \omega_{I'J}^{\ominus} = 0 \\ \bigcup_{\forall \omega_{I''I'}^{\flat} = 1} \prec (I'', J) & \text{if } \omega_{I'J}^{\ominus} = 1 \\ \emptyset & \text{otherwise} \end{cases}$$

or alternatively for the second rule (36)

$$\begin{aligned} &\text{for } \forall J' \in \omega_{I'I}^{\flat} > 0 \\ &\quad \text{for } \forall J'' \in \succ (I, J') \\ &\quad\quad \omega_{IJ''}^{\flat} = \omega_{JJ''}^{\flat} \cdot \omega_{IJ}^{\flat} \end{aligned} \tag{38}$$

where

$$\succ (I, J') = \begin{cases} \{J'\} & \text{if } \omega_{IJ'}^{\ominus} = 0 \\ \bigcup_{\forall \omega_{J'J''}^{\flat} = 1} \succ (I, J'') & \text{if } \omega_{IJ'}^{\ominus} = 1 \\ \emptyset & \text{otherwise} \end{cases}$$

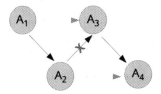
Fig. 12 Detected Corruption

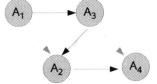
Fig. 13 Solving a Corruption

Note that the maximal recursion depth is bounded by $\mathcal{O}(|\mathscr{A}|)$ because of $\omega_{II}^{\ominus} = 0$.

Using the same input data (Table 1) as for the chain model, the skeleton is initialized as given in Figure 3. The functional dependency $A_4 \to A_1$ is detected as corrupted (Figure 14), which leads to swapping attributes A_1 and A_4, i.e. functional dependency between the attributes will be oriented in the inverse direction. In the next step (Figure 15), the corruption of $A_3 \to A_1$ is detected, and the attributes are analogically swapped.

Detecting $A_3 \to A_2$ to be corrupted (Figure 16) provides the attribute swapping leading to the activation of functional dependency $A_3 \to A_1$, which is already marked as invalid. In order to save cycles, the next valid functional dependency has to be used to make a bridge (Figure 17). As shown in Figure 18, these are $A_3 \to A_4$ and $A_2 \to A_1$, which cause the original cycle

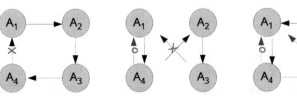

Fig. 14 $A_4 \nrightarrow A_1$ Fig. 15 $A_3 \nrightarrow A_1$ Fig. 16 $A_3 \nrightarrow A_2$ (a)

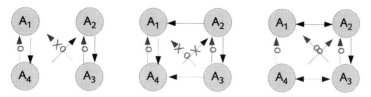

Fig. 17 $A_3 \nrightarrow A_2$ (b)   **Fig. 18** $A_3 \nrightarrow A_2$ (b)   **Fig. 19** $A_4 \nrightarrow A_2$

division into two cycles $A_1 \rightarrow A_4 \rightarrow A_2 \rightarrow A_1$ and $A_3 \rightarrow A_4 \rightarrow A_2 \rightarrow A_3$. Finally, $A_4 \rightarrow A_2$ corruption is detected (Figure 19). The final model is the same as in Figure 11.

2.3 Model Uncertainty

Since it is not possible to decide whether the estimated model (13) is correct, the functional dependencies should be weighted. The recent approaches [15] mostly measure functional dependencies by amount of instances passing injection test (15). The main benefit is error proneness in the input data (principally implementing Ocamm razor). Such measures often divide the active domain $\mathscr{D}_\alpha^\mathscr{R}(A_I)$ into two parts. The first part $\mathscr{D}_{\alpha\oplus(IJ)}^\mathscr{R}(A_I)$ satisfies functional dependency condition $A_I \rightarrow A_J$, and the second $\mathscr{D}_{\alpha\ominus(IJ)}^\mathscr{R}(A_I)$ concentrates on counterexamples (corrupting the functional dependency). Consequently the *(fuzzy) functional dependency weight* can be evaluated by

$$\omega_{IJ}^\natural = \frac{|\mathscr{D}_{\alpha\oplus(IJ)}^\mathscr{R}(A_I)|}{|\mathscr{D}_\alpha^\mathscr{R}(A_I)|} \tag{39}$$

Using such kind of the weights requires full form matrices (the transitivity feature is not satisfied in general).

Alternatively, the uncertainty coming from within the estimation process can be weighted by a minimal change leading to the modification of functional dependency system. For each valid functional dependency, it can be established by a *upper bound of probability* that the next inserted tuple will cover a counterexample to a given functional dependency:

$$\omega_{IJ}^{\natural\natural} = \omega_{IJ} \cdot \frac{|\mathscr{D}_\alpha^\mathscr{R}(A_I)|}{|\mathscr{D}_\alpha^\mathscr{R}(A_I)| + 1} \tag{40}$$

3 Integrating Repository

The repository described above represents one data source. Data on the web are often redundantly provided by various sources. These sources are constructed independently from each other; their schemas are designed by different designers, for different purposes or from a different point of view. This leads to generally heterogeneous schemas. In order to enable advanced co-processing between data sources, the correspondences between the source schemas have to be established.

The problem of finding such correspondences can be solved by *schema matching* [16, 17]. The matching operation takes two schemas as an input and produces mappings describing relationships between schemas as an output. The mostly used architecture for data integration is a mediator system, connecting the sources by mediator views. Using the matrix formalism, the mediators connecting repositories can be expressed by a *mediator matrix*.

Centralized Solution

A classical approaches use a centralized solution of the data integration; a global schema is designed and correspondence to local schemas is provided by views. Using the repository notation, a set of global elements $\mathscr{E}_{\mathscr{S}}$ is defined as an intersection over all local ones

$$\mathscr{E}_{\mathscr{S}} = \bigcup_{\forall S \in \mathscr{S}} \mathscr{E}_S \tag{41}$$

and integration process is driven by local to global (element/attribute indeces) mappings. The mappings between the local e_i and global e_j elements can be expressed by a binary mediator matrix

$$\Gamma_S = [\gamma_{ij}^S], \; \gamma_{ij}^S = \begin{cases} 1 & \text{if } e_i \sim e_j : e_i \in \mathscr{E}_S, e_j \in \mathscr{E}_{\mathscr{S}} \\ 0 & \text{otherwise} \end{cases} \tag{42}$$

The virtual repository matrix, representing all the relationship from the particular sources $S \in \mathscr{S}$, can be expressed as

$$\Phi_{\mathscr{S}} = \sum_{\forall S \in \mathscr{S}} \Gamma_S^T \Phi_S \Gamma_S \tag{43}$$

Each source S is represented by its repository matrix Φ_S, which is via mappings Γ_S transformed into a virtual repository layer. A virtual repository $\Phi_{\mathscr{S}}$ merges all layers into virtual one.

Since the mediators can be given as real numbers, a value of the position in the repository matrix can express a weight (sureness) assigned to the instance.

Decentralized Solution

The web has a decentralized architecture, where independent sources are connected by links. This architecture is shown in Figure 21, the sources are linked together via local to local mediators:

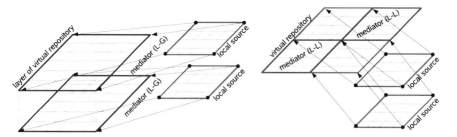

Fig. 20 Centralized Data Integration   **Fig. 21** Decentralized Data Integration

$$\Psi_{IJ} = [\psi_{ij}^{IJ}], \; \psi_{ij}^{IJ} = \begin{cases} 1 & \text{if } e_i \sim e_j : e_i \in \mathcal{E}_{S_I}, e_j \in \mathcal{E}_{S_J} \\ 0 & \text{otherwise} \end{cases} \tag{44}$$

Using such mappings, the virtual repository matrix can be evaluated by

$$\Phi_{\mathscr{S}} = \begin{bmatrix} \Phi_1 & \Psi_{12} & \cdots & \Psi_{1|\mathscr{S}|} \\ \Psi_{21} & \Phi_2 & \cdots & \Psi_{2|\mathscr{S}|} \\ \vdots & & \ddots & \\ \Psi_{|\mathscr{S}|1} & \cdots & \cdots & \Phi_{|\mathscr{S}|} \end{bmatrix} \tag{45}$$

The local to local mediator can be established from global to local mediators of both sources:

$$\Psi_{IJ} = \Gamma_J^T \Gamma_I \tag{46}$$

3.1 Semiautomatic Data Integration

The data integration is currently mainly solved manually with significant limitations (e.g. time consumption, errors proneness and expensiveness). Therefore a need for automatization of the task is obvious. (Semi)automated approaches [18] aim at using knowledge from all the data sources together with an additional external knowledge (such as an ontological description) in order to obtain the mappings between source schemas. The obtained result of such automatic processes is a set of *matching candidates*, and the designer needs to (manually) tune the returned mappings to warrant their correctness. This phase is required because the schemas have often some non-expressed semantics affecting the mappings or it is not possible to fully automatically make unique decision about the integration rule.

Approaches for searching the schema correspondences [17, 19] can be basically distinguished according to the information level [20] at which the schemas are compared:

- At the *instance level*: instances satisfying considered schemas are compared.

- At the *element level*: proposed techniques consider elements as being isolated (without any relationships to other elements in the schema). These are further divided into:

  - *Name-based methods* cover lexical or linguistic techniques. For example prefix, suffix, string distance are analyzed. Also natural language processing methods or some external information sources (e.g. domain ontologies) can be used (e.g. analyzing synonyms).
  - *Type-based methods* considers data type of the element value through an analysis driven by a specification of a data type similarity distance. Methods can consider not only domains or active domains, but also other constraints as unique values, cardinalities etc.

- At the *structure level*: techniques search for mappings using structural relationships of elements. For example, elements are compared to their positions in the structure (inner nodes or leaves) and similarity of their parents and children are considered.

All found correspondences can be extended by their preferences. Matching preference is often expressed using some *similarity function*. It can be based on probability [21], on cosine measure of particular element feature vectors [22], or some other measure describing the number of explored aspects in which they correspond [23]. Sometimes, some additional techniques like candidates refinement [24] or machine learning [25] are used.

The scale of matching methods is very large and also their complexity is various. Therefore a selection of the method and its relevant measure to describe matching preference is strongly important, and may significantly affect a whole mapping result.

3.2 Trivial Element Mapping

The simplest automatically estimated mappings at the element level are based on making pairs of elements with the same values, i.e. the elements from different sources having assigned the same value will be marked as equivalent. These pairs can be expressed by a binary matrix:

$$\xi^{kl}(e_i, e_j) = \begin{cases} 1 & \text{if } \begin{matrix} e_i = (A_I, v) \in \mathscr{E}_{S_k} \\ e_j = (A_J, v) \in \mathscr{E}_{S_l} \\ v \in \mathscr{D}_\alpha^{S_l}(A_I) \cap \mathscr{D}_\alpha^{S_k}(A_J) \end{matrix} \\ 0 & \text{otherwise} \end{cases} \quad (47)$$

The local to local mediator (44) between sources $S_k, S_l \in \mathscr{S}$ can be directly evaluated using the measure by

$$\hat{\Psi}_{kl} = [\hat{\psi}_{ij}^{kl}], \ \hat{\psi}_{ij}^{kl} = \xi^{kl}(e_i, e_j) \quad (48)$$

The amount of possible mappings is $\mathscr{O}(|\mathscr{E}_{S_k}| \cdot |\mathscr{E}_{S_l}|)$, but once created mappings can be incrementally updated after inserting new tuples. Unfortunately, using such trivial mappings is usually not well conditioned, because the same values very often occur with different meaning. This is the reason, why these mappings are further restricted by additional criteria (see next subsection).

3.3 Attribute Equivalence Mappings

Such an additional criterion can be an aggregation of the element equivalence onto the attribute level. The measure handling the similarity between attributes by similarity of their active domains can be defined as:

$$\xi^{kl}(A_I, A_J) = \frac{|\mathscr{D}_\alpha^{S_k}(A_I) \cap \mathscr{D}_\alpha^{S_l}(A_J)|}{|\mathscr{D}_\alpha^{S_k}(A_I) \cup \mathscr{D}_\alpha^{S_l}(A_J)|} \ , \ A_I \in \mathscr{A}_{S_l}, A_J \in \mathscr{A}_{S_k} \qquad (49)$$

where

$$|\mathscr{D}_\alpha^{S_k}(A_I) \cap \mathscr{D}_\alpha^{S_l}(A_J)| = ||\Delta_{S_k} \hat{\Psi}_{kl} \Delta_{S_l}^T|| \qquad (50)$$

The measure tries to reflect the fact that similar attributes should have similar (active) domains.

The attribute mapping candidates can be expressed by a matrix

$$\tilde{\Pi}_{kl} = [\pi_{ij}^{kl}]; \pi_{ij}^{kl} = \xi^{kl}(A_I, A_J) \qquad (51)$$

Candidates bellow a given preference threshold τ can be ignored

$$\tilde{\Pi}_{kl}^\tau = [\dot{\pi}_{ij}^{kl}], \dot{\pi}_{ij}^{kl} = \begin{cases} \xi^{kl}(A_I, A_J) & \text{if } \xi^{kl}(A_I, A_J) \geq \tau \\ 0 & \text{otherwise} \end{cases} \qquad (52)$$

For the automatic design, only rules with the uniquely maximal support are allowed. Formally

$$\tilde{\Pi}_{kl}^\tau = [\ddot{\pi}_{ij}^{kl}], \forall \ddot{\pi}_{ij}^{kl} > 0 : \forall j' \dot{\pi}_{ij}^{kl} > \dot{\pi}_{ij'}^{kl} \wedge \forall i' \dot{\pi}_{ij}^{kl} > \dot{\pi}_{i'j}^{kl} \qquad (53)$$

Analogically to Formula (4) only element equivalence given by (48) satisfying also the relationship at the higher (attribute) level is considered.

$$\Psi_{kl} = \hat{\Psi}_{kl} \odot \Delta_{S_l}^T \tilde{\Pi}_{kl}^\tau \Delta_{S_k} \qquad (54)$$

The complexity of finding the attribute equivalence rules is $\mathscr{O}(|\mathscr{E}_{S_k}| \cdot |\mathscr{E}_{S_l}|)$, the complexity of achieving Condition (53) is $\mathscr{O}(|\mathscr{A}_{S_k}| \cdot |\mathscr{A}_{S_l}|)$. Because the equivalence is symmetric,

$$\Pi_{kl} = \Pi_{lk}^T \qquad (55)$$

only the items bellow or at the diagonal ($i \leq j$) can be evaluated.

The method will give sound results, when the data provided by the sources S_k, S_l are redundant, and the active domains of integrated attributes are disjunct:

$$\forall A_I, A_J \in \mathscr{A}_{S_k}, \quad \forall A_{i'}, A_{j'} \in \mathscr{A}_{S_l} : \tag{56}$$
$$(\mathscr{D}_\alpha^{S_k}(A_I) \cup \mathscr{D}_\alpha^{S_l}(A_{i'})) \cap \mathscr{D}_\alpha^{S_k}(A_J) \cup \mathscr{D}_\alpha^{S_l}(A_{j'})) = \emptyset$$

When the condition of disjunct domains is not satisfied, rules can be assigned ambiguously. Such situations have to be eliminated by Condition (53) in the automatic mode and in the case of conflicting rules with the same support $\pi_{ii'}^{kl} = \pi_{jj'}^{kl}$, no rule will be selected.

Attribute Equivalence Rule Example

Let us illustrate the matching process on the repositories of two sources:

$$\Phi_1 = \begin{bmatrix} 1 & 0 & 0 & 0 & 0 & 0 & 0 \\ 0 & 1 & 0 & 0 & 0 & 0 & 0 \\ 0 & 0 & 1 & 0 & 0 & 0 & 0 \\ 1 & 0 & 0 & 1 & 0 & 0 & 0 \\ 0 & 1 & 1 & 0 & 1 & 0 & 0 \\ 1 & 1 & 0 & 1 & 0 & 1 & 0 \\ 0 & 0 & 1 & 0 & 1 & 0 & 1 \end{bmatrix} \begin{matrix} \text{Town, Praha} \\ \text{Town, Košice} \\ \text{Town, Bratislava} \\ \text{State, Czechia} \\ \text{State, Slovakia} \\ \text{Currency, CZK} \\ \text{Currency, SSK} \end{matrix} \tag{57}$$

and

$$\Phi_2 = \begin{bmatrix} 1 & 0 & 1 & 0 \\ 0 & 1 & 0 & 1 \\ 1 & 0 & 1 & 0 \\ 0 & 1 & 0 & 1 \end{bmatrix} \begin{matrix} \text{State, Czech Republic} \\ \text{State, Slovakia} \\ \text{Capital, Praha} \\ \text{Capital, Bratislava} \end{matrix} \tag{58}$$

Similarity between attributes can be evaluated by measure (51):

$$\tilde{\Pi}_{12} = \begin{bmatrix} 0 & \frac{2}{3} \\ \frac{1}{3} & 0 \\ 0 & 0 \end{bmatrix} \tag{59}$$

Because Condition (53) is met, the final attribute and (corresponding) element mappings can be established as:

$$\Pi_{12} = \Pi_{21}^T = \begin{bmatrix} 0 & \frac{2}{3} \\ \frac{1}{3} & 0 \\ 0 & 0 \end{bmatrix} \qquad \Psi_{12} = \Psi_{21}^T = \begin{bmatrix} 0 & 0 & \frac{2}{3} & 0 \\ 0 & 0 & 0 & 0 \\ 0 & 0 & 0 & \frac{2}{3} \\ 0 & 0 & 0 & 0 \\ 0 & \frac{1}{3} & 0 & 0 \\ 0 & 0 & 0 & 0 \\ 0 & 0 & 0 & 0 \end{bmatrix} \tag{60}$$

The elements of equivalent attributes are aggregated during a virtual repository is queried and element activations can be summed (weights can be given by the cosine measures) to evaluate the support for the assertion of the element to the query (activating element $(Town, Prague)$ in the query vector):

Attribute	Value	S_1	S_2	$\sum_{S_*}$
Town = Capital	Prague	1	0.66	1.66
State = State	Czechia	1		1
	Czech Republic		0.66	0.66
Currency	CZK	1		1

$\qquad\qquad(61)$

As shown the weights assigned to an element activation may be used as a clue for finding data inconsistency. The methods above are not responsible for identifying $Czechia$ and $CzechRepublic$ as synonyms. Therefore the repository returns all activated (possible opposing) elements and a final interpretation is up to the end user. Note that the resulted weights depend on the query vector (which source(s) is/are primary chosen for an answer).

3.4 Attribute Hierarchy Mappings

Aggregating elements of attributes marked as equivalent may not be sound (follow the activations of Košice $\rightarrow$ Slovakia $\rightarrow$ Bratislava causing inconsistency at integrated attribute Capital $\sim$ Town in the example table above. Such situations can be solved by using the attribute hierarchy rule instead of equivalence which allows activation of the element only in one direction and does not aggregate the elements.

Similarly to the measure for attribute equivalence, attribute hierarchy can be measured by

$$\xi_{\sqsubset}^{kl}(A_I, A_J) = \frac{|\mathscr{D}_\alpha^{S_k}(A_I) \cap \mathscr{D}_\alpha^{S_l}(A_J)|}{|\mathscr{D}_\alpha^{S_k}(A_I)|} \ , \ A_I \in \mathscr{A}_{S_l}, A_J \in \mathscr{A}_{S_k} \qquad (62)$$

The complexity of the measure evaluation is asymptotically the same as in the previous case $\mathcal{O}(|\mathscr{E}_{S_k}| \cdot |\mathscr{E}_{S_l}|)$, but note the matrices Π_{kl} and Π_{lk} have to be established independently (hierarchy rules are not symmetric).

The final mapping can be established using the same steps (51 -54) as in the previous case, or the hierarchical rules can be combined with equivalence ones. A preference to hierarchical or equivalence rules can be driven by a symmetry: if the attribute domain measures are symmetric (i.e. domains are the same), the equivalence will be preferred; if the measures are asymmetric, a hierarchical rule is used. Formally, the measures can be rewritten as:

$$\sigma^\sim(A_I, A_J) = \xi_{\sqsubset}(A_I, A_J) \cdot \xi_{\sqsubset}(A_J, A_I) \text{ for equivalence rule}$$
$$\sigma^{\sqsubset}(A_I, A_J) = \xi_{\sqsubset}(A_I, A_J) \cdot (1 - \xi_{\sqsubset}(A_J, A_I)) \text{ for hierarchical rule}$$

If the measures are combined, comparable measures $\sigma^\sim(A_I, A_J), \sigma^{\sqsubset}(A_I, A_J)$ are used to find the maximal support.

A final selection of hierarchical rules has to avoid situations where chosen rules make a "cycle":

$$A_J \sqsubset A_k \sqsubset A_I : A_I, A_J \in \mathscr{A}_{S_k}, A_I \neq A_J, A_k \in \mathscr{A}_{S_l} \tag{63}$$

The complexity of the test is $\mathscr{O}(|\mathscr{A}_{S_k}| + |\mathscr{A}_{S_l}|)$.

Attribute Hierarchy Rule Example

Using the same example sources (57,58) a support for hierarchical is:

$$\Pi_{12}^{\sqsubset} = \begin{bmatrix} 0 & \frac{1}{2} \\ \frac{2}{2} & 0 \\ 0 & 0 \end{bmatrix} \quad \Pi_{21}^{\sqsubset} = \begin{bmatrix} 0 & \frac{1}{2} & 0 \\ \frac{2}{3} & 0 & 0 \end{bmatrix} \tag{64}$$

Using the measures $\sigma^{\sim}(A_I, A_J), \sigma^{\sqsubset}(A_I, A_J)$, the following rules will be selected:

Capital $\sqsubset$ Town	$\frac{2}{3}$	$\oplus$
Capital $\sim$ Town	$\frac{2}{9}$	$\ominus$
State $\sim$ State	$\frac{1}{4}$	$\oplus$
State $\sqsubset$ State	$\frac{1}{4}$	$\ominus$
State $\sqsubset$ State	$\frac{1}{4}$	$\ominus$
Town $\sqsubset$ Capital	$\frac{1}{9}$	$\ominus$

Finally, the selected rules can be rewritten into the mediator matrices:

$$\Pi_{12} = \begin{bmatrix} 0 & \frac{2}{3} \\ \frac{1}{3} & 0 \\ 0 & 0 \end{bmatrix} \quad \Pi_{21}^{T} = \begin{bmatrix} 0 & 0 \\ \frac{1}{3} & 0 \\ 0 & 0 \end{bmatrix} \quad \Psi_{12} = \begin{bmatrix} 0 & 0 & \frac{2}{3} & 0 \\ 0 & 0 & 0 & 0 \\ 0 & 0 & 0 & \frac{2}{3} \\ 0 & 0 & 0 & 0 \\ 0 & \frac{1}{3} & 0 & 0 \\ 0 & 0 & 0 & 0 \\ 0 & 0 & 0 & 0 \end{bmatrix} \quad \Psi_{21}^{T} = \begin{bmatrix} 0 & 0 & 0 & 0 \\ 0 & 0 & 0 & 0 \\ 0 & 0 & 0 & 0 \\ 0 & 0 & 0 & 0 \\ 0 & \frac{1}{3} & 0 & 0 \\ 0 & 0 & 0 & 0 \\ 0 & 0 & 0 & 0 \end{bmatrix}$$

$$\tag{65}$$

Subsequently evaluating all available information about Košice leads to

Attribute	Value	S_1	S_2	$\sum_{S_*}$
Town	Košice	1		1
State = State	Slovakia	1	0.33	1.33
Currency	SKK	1		1
Capital	Bratislava	0.33		0.33

$$\tag{66}$$

e.g. the relationship asserting the capital to each town in the state, which has not been covered by any local source, is discovered.

4   Source Reputation

The weight of the integration rule express established soundness of the rule, which is given indirectly by the measures analyzing structural relationships over data. As shown on the example (61), the weights can be used to scale the element activations provided by other sources, and in this way to protect (by lower activations) the results of the sources being queried.

From this point of view, the weights can also be used to express *quality of the source*. The quality of data presented by a source S_a can be given by their *freshness* and *consistency*.

Evaluation of the source quality is not an easy task in a totally distributed environment (e.g. the Web) with constantly changing topology. A similar issue is solved by *trust management systems* being responsible for building trust between communicating entities.

4.1   Trust Management Systems

Trust management systems can be categorized according to the way adopted for establishing and evaluating trust as follows:

- *credential* and *policy based trust management*;
- *reputation based trust management*, and;
- *social network based trust management*.

Policy based approach has been proposed in the context of open and distributed service architectures [26] as well as in the context of Grids [27] as the solution to the problem of authorization and an access control in open systems. Its focus is on the trust management mechanisms employing different policy languages and engines for specifying and reasoning on rules for the trust establishment. Since the primary aim of such systems is to enable access control, the trust management is limited to verification of credentials and restricting an access to sources according to policies defined by a required source owner [28].

Reputation based trust management systems provide entities with tools for evaluating and building a trust relationships. This approach has emerged in the context of electronic commerce systems, e.g. eBay. In distributed settings, reputation-based approaches have been proposed for managing trust in public key certificates, P2P systems XREP [29], mobile ad-hoc networks, and recently, also in the Semantic Web [30], NICE [31], DCRC/CORC [32], EigenTrust [33], EigenRep[34].

Social network based trust management systems utilize, in addition, social relationships between entities to infer trust. In particular, the social network based system views the whole structure as a social network with relationships defined amongst entities. Examples of such trust management systems include Regret [35], NodeRanking [36].

4.2 Model for Managing Reputations

The main effort behind the proposal of the reputation system for the presented global repository is to enable evaluation of the quality of data presented by sources. A proposed model for managing source reputation will use the hypergraph model. The hypergraph model is able to express arbitrarily complicated structure of sources and their instances in very straightforward and efficient way [37].

A hypergraph $\mathbb{H} = (\mathcal{U}, \mathcal{N})$ is defined [38] as a set of vertices $\mathcal{U}$ and a set of hyperedges $\mathcal{N} \subseteq 2^{\mathcal{U}}$. A hyperedge, unlike edges in graphs, can connect arbitrary many vertices and one vertex can be included in more hyperedges (be a *pin* of more hyperedges).

In the hypergraph model for reputation holds:

- vertices $u \in \mathcal{U} \subseteq \mathcal{E}_{\mathcal{S}} \times \mathcal{E}_{\mathcal{S}}$ represent virtual instances $\phi_{ij}^{\mathcal{S}} > 0$ from the integrated repository matrix $\Phi_{\mathcal{S}}$.
- hyperedges $S_l \in \mathcal{N} = \mathcal{S}$ represent sources
- pins of a hyperedge represent all the instances provided by source S_l:

$$\mathrm{pins}(S_l) = \{[e_i \to e_j] \subseteq \mathcal{U} : \phi_{ij}^{S_l} > 0\}$$

- nets of a vertex represent all sources providing the instance $\phi_{ij}^{\mathcal{S}}$.

$$\mathrm{nets}(\phi_{ij}^{\mathcal{S}}) = \{S_l \subseteq \mathcal{S} : \phi_{ij}^{S_l} > 0\}$$

In Figure 22, an example situation of 3 sources presenting several instances in the hypergraph model is shown. If a source asserts an instance, this instance is included in the oval representing the source. If the instance is asserted by more sources (e.g. $\phi_{3,1}$) it is included in more ovals.

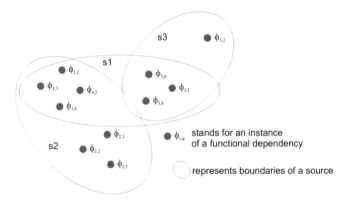

Fig. 22 A hypergraph representing instances at sources

4.3 Dynamic Reputation of Sources

Because the sources dynamically update data, managing reputation is strongly dynamic process. Therefore all factors affecting a source reputation depend on the changes occurring during the k-th step.

The following list gives components of the proposed reputation system influencing the source reputation:

- $a_k^{S_l}$ to prefer sources providing verified data,
- $b_k^{S_l}$ to prefer sources providing new (actual) data,
- $c_k^{S_l}$ to penalize sources providing inconsistencies.

The overall reputation $\rho_k^{S_l}$ of the source S_l can be composed from all these components as:

$$\rho_k^{S_l} = \frac{a_k^{S_l} \cdot (1 + b_k^{S_l}) \cdot (1 - c_k^{S_l})}{2} \in \langle 0, 1 \rangle \tag{67}$$

This reputation can be put into effect during a integration process by extending formula (43):

$$\Phi_{\mathscr{S}} = \sum_{\forall S_l \in \mathscr{S}} \rho_k^{S_l} \cdot \Gamma_{S_l} \Phi_{S_l^k} \Gamma_{S_l}^T \tag{68}$$

Confirmation Component $a_k^{S_l}$

A confirmation component $a_k^{S_l}$ represents the fact that the instances confirmed by more sources are more likely true. The confirmation component is calculated for each source S_l as

$$a_k^{S_l} = \sum_{u \in pins(S_l^k)} \frac{(|\mathrm{nets}(u)| - 1)}{|\mathscr{U}|} \tag{69}$$

where $u \in pins(S_l^k)$ are instances presented by source S_l in k-th step and $|\mathrm{nets}(u)|$ is the total amount of sources presenting instance u. In order to scale the factor, the amount is normalized by the total amount $|\mathscr{U}|$ of unique instances presented by all the sources.

In the matrix notation,

$$a_k^{S_l} = \frac{\Phi_{S_l(k)} \odot (\Phi_{\mathscr{S}(k)} - 1)}{||\Phi^{\mathscr{S}(k)}||} \tag{70}$$

Novelty Component $b_k^{S_l}$

A novelty component represents the fact that sources S_l providing new data are appreciated. The factor is evaluated by the amount of new instances added during last r steps; i.e. instances, which are present in the k-th step but not in the $k - r$ step. Formally in the hypergraph notion, the factor increment $\Delta b_k^{S_l}$ is given as

$$\Delta b_k^{S_l} = \sum_{\substack{u \in \text{pins}(S_l(k)) \\ u \notin \text{pins}(S_l(k-r))}} \frac{|nets(u)| - 1}{|\mathscr{U}|} \tag{71}$$

Alternatively, in the matrix notion

$$\Delta b_k^{S_l} = \frac{(\Phi_{S_l(k)} - \Phi_{S_l(k-r)}) \odot (\Phi_{\mathscr{S}(k)} - 1)}{||\Phi^{\mathscr{S}(k)}||} \tag{72}$$

While the confirmation component (69/70) stands for a long-term characteristics, the novelty component represents a short term one; it is selected by the maximal history radius r and the dynamic coefficient ϵ. The current value of the factor is recursively established by

$$b_k^{S_l} = \epsilon b_{k-1}^{S_l} + (1 - \epsilon)\Delta b_k^{S_l} \tag{73}$$

Inconsistency Component $c_k^{S_l}$

The inconsistency component deals with sources that provide inconsistent data. For example, assume a pair of instances $\phi_{ij}^{\mathscr{S}}$ and $\phi_{ij'}^{\mathscr{S}}$ having the same element e_i on the left side but different element $e_j \neq e_j'$ of the same integrated attribute (55) on the right side. The inconsistency component $\delta_{ij}^{S_l}$ is calculated as the weight of the instance causing the inconsistency $\phi_{ij}^{S_l}$ normalized by the sum of weights of all relevant (all of equivalent attributes) elements:

$$\delta_{ij}^{S_l} = 1 - \frac{\phi_{ij}^{S_l} \cdot \phi_{ij}^{\mathscr{S}}}{\sum_{\forall j':e_j'=(A',\star)\in\{e_i=(A,\star):\forall A\sim A'\}} \phi_{ij'}^{\mathscr{S}}} \tag{74}$$

If there are no inconsistencies then

$$\delta_{ij}^{S_l} = 1 - \frac{1 \cdot \phi_{ij}^{\mathscr{S}}}{\phi_{ij}^{\mathscr{S}}} = 0 \tag{75}$$

All the inconsistencies caused to the virtual repository by instances of the source S_l can be aggregated into the inconsistency component as

$$c_k^{S_l} = \frac{\sum \delta_{ij}^{S_l}}{||\Phi^{\mathscr{S}(k)}||} \tag{76}$$

Details on Dynamics of the Source Reputation

In the following set of figures an example scenario of three sources (S_1, S_2, S_3) is shown. In Figure 23, source S_1 introduces a new instance. In the next step (Figure 24) the source S_2 discovers the new instance and adds it into its

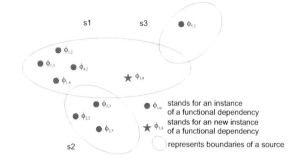

Fig. 23 A new instance is introduced by source S_1

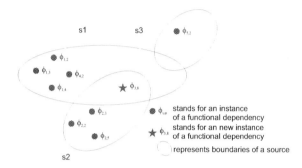

Fig. 24 Source S_2 confirms the instance introduced by S_1

Fig. 25 Source S_3 also confirms the new instance

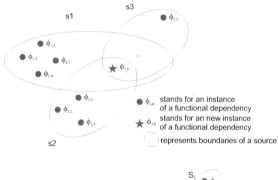

Fig. 26 Source S_1 presents inconsistency $\phi_{2,5'}$ against S_3

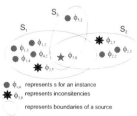

set of the instances. In the following step even source S_3 takes the instance over (Figure 25). In Figure 26 source S_1 introduces an inconsistency again source S_2.

Table 2 Evolution of Reputation Components ($\epsilon = 0.8, r = 2$)

k	Fig.	$|\mathcal{U}|$	a^{S_1}	Δb^{S_1}	b^{S_1}	c^{S_1}	a^{S_2}	Δb^{S_2}	b^{S_2}	c^{S_2}	a^{S_3}	Δb^{S_3}	b^{S_3}	c^{S_3}
0	-	8	0	0	0	1	0	0	0	1	0	0	0	1
1	23	9	0	0	0	1	0	0	0	1	0	0	0	1
2	24	9	0.11	0.11	0.02	1	0.11	0	0	1	0	0	0	1
3	25	9	0.22	0.22	0.06	1	0.22	0	0	1	0.22	0	0	1
4	-	9	0.22	0.11	0.07	1	0,22	0	0	1	0.22	0	0	1
5	-	9	0.22	0	0.06	1	0.22	0	0	1	0.22	0	0	1
6	26	10	0.20	0	0.05	0.95	0.20	0	0	0.95	0.20	0	0	1

Table 2 shows an evolution of the components. At the beginning all sources have the same component values. Afterwards source S_1 introduces a new instance $\phi_{3,8}$ at time $k = 1$. In the following two steps both sources S_2 and S_3 took the instance over causing increments in component b^{S_1} and components a^{S_2} and a^{S_3}. The important point comes in during the step $k = 4$. For parameter $r = 2$ the instance $\phi_{3,8}$ is no loner consider as a new implying decrease of component b^{S_1} in the subsequent steps. The inconsistency introduced by S_1 against S_2 causes decrease of components c^{S_2} and c^{S_1}.

4.4　Analysis of Attacks

Even thought the world with only honest sources would be a wonderful place to live, it is not the case of the current and very probably the future Web. Currently running computer systems are often under attack of hackers, viruses, etc. In our proposal we introduced, besides the other attributes, the reputation of a source as an additional parameter influencing query processing. In this scenario, the reputation of a sources has direct impact on result of a given query. Therefore the reputation system managing trust must be able to cope with certain threats. In the following paragraphs we put our reputation system under investigation against well known attacks.

Fake transaction

Malicious colluding peers always cooperate with others in order to receive strong reputation. They then provide misinformation to promote actively malicious peers [39].

In our system, the responsibility for creating transaction and evaluation a feed-back (whether transaction was done correctly with demanded results or otherwise) lays upon our framework. In other words, the framework presented does not require any feed-back from sources, since it can generate feed-backs automatically for each integration step.

On the other hand, there is another type of fake transaction attack, which may cause damage to a source reputation. Assume that there are two sources

(S_1 and S_2) having the very similar overall reputation. Under this configuration queries as well as users are quite uniformly redirected between these sources. Assume that source S_2 is willing to gain on number of users being redirected to it. Therefore S_2 creates a new shadow source $S_{2'}$ which main aim is to destroy reputation of source S_1 by introducing new instances of functional dependencies being inconsistent with source S_1. This will cause increase in inconsistencies of source S_1 followed by decrease in S_1 reputation.

To see how our reputation system works such scenario out assume two sources S_1 and S_2 with similar reputation ($\rho_{S_1}^k = 0.998$ and $\rho_{S_2}^k = 0.989$). S_2 introduces a new source $S_{2''}$ which reputation is naturally low at the beginning ($\rho_{S_{2''}}^k = 0.001$). $S_{2''}$ wittingly introduces inconsistent instances with S_1. This will cause increase of parameter $c_k^{S_\star}$ of both sources.

As the parameter c weights the number of inconsistencies by reputation of the inconsistent source, it results in situation where S_{new} has its reputation decreased much more (while parameter $c_k^{S_{new}}$ is multiplied by reputation of source S_1 which is high) than the reputation of source S_1. With more instances introduces the gap is getting bigger for reputation of S_{new} is continually decreasing causing smaller and smaller decreases in reputation of S_1.

In this way such a kind of attack has very limited impact on overall reputation of a source.

In other words, having a lot of inconsistencies with a source with high reputation is risky.

Collusion

Multiple malicious peers cooperating together to cause more damage [40],[41].

Figure 27 shows an example situation of 3 sources sharing the same fake functional dependencies to boost their reputations. If the overall reputation given in (67) had included only component $a_k^{S_i}$, the reputation system would have been affected very easily by collusion. In our proposal reputation of source includes also components $b_k^{S_i}$ and $c_k^{S_i}$. Component $b_k^{S_i}$ prices newly introduced instances of functional dependencies, thereby giving more reputation to sources that introduce new instances. Parameter $c_k^{S_i}$, on the other hand, gives more reputation to sources providing instances often included in results to queries. In collision, the sources providing fake or meaningless

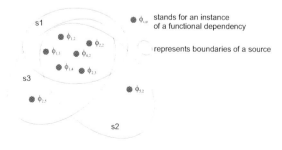

Fig. 27 An example situation of collision attack on reputation system. Sources co-operate in generation possible false dependencies that are shared to boost parameter a^k

data are punished by component $c_k^{S_i}$. Moreover sources in collision would unlikely provide a new data that is accepted by a source with hight reputation (component $b_k^{S_i}$).

Whitewashers

Peers purposefully leave and rejoin the system with a new identity in an attempt to shed any bad reputation they have accumulated previously [42].

In our framework, entities are not anonymous peers but rather real servers with at least IP addresses. Under such circumstances, it is unlikely to assume such type of attack as making a new identity might be expensive and reputation is at the beginning given only by parameter $a_k^{S_i}$.

5 Conclusion

The chapter has proposed an automatic incremental algorithm for building a Semantic Web repository, which has been inspired by a functional dependency discovery approach. This approach has been used to define a data structure estimated from input data. As shown, the incremental algorithm has a polynomial complexity. Moreover, the optimization of the repository has been described in a detail: only ground relationships have been stored in the repository, which are used to derive other by a transitivity rule. The algorithms handling directly a reduced form of the repository have been proposed and their complexity analyzed.

Further, the automatic data integration process has been described, and measures evaluating possible valid equivalence and hierarchical rules have been given. Since the (semi)automatic methods generally return a weight (driven by a support) of an obtained rule, these weights can be used for querying the repository - the rule is activated only in the rate given by the measure. Finally, these weights can be influenced by a source reputation managed by the reputation system, analyzing a former source behavior. In this way, the returned result to the query can be extended by a weight expressing soundness of each result part

References

1. Raghavan, P.: Information retrieval algorithms: a survey. In: SODA 1997: Proceedings of the eighth annual ACM-SIAM symposium on Discrete algorithms, Philadelphia, PA, USA, pp. 11–18. Society for Industrial and Applied Mathematics (1997)
2. Langville, A.N., Meyer, C.D.: Google's PageRank and Beyond: The Science of Search Engine Rankings. Princeton University Press, Princeton (2006)
3. Antoniou, G., van Harmelen, F.: A Semantic Web Primer (Cooperative Information Systems). MIT Press, Cambridge (2004)

4. Pivk, A.: Automatic ontology generation from web tabular structures. AI Communications (2005)
5. Flach, P.A., Savnik, I.: Database dependency discovery: A machine learning approach. AI Communications 12(3), 139–160 (1999)
6. Date, C.J.: An Introduction to Database Systems. Addison Wesley Longman (October 1999)
7. Řimnáč, M.: Data structure estimation for rdf oriented repository building. In: Proceedings of the CISIS 2007, pp. 147–154. IEEE Computer Society, Los Alamitos (2007)
8. Quilitz, B., Leser, U.: Querying distributed rdf data sources with sparql. In: Bechhofer, S., Hauswirth, M., Hoffmann, J., Koubarakis, M. (eds.) ESWC 2008. LNCS, vol. 5021, pp. 524–538. Springer, Heidelberg (2008)
9. Mannila, H., Räithä, K.J.: Design by example: An applications of armstrong relations. Journal of computer and system sciences 33, 129–141 (1986)
10. Mannila, H., Räihä, K.J.: Dependency inference. In: Proc. of VLDB, pp. 155–158 (1987)
11. Mannila, H., Räithä, K.J.: Algorithms for inferring functional dependencies from relations. Data & Knowledge Engineering 12, 83–99 (1994)
12. Akutsu, T., Miyano, S., Kuhara, S.: A simple greedy algorithm for finding functional relations: efficient implementation and average case analysis. Theoretical Computer Science 292, 481–495 (2003)
13. Simon, K.: On minimum flow and transitive reduction. In: Lepistö, T., Salomaa, A. (eds.) ICALP 1988. LNCS, vol. 317, pp. 535–546. Springer, Heidelberg (1988)
14. Simon, K.: Finding a minimal transitive reduction in a strongly connected digraph within linear time. In: Nagl, M. (ed.) WG 1989. LNCS, vol. 411, pp. 245–259. Springer, Heidelberg (1990)
15. Giannella, C., Robertson, E.: On approximation measures for functional dependencies. In: ADBIS 2002: Advances in databases and information systems, pp. 483–507. Elsevier, Amsterdam (2004)
16. Rahm, E., Bernstein, P.A.: A survey of approaches to automatic schema matching. VLDB Journal: Very Large Data Bases 10(4), 334–350 (2001)
17. Shvaiko, P., Euzenat, J.: A survey of schema-based matching approaches. In: Spaccapietra, S. (ed.) Journal on Data Semantics IV. LNCS, vol. 3730, pp. 146–171. Springer, Heidelberg (2005)
18. Mitra, P., Wiederhold, G., Jannink, J.: Semi-automatic integration of knowledge sources. In: Proc. of the 2nd Int. Conf. On Information FUSION 1999 (1999)
19. Do, H.-H., Melnik, S., Rahm, E.: Comparison of schema matching evaluations. In: Chaudhri, A.B., Jeckle, M., Rahm, E., Unland, R. (eds.) NODe-WS 2002. LNCS, vol. 2593, pp. 221–237. Springer, Heidelberg (2003)
20. Lenzerini, M.: Data integration: a theoretical perspective. In: PODS 2002: Proceedings of the twenty-first ACM SIGMOD-SIGACT-SIGART symposium on Principles of Database Systems, pp. 233–246. ACM Press, New York (2002)
21. Nottelmann, H., Straccia, U.: Information retrieval and machine learning for probabilistic schema matching. Information Processing and Management 43, 552–576 (2007)
22. Su, X., Gulla, J.A.: An information retrieval approach to ontology mapping. Data Knowl. Eng. 58(1), 47–69 (2006)
23. Yi, S., Huang, B., Chan, W.T.: Xml application schema matching using similarity measure and relaxation labeling. Inf. Sci. 169(1-2), 27–46 (2005)

24. Do, H.-H., Rahm, E.: Matching large schemas: Approaches and evaluation. Inf. Syst. 32(6), 857–885 (2007)
25. Xu, L., Embley, D.W.: A composite approach to automating direct and indirect schema mappings. Inf. Syst. 31(8), 697–732 (2006)
26. Li, N., Mitchell, J.: A role-based trust-management framework. In: DARPA Information Survivability Conference and Exposition (DISCEX), Washington, D.C. (April 2003)
27. Basney, J., Nejdl, W., Olmedilla, D., Welch, V., Winslett, M.: Negotiating trust on the grid. In: 2nd WWW Workshop on Semantics in P2P and Grid Computing, New York, USA (May 2004)
28. Grandison, T., Sloman, M.: Survey of trust in internet applications. IEEE Communications Surveys 3(4) (2000)
29. Damiani, E., di Vimercati, S.D.C., Paraboschi, S., Samarati, P., Violante, F.: A reputation-based approach for choosing reliable resources in peer-to-peer networks. In: Proceedings of ACM Conference on Computer and Communications Security, pp. 202–216 (2002)
30. Teacy, W.T.L., Patel, J., Jennings, N.R., Luck, M.: Travos: Trust and reputation in the context of inaccurate information sources. Autonomous Agents and Multi-Agent Systems 12(2), 183–198 (2006)
31. Lee, S., Sherwood, R., et al.: Cooperative peer groups in nice. In: IEEE Infocom, San Francisco, USA (2003)
32. Gupta, M., Judge, P., et al.: A reputation system for peer-to-peer networks. In: Thirteenth ACM International Workshop on Network and Operating Systems Support for Digital Audio and Video, Monterey, California (2003)
33. Kamvar, S., Schlosser, M., et al.: The eigentrust algorithm for reputation management in p2p networks. In: WWW, Budapest, Hungary (2003)
34. Kamvar, S.D., Schlosser, M.T., Garcia-Molina, H.: Eigenrep: Reputation management in p2p networks. In: Proceedings of 12th International WWW Conference, pp. 640–651 (2003)
35. Sabater, J., Sierra, C.: Regret: A reputation model for gregarious societies. In: 4th Workshop on Deception, Fraud and Trust in Agetn Societies, Montreal, Canada (2001)
36. Pujol, J., Sanguesa, R., et al.: Extracting reputation in multi agent systems by means of social network topology. In: First International Joint Conference on Autonomous Agents and Multi-Agent Systems, Bologna, Italy (2002)
37. Catalyurek, U.V., Aykanat, C.: Hypergraph-partitioning-based decomposition for parallel sparse-matrix vector multiplication. IEEE Transactions on Parallel and Distributed Systems 10(7), 673–693 (1999)
38. Golumbic, M.C.: Algorithmic graph theory and perfect graphs. Academic Press, London (1980)
39. Feldman, M., Lai, K., Stoica, I., Chuang, J.: Robust incentive techniques for peer-to-peer networks. ACM Press, New York (2004)
40. Marti, S., Garcia-Molina, H.: Limited reputation sharing in P2P systems. ACM Press, New York (2004)
41. Maniatis, P., Roussopoulos, M., Giuli, T., Rosenthal, D., Baker, M., Muliadi, Y.: Preserving peer replicas by rate-limited sampled voting. Technical Report arXiv:cs.CR/0303026, Stanford University (2003)
42. Lai, K., Feldman, M., Stoica, I., Chuang, J.: Incentives for cooperation in peer-to-peer networks. In: Workshop on Economics of Peer-toPeer Systems (2003)

A Data Mining Approach for Adaptive Path Planning on Large Road Networks

A. Awasthi, S.S. Chauhan, M. Parent, Y. Lechevallier, and J.M. Proth

Abstract. Estimating fastest paths on large networks is a crucial problem for dynamic route guidance systems. This chapter presents a data mining based approach for approximating fastest paths on urban road networks. The traffic data consists of input flows at the entry nodes, system state of the network or the number of cars present in various roads of the roads, and the paths joining the various origins and the destinations of the network. To find out the relationship between the input flows, arc states and the fastest paths of the network, we developed a data mining approach called hybrid clustering. The objective of hybrid clustering is to develop IF-THEN based decision rules for determining fastest paths. Whenever a driver wants to know the fastest path between a given origin-destination pair, he/she sends a query into the path database indicating his/her current position and the destination. The database then matches the query data against the database parameters. If matching is found, then the database provides the fastest path to the driver

A. Awasthi
CIISE, Concordia University, Montreal, Canada
e-mail: awasthi@ciise.concordia.ca

S.S. Chauhan
JMSB, Concordia University, Montreal, Canada
e-mail: sschauha@jmsb.concordia.ca

M. Parent
AXIS, INRIA Rocquencourt, Le Chesnay, France
e-mail: yves.lechevallier@inria.fr

Y. Lechevallier
IMARA, INRIA Rocquencourt, Le Chesnay, France
e-mail: michel.parent@inria.fr

J.M. Proth
INRIA Lorraine, Ile du Saulcy, Metz, France
e-mail: proth.jm@wanadoo.fr

A. Abraham et al. (Eds.): Foundations of Comput. Intel. Vol. 6, SCI 206, pp. 297–320.
springerlink.com   © Springer-Verlag Berlin Heidelberg 2009

using the corresponding decision rule, otherwise, the shortest path is provided as the fastest path.

A numerical experiment is provided to demonstrate the utility of our approach.

1 Introduction

The problem of finding fastest path between two points on a network is same as finding the shortest path on a network provided the state of the network or the number of vehicles present in various arcs of the network does not evolves. The solution lies in simply replacing the arc travel times by the arc lengths and usage of any standard shortest path algorithm for computing the fastest paths. Unfortunately, the state of the network does not always remains constant due to the randomly varying input flows and arc states and this leads to continuously changing fastest paths inside the network. This is the reason why a standard shortest path computation algorithm cannot be used for finding fastest paths on urban networks.

A number of methodologies for finding out the shortest paths have been proposed by researchers over recent years. Zhan [24] presents a set of three shortest path algorithms that run fastest on real road networks. These are the graph growth algorithm implemented with two queues, the Dijkstra algorithm implemented with approximate buckets and the Dijkstra algorithm implemented with double buckets. Buckets are sets arranged in a sorted fashion. The search of an element in a bucket data structure is less time consuming than a non-ordered list. Smith [20] introduced a hierarchical approach for solving shortest paths in a large scale network. Their approach consists of two steps. The first step consists of decomposing the network into several smaller (low level) sub-networks and the second step consists of:

- Finding the sequence of the sub-networks to visit for computing the shortest path.
- Optimization of travel time inside each sub-network.

Data analysis techniques Cutting [4], Marcotorchino [16], Diday [5] have been applied in the domain of transportation by the following researchers:

- Kohonen Self Organizing Feature Maps (SOFM) [Murtagh [18], Kohonen [13], Ambroise [1] have been used with advanced neural networks for freeway travel time prediction. Park and Rillett [19], Dougherty [8]use SOFM to transform an incoming signal pattern of multiple dimensions into a one- or two- dimensional discrete map and to perform this transformation adaptively in a topologically ordered fashion. Kisgyorgy and Rillett [12] use Kohonen SOFM to classify the input vectors into different clusters where the vectors associated with each cluster have similar features.
- Jain et al [11]present an overview of pattern clustering methods from a statistical pattern recognition perspective.

Our approach uses hybrid clustering followed by canonical correlation analysis for prediction of fastest paths on dynamically changing urban networks. Hybrid clustering is conducted in four steps. The first steps consists of subjecting the traffic data to multiple correspondence analysis. This stage identifies the principal factors containing maximum amount of information. In the second phase, the dimension space of the data points is reduced. This gives us the most significant data points in the reduced dimension space. The third phase consists of classifying the data points obtained from phase 2 through k-means clustering [Everitt [9], Jain [11], Meila [17], Diday [6] [7], Tomassone [21], Roux [22], Hansen [10], Klein [14]. In the fourth phase, we group the classes obtained from phase 3 through a hierarchical agglomerative clustering method called Ward's method [23]. The elements of the classes obtained at the end of phase 4 are subjected to canonical correlation analysis in order to find out the relationship among input flows, arc states and paths, which is the objective of our study.

2 Dynamic Route Guidance Method

The route guidance method developed by us is hierarchical in nature. Our approach for predictive route guidance on large scale urban networks consists of the following steps:

1. Decomposition of the urban networks into sub-networks of reasonable size that would be as far as possible independent from each other. In other words, we are trying to minimize the number of boundary nodes of subnetworks.
2. Finding the sequence of sub-networks to visit in order to reach the destination.
3. Optimization of travel time inside the sub-networks.

Assuming that the sub-networks have already been obtained, we will concentrate mainly on the second and the third steps. The second step involves dynamic programming for finding out the sequence of subnetworks to visit in order to reach a destination. The dynamic programming will provide us a global path that will be used for determining the input and output nodes of the subnetworks to be visited. The third step involves local path computation. This path will be computed using the decision rules obtained from hybrid clustering of traffic data. This paper mainly focuses on local path computation approach.

Note that before subjecting the traffic data to data analysis, we classify the traffic data into ranks. The classification is done in such a way that traffic elements having the same rank will lead us to approximately the same results. For example, two arcs having same arc state ranks will lead to identical travel times for the paths (of which they are a part) between a given origin and the exit node. Likewise, we also assign ranks to input flows and paths.

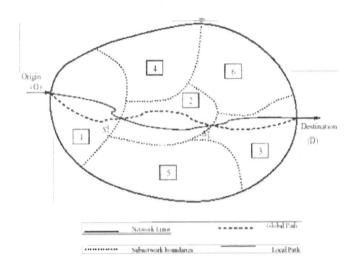

Fig. 1 Route guidance approach

Note that the global path computation assumes that the characteristics of the state of the system remain constant throughout the journey. This is the only characteristic that differentiates the global approach from the local approach for fastest path computation which assumes that characteristics of the network continuously vary with time. The local approach computes the travel time inside the current sub-network and the next subnetworks depending upon the local traffic conditions prevailing at that instant. Thus, our hierarchical guidance approach involves both global and local path computation. The objective of global path computation is to find out the sequence of the sub-networks to be visited in order to reach the destination and the local path computation is done in order to find out the fastest path between the entry and exit point of each sub-network taking into account the real time state of the system. Figure 1 schematises this approach.

In the example presented in Figure 1, the network is decomposed into 6 subnetworks namely, 1,2,3,4,5 and 6. The vehicle enters the network at origin O and wants to go to the destination D. When the vehicle enters inside subnetwork 1 (point O), the global path linking O to D is calculated. This is the path O-S_2^1-S_3^2-D. From the global path we find that the exit node of the sub-network 1 is S_2^1, therefore we will now calculate the local path between origin O and the exit point S_2^1 of the sub-network 1. This path is computed using the rules obtained from hybrid clustering. When the vehicle exits S_2^1, then the next step is to find out the local path for the next sub-network to be visited i.e. sub-network 2. The local path for subnetwork 2 is computed between the entry point S_2^1 and exit point S_3^2. After exiting from the subnetwork 2, the vehicle enters the subnetwork 3. The entry point for subnetwork 3 is S_3^2 and the exit point is D. Now the local path between these two points will be calculated. It can be seen that the vehicle leaves the subnetwork 3 at the exit point D which is the final destination of the vehicle. It is evident from this example that the local path computation inside the subnetworks is done over a rolling horizon.

3   Local Path Computation: Methodology

This section is dedicated to the search of the best local path inside the subnetworks. When a vehicle enters a subnetwork, we are interested in not only knowing the path that will connect the point of entry to the point of the exit (or at the end of the path if it appears in the subnetwork), but also the optimisation of travel time on this path taking into account the real time state of the system.

When a driver enters into a sub-network, we know:

- The input flows at the entry points of the subnetwork.
- The number of vehicles inside the arcs of the subnetwork.

The input flows being stepwise constant, we can assume that they are going to remain stable at least for few minutes and will sufficiently allow a vehicle to traverse this sub-network. These flows modify the initial state of the system and therefore influence the decisions of the drivers who want to reach the exits of the sub-networks as fast as possible. Hybrid clustering is used to compute the local paths which will be used by the drivers in order to reach the exits in minimum time. The traffic data for hybrid clustering is generated using a traffic flow simulation software developed by us. The inputs to the software are the input flows and the arc states and the outputs from the software are the different paths connecting the origins and the destinations of the network. The traffic data before conducting the hybrid clustering is prepared in the following manner:

- We generate at random a large number of input flows and arc states within a fixed range. Two types of preparatory activities are done before starting the simulation:

  - Finding the domain or the range in which each of the variables will take an important percentage (95% for example) of the observed values.
  - Finding the correlation between the variables arc system state and the input flow.

  For example, it is less probable that an arc will be saturated if its predecessors arcs are empty. Therefore, there exists certainly a strong correlation certainly exists between the number of cars in arc l_1 and the number of cars in arc l_2. To cope up with this, we will perform random generation of system states.

  The preliminary study has not been conducted in the paper. We have simply assumed that the arc state and the input flow variables are non-correlated and that they take their values between 0 and the maximum value generated from a uniform probability function.

- For each of these pairs (input flow, system state), we compute the paths and their corresponding travel times for all origin-destination pairs of the network. This calculation is done using simulation.

We use the results obtained from the numerous simulations for the construction of a memory or database that will inform the user the best way to traverse the sub-network in which he wants to enter. This requires the knowledge of the pair (arc system state, input flows) to extract the corresponding best path from the memory. The best paths are stored in the memory in the form of rules. These rules provide the fastest path as a function of the input flow and the system state of the arcs.

Remark: For developing a hierarchical route guidance system, we also developed algorithms for decomposing large scale urban networks into small sub-networks such that they are as far as possible independent from each other (Awasthi [2]). In other words, we tried to obtain sub-networks in a way such that the total number of input and output nodes or the boundary nodes for all the sub-networks is minimized. This also leads to the minimisation of the volume of the traffic data for each subnetwork.

4 Classification of Traffic Data

The classification of traffic data consists of transforming the input data (input flow, arc states) and the output data (paths) obtained from the simulation software into classes. The inputs of the simulation software i.e. the input flow and the number of vehicles in each arc of the system are divided into three classes. We assume that the values of input flow or the number of vehicles in arcs have same impact on path travel times as their ranks, therefore the division of traffic data into classes or ranks is justified. Each traffic variable is provided a class out of the three classes as follows:-

- It belongs to class 1 or has rank 1 if its value lies between 0 and 60% of the maximum value.
- It belongs to class 2 or has rank 2 if its value lies between 60% and 90% of the maximum value.
- It belongs to class 3 or has rank 3 if its value lies between 90% and 100% of the maximum value.

For instance, if the input flow at an entry point is equal to 90% and 100% of the maximum flow permitted at that entry point, then we will say that the input flow belongs to class 3 or has rank 3. Likewise, if an arc state has rank 2, this means that the number of vehicles present in the arc at that instant lie between 60% and 90% of the maximum capacity of the arc. Note that the classification of traffic variables is done into 3 classes only because these ranges allow us to sufficiently view the impact of input flows and arc states

Table 1 Simulation Results

No.	Input 1		Input 2		-	$Arc(i,j)$		$Arc(k,l)$		-	$PathC1$		$PathC2$		-
1	X		X		-	X		X		-	X			X		-
2		X	X		-	X			X	-		X	X			-
3		X		X	-	X			X	-	X		X			-
7		X		X	-	X		X		-	X			X		-
-	-	-	-	-	-	-	-	-	-	-	-	-	-	-	-
-	-	-	-	-	-	-	-	-	-	-	-	-	-	-	-
-	-	-	-	-	-	-	-	-	-	-	-	-	-	-	-

on travel times. However, other users can choose different classes or ranges depending upon their utility.

The classification or ranking of the paths is done on the basis of travel times. They are classified in three ranks as follows:

- The fastest path is assigned rank 1.
- The second fastest path is assigned rank 2.
- The third fastest path is assigned rank 3.

Each simulation result i.e. input flow, arc state and path is thus transformed into ranked data in the form of 1,2 or 3 for all the origin nodes, arcs and paths of the network. If n_1 is the total number of origin nodes, m is the total number of arcs and p_1 is the total number of paths of the network, then the number of input parameters are $n_1 + m + p_1$. Since each parameter can take 3 ranks, therefore the total number of traffic parameter combinations available for hybrid clustering is equal to $3^{n_1+m+p_1}$. Note that all the traffic parameters do not have the same importance and the objective of the method presented in the next paragraph is only to extract the parameters that are most significant from the point of view of decision making. We will therefore observe the importance of these parameters by keeping a limit on the size of the network i.e. by reducing m.

Table 1 presents the various traffic parameters (input flow, arc state, path) obtained from the simulation. It can be seen in Table 1 that the arc states, input flows and paths are divided into 3 classes i.e. rank 1, 2 and 3 and that one simulation experiment comprises of only one X allotted to the triplet of columns assigned for the different traffic parameters. This means that during one simulation run, the impact of one class of input flows (at different entry points) and one class of arc states (for different arcs) is observed on the paths (between different origin-destination pairs) of the network.

5 Hybrid Clustering

Hybrid Clustering is performed on the ranked data obtained from traffic simulation. The statistical software used for conducting the hybrid clustering

Fig. 2 Four Methods in
SPAD

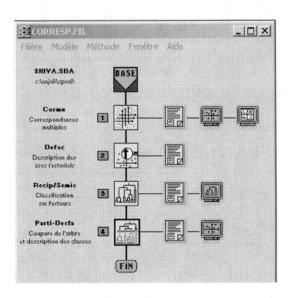

is SPAD [3]. The four steps of the hybrid clustering (described in section 1)
were performed by four modules of SPAD namely CORMU, DEFAC, SEMIS
and PARTI/DECLA. Figure 2 depicts the four modules of SPAD.

5.1 Multiple Correspondence Analysis

The first step of hybrid clustering consists of multiple correspondence analysis
(MCA). This step was achieved through CORMU module of SPAD. Before
commencing MCA, the data's are filtered in the following manner:-

Let us consider Table 1. We count the number of crosses (X) in each column
and divide it by the total number of simulations or the total number of rows
of the table in order to compute a ratio. This ratio gives us proportion in
which the variable of the column appears in the total number of simulations.
If this ratio is inferior than a threshold value r, then we reject that column
from further study because it is considered to be contributing insignificantly
to the study of the problem. Note that the value of r is provided by the user
and we will see later in the numerical example section that $r = 0.02$.

The filtered data is subjected to multiple correspondence analysis. The
results are the projection of data points (each point being obtained from the
simulation) on the factor space. The MCA graphs represent the projections,
i.e. two points that are close on this graph are not necessarily close on the
complete dimensions (even for a small network like that presented in the
numerical example section, the number of dimensions needed to explain the
total variance are 59). It is therefore important to correctly interpret the
coordinates of these projected data points in the space defined by the factors.

Treatment 1: *The multiple correspondence analysis allows us to detect the relationship between the input flow ranks, arc state ranks and the paths of rank 1 or the fastest paths. Besides, it also allows us to identify the principal factors containing the maximum information i.e. which contribute maximum to the total inertia.*

5.2   Reduction of Variable Space

Let d be the number of factors used to define the traffic variables. This means that each traffic data i.e. input flow, arc state and fastest path can be represented by a point in d dimension space. We then select the factors that contain the maximum amount of information. Let f denote the number of these factors and $f < d$. In the numerical example presented in section 8, $d = 59$ and $f = 10$.

The DEFAC module of SPAD allows us to project the group of points (the simulation data) initially situated in a space of dimension d into a reduced space of dimension f.

Treatment 2: *The module DEFAC allows us to represent the data points obtained from MCA in a reduced dimension space. The data points are represented in the new coordinates defined by the factors that contain the maximum amount of information(i.e. which contribute maximum to the total inertia).*

5.3   Classification of Points in the New Space

The third method consists of applying k-means & Ward's clustering method on the data points obtained in the reduced factor space. The k-means clustering (Legendre [15]) is performed by the SEMIS module of SPAD. Let N be the total number of points to be classified in the dimension space f. The k-means clustering method can be explained as follows:-

5.3.1   k-Means Clustering

Step 1: Initialization
Let $N = \{1, 2, 3...n\}$. We select $k^o < n$ elements from N at random. These elements represent the initial centres and are denoted by U_z^o. The set containing the centres $U_1^o, U_2^o, ...$ etc. is represented by S^o. In other words $S^o = \{U_1^o, U_2^o, U_3^o, ..U_{k^o}^o\}$ where $S^o \subset N$.
Step 2: Assignment
For $i = 1, 2, ..n$ we assign an element i to cluster C_z^o if

$$d(i, U_z^o) = Min_{\{j=1,2,..k^o\}} d(i, U_j^o)$$

Step 3: Computation of S^1
For each C_z^o, we compute the mean value of the elements of C_z^o. Let U_z^1 be this value and $S^1 = \{U_1^1, ...U_{k^1}^1\}$. Note that we may have $k^1 < k^o$.

Step 4: Test

If $S^1 \equiv S^o$ stop, otherwise set $k^o = k^1, S^o = S^1$ and go to step 2.

It has been proved that this algorithm converges. In practice, this method is conducted many times in order to obtain the stable classes i.e. classes which are formed always irrespective of the initial choice of random centers at each experiment.

Treatment 3: *The k-means clustering method is run multiple times using randomly chosen initial centers in order to obtain the stable classes. We know that the data points contained in a class are very close and therefore a class can be replaced merely by a unique point. In fact, the last point that joins a class also be used to represent that class.*

5.4   Regrouping of Classes Closest to Each Other

This step performs hierarchical agglomerative clustering using Ward's method. An agglomerative hierarchical clustering procedure produces a series of partitions of the data, P_n, P_{n-1}, ... , P_1. The first P_n consists of n single object 'clusters', the last P_1, consists of single group containing all n cases. At each particular stage the method joins together two clusters which are closest to each other (most similar). In the Ward's method, the union of those two clusters occur at each step whose fusion results in minimum increase in 'information loss'. Information loss is defined by Ward in terms of an error sum-of-squares criterion (Subsection 5.4.1).

PARTI/DECLA module of SPAD is used to perform this step (see Figure 2). The classes which are close to each other are consecutively grouped together until a predefined number of classes (i.e. $\geq$ threshold value provided by the user) have been obtained. For example, In Figure 3, 7 classes have been obtained from 10 initial classes using Ward's method. It can be seen that these 7 classes explain 29%, 18%, 11%, 17%, 3%, 9% and 12% of the total inertia.

Note that the higher is the threshold value, the greater are the number of classes obtained at the end of Ward's clustering. However, higher number of classes may also yield results that are less precise i.e. there is less possibility that the elements of one class are present in other classes. Ward's method is explained as follows:

5.4.1   Ward's Method

Ward [1963] considered that any stage of regrouping of elements, the loss of information which results from the grouping of elements into clusters or classes can be measured by the total sum of squared deviations of every point from the mean of the cluster to which it belongs. At each step, union of

Classification mixte

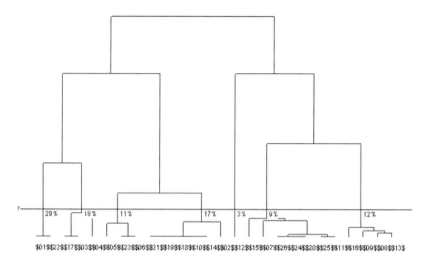

Fig. 3 Ward's Hierarchical Tree/Dendogram

those two possible clusters which cause minimum increase in the Error Sum of Squares (E.S.S) is considered. If x_{ij} denotes the jth element of cluster j, and the total number of clusters is k, the number of elements in each cluster being n_j , then the E.S.S for cluster c represented by v_c is given by:

$$E.S.S.(v_c) = \sum_{j=1}^{k} \sum_{i=1}^{n_j} (x_{ij} - \overline{x_j})^2$$

where

$$x_j = (1/n_j) \sum_{i=1}^{n_j} x_{ij}$$

Initially, each cluster consists of a single element, therefore E.S.S = 0 for all clusters.

Treatment 4: *The data points clustered together at the fourth stage lead to classes (also called groups or clusters) each of which contains elements that we considered at the start of the analysis i.e. ranked input flows, arc system states and paths.*

It is likely that in certain classes, we will observe paths of rank 1 along with different arc states and input flows. This implies that the ranks of these input flows and arc states govern the fastest path. This approach will be adopted for interpreting the results. Later on, we will also find the degree of correlation between the input flow ranks, arc state ranks and the path ranks in each class.

5.5   Canonical Correlation Analysis

Canonical correlation analysis is used to find out the correlation between the input flow ranks, arc state ranks and the path ranks of each class. A threshold value is set for the correlation coefficient in order to identify the strength of relationship among the traffic elements.

Treatment 5: *The canonical correlation analysis allows us to find out the input flow and arc state ranks which are correlated with the paths of rank 1 in each class. The strength of co-relationship between the input flow ranks, arc system state ranks with paths of rank 1 is determined by comparison against a threshold value.*

6   Results

The results of hybrid clustering are presented in the form of decision rules that will permit the driver to decide the fastest path in real time in order to reach the destination. The decision rules can be of the following type:-

   IF $(r(i_1, j_1) = 1)$ **AND** $(r(i_2, j_2) = 2)$ **AND** $(r(e_1) = 3)$ **AND** $(r(e_2) = 1)$ **THEN** *the fastest path is (a1, a2, ..as)*.
In this rule:

- $r(i_1, j_1)$ represents the state of the arc joining node i_1 to node j_1. The condition $r(i_1, j_1) = 1$ indicates that the rank of the state of arc (i_1, j_1) is 1 or in other words the number of vehicles in the arc (i_1, j_1) lies between 0% and 60% of its maximum capacity. The reverse of this condition would be $r(i_1, j_1) \neq 1$. Let us assume that in a rule we observe $r(i_1, j_1) \neq 1$. The interpretation of this condition would be that the system state of the arc (i_1, j_1) does not lies between 0% - 60% of its maximum capacity. This means that either it has rank 2 (60% - 90% of the maximum capacity) or rank 3 (90% - 100% of the maximum capacity), but never rank 1.
- The condition $(r(e_1) = 1)$ indicates that the input flow at the entry node e_1 has rank 1 or it lies between 0-60% of the maximum flow. As illustrated before, we may across a condition of the type $(r(e_1) \neq 1)$ which would mean than input flow never possesses rank 1 though other ranks i.e. 2 or 3 are possible.

Therefore, decisions for predicting the fastest paths are done using rules. Whenever a driver enters the network, the current ranks of the input flows and the arc states are searched inside the database to find out the decision rule which will provide the fastest path under these traffic conditions. Following cases are possible:

1. There is no rule in the database for certain pairs of input flow ranks and arc state ranks. This signifies that we have not succeeded in identifying the fastest path under these conditions. This situation may arise due to two

cases. The first case occurs when all the initial arc states and the desired input flows have not been simulated i.e. they have not been considered while performing the simulations (result of a badly conducted study). In the second case, insufficient number of simulations are conducted on the initial data. In both of these cases, no rule should be provided and the shortest path is predicted as the fastest path.

2. There are many rules in the database for fastest path between a given origin-destination pair. This would mean that there exist various combinations of arc state and input flow ranks that lead to the selection of the fastest path.

7 Important Remarks

- It is clear from the previous sections that the quality of the results depends on the traffic data (i.e. input flow ranks, arc state ranks and path ranks) and the number of experiments chosen for simulation. If the number of simulation experiments conducted do not correspond to reality, then the rules obtained are not efficient and in fact they can be counter-productive also.

- It is absolutely important that the route guidance system should be aware of the input flows and arc states on the network on a real time basis. However, this information can be limited to only those that appear in the rules (which are in most of the cases less than the information needed for all arc states and input flows for all the origin nodes). Finally, when a vehicle enters a sub-network, it knows its origin and destination (obtained from global path). We therefore need only a fixed number of rules corresponding to the required origin and destination for finding the fastest path inside the sub-network (local path).

8 Numerical Example

8.1 Input Data

Let us consider the network depicted in Figure 4.

The traffic simulation on this network was conducted under varying input flows and arc states. The input flows and arc states were generated at random between 0 and their maximum value. Ranks were allotted to the input variables i.e. input flows and arc states as following:

- The variable has rank 1 if its value lies between 0 and 60% of the maximum value.
- The variable has rank 2 if its value lies between 60% and 90% of the maximum value.

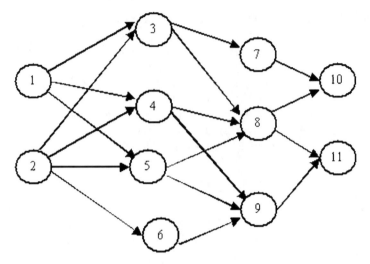

Fig. 4 A Multiple Origin Destination Network

- The variable has rank 3 if its value lies between 90% and 100% of the maximum value.

The classification of various paths of the network was done as following :

- The fastest path is assigned rank 1.
- The second fastest path is assigned rank 2.
- The third fastest path is assigned rank 3.

The maximum permitted value of input flows at the entry points of the network was set equal to 70. These flows remained constant during a time interval of 200 time units. The input flows arriving at the two entry points of the network varied between 0 and the maximum permitted value i.e. 70. Different ids were assigned to the input flows depending upon their value. For example, for the origin node 1:

- The input flow was allotted an id I1 if its rank was 1.
- The input flow was allotted an id I2 if its rank was 2.
- The input flow was allotted an id I3 if its rank was 3.

Likewise, I4, I5 and I6 represent the ids of the ranked input flows for the origin node 2.

Table 2 presents the characteristics of the network. The first column represents the arc ids. An arc id (i, j) indicates i is the tail and j is the head of the arc. The second column represents the traffic accommodation capacity of the arc or in other words the maximum number of vehicles that can be present inside the arc (i, j). Finally, the last column represents the identities assigned to various arcs of the network on the basis of their ranks. For instance,

Table 2 Network Parameters at time t_o

Arc Id (i,j)	Maximum Capacity	Arc identities as per ranks 1,2 & 3
(1,3)	175	S1,S2,S3
(1,4)	210	S4,S5,S6
(1,5)	240	S7,S8,S9
(2,3)	200	S10,S11,S12
(2,4)	160	S13,S14,S15
(2,5)	150	S16,S17,S18
(2,6)	180	S19,S20,S21
(3,7)	260	S22,S23,S24
(3,8)	180	S25,S26,S27
(4,8)	165	S28,S29,S30
(4,9)	132	S31,S32,S33
(5,8)	192	S34,S35,S36
(5,9)	180	S37,S38,S39
(6,9)	140	S40,S41,S42
(7,10)	119	S43,S44,S45
(8,10)	120	S46,S47,S48
(8,11)	130	S49,S50,S51
(9,11)	132	S52,S53,S54

- S1 represents the id of arc (1,3) when its state lies between 0 and 60% of its maximum capacity 175, i.e. between 0 and 105 vehicles.
- S2 represents the id of arc (1,3) when its state lies between 60% and 90% of its maximum capacity 175, i.e. between 105 and 157.5 vehicles.
- S3 represents the id of arc (1,3) when its state lies between 90% and 100% of its maximum capacity 175, i.e. between 157.5 and 175 vehicles.

Table 3 presents the rank ids for different paths of the network. The first column represents the origin-destination pair. For each of these pairs:

- The column 2 represents the various paths that join this origin-destination pair.
- The column 3 represents the lengths of these paths.
- The column 4 represents the ids of the paths. For example, path 1-3-7-10 has an id P1 if it is the fastest path, an id P2 if it is the second fastest path and an id P3 if it is the third shortest path.

8.2 Application of the Method

The total number of simulations carried out were 10,000 where the number of iterations per simulation were 200. During each simulation experiment, we generated input flows and arc system states at random (as explained in section 4). SPAD statistical software [CISIA, 1997] was used to perform hybrid clustering on the traffic data obtained from simulation. The four methods

Table 3 Path Details

O-D Pair	Paths	Length	Path identities for ranks 1,2 & 3
1-10	1-3-7-10	27	P1,P2,P3
	1-3-8-10	25	P4,P5,P6
	1-4-8-10	26	P7,P8,P9
	1-5-8-10	28	P10,P11,P12
1-11	1-3-8-11	27	P13,P14,P15
	1-4-8-11	28	P16,P17,P18
	1-4-9-11	29	P19,P20,P21
	1-5-8-11	30	P22,P23,P24
	1-5-9-11	28	P25,P26,P27
2-10	2-3-7-10	30	P28,P29,P30
	2-3-8-10	28	P31,P32,P33
	2-4-8-10	27	P34,P35,P36
	2-5-8-10	26	P37,P38,P39
2-11	2-3-8-11	30	P40,P41,P42
	2-4-8-11	29	P43,P44,P45
	2-4-9-11	30	P46,P47,P48
	2-5-8-11	28	P49,P50,P51
	2-5-9-11	26	P52,P53,P54
	2-6-9-11	27	P55,P56,P57

of hybrid clustering were performed by the CORMU, DEFAC, SEMIS and PARTI/DECLA modules of SPAD.

Multiple correspondence analysis (MCA) was done on the traffic data using CORMU. Before subjecting to MCA all the traffic variables were classified as nominal active and variables that contributed less than 2% to the simulation results were discarded. Figure 5 presents the results obtained from MCA. In the Figure 5, the paths, arc states and input flow ranks are represented by their ids.

It can be seen in Figure 5 that the paths and arcs are well-separated with most of the arcs and input flows located around the centre and the paths spread far away. However, few arcs are situated close to the paths and can be seen around the centre. For instance P39 and S39. The path id P39 (path 2-5-8-10 with rank 3) is close to the arc id S39 (arc (5,9) with rank 3). The MCA findings indicate that a total of 59 factors or axes are required for explaining 100% of the total inertia.

The second method of hybrid clustering namely reduction of factors was done using DEFAC. The outcome was 10 principal factors that contribute maximum to the total inertia. The variances explained by these factors are 7.55%,7.35%, 6.33%,3.59%, 2.59%, 2.01%, 2.0%, 1.86%, 1.85% and 1.84%. (= 36.96% of the total inertia). At this stage, the number of factors reduce from 59 to 10.

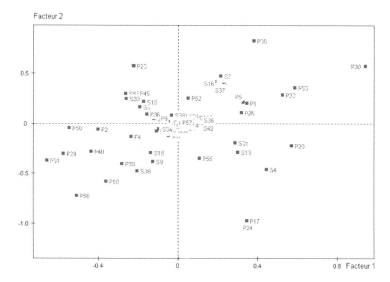

Fig. 5 Multiple Correspondence Analysis

The third method namely k-means clustering was performed using SEMIS. For generating the clusters in the traffic data, 10 centres were randomly chosen and 20 iterations of clustering were done. In this way 3 clustering experiments were done with randomly chosen initial centres. At the end of each experiment 10 clusters were obtained. These clusters where then intersected to find out the stable classes. 26 stable classes were obtained at the end of the experiment.

These 26 stable classes obtained from k-means clustering were subjected to Ward's agglomerative hierarchical clustering. The PARTI/DECLA module of SPAD was used at this stage. The output of the Ward's clustering are 7 classes or clusters that can be seen in Figure 6. Each cluster contains individuals called parangons that are normally the closest points to the centre of each class and best represent the cluster. The parangons can be seen at the centre of each cluster in Figure 6.

Table 4 presents the constituent elements of the 7 classes obtained from Ward's clustering. Each class contains few traffic elements namely input flows, arc states and paths. The identities of the paths, arcs and input flows are present in column 3 of Table 4. It can be seen in Table 4 that all the classes contain paths and arc states while input flows are present only in class number 2,4,5,6 and 7. The reason being the topology of the network and the short time period used for conducting the simulation which was not large enough for significantly affecting the paths.

Each of the classes obtained from hybrid clustering were subjected to canonical correlation analysis for measuring the correlation among the traffic elements i.e. input flows, arc states and paths. The canonical R value for each

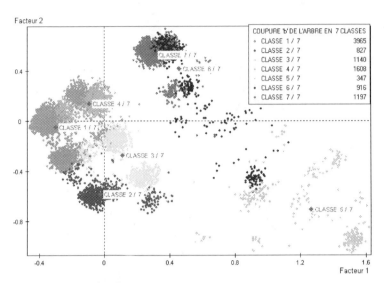

Fig. 6 Cluster Analysis

Table 4 Seven Clusters

Cluster	Frequency	Elements
1	.2868	P31,P2,P4,P18,P23,P50,P29,P40,P45,P36 S33,S34,P39,S28,S25,P9,S6,P38,S15,S5,S10,P13 S1,S14,S38,P12,S40
2	.1848	P17,P24,P31,P29,P2,P4,P40,P44 P51,P39,S4,S39,S25,S13,S34,S9,P9,P18 S28,S31,S10,S1,S32,P35,P13,I1,S8,S19,S17,P12
3	.1092	P17,P24,P28,P5,P1,P32,P39,P44,P40 P51,S31,S36,S27,S9,S30,S39,P36,S4,S18,P6 S3,S13,S29,S26,S42,S12
4	.1732	P5,P1,P28,P50,P32,P23,P18,P39,P40,S12 P45,S36,S30,S27,P36,P9,S6,S33,S26,S5,S15,S3 S29,S14,P13,S21,S9,S37,S32,S35,S8,P12,I2,P57
5	.0347	P34,P8,P3,P43,P24,P30,P51,S4,P17 P41,S13,P35,P33,P32,P4,S28,S25,S34,P44,S31 S42,S49,S32,S39,I1
6	.0917	P33,P49,P41,P1,P5,P38,P37,P28 P35,S37,S7,S27,S16,S36,S30,S31,P18,P43,P23 S12,S3,P57,P9,S13,S42,I3
7	.1196	P37,P49,P30,P41,P32,P18,P23,S7,P2 P4,P45,S16,S37,S34,S25,P36,S28,P9,P57,S15 S33,S1,S21,S52,I1

Table 5 Fastest Paths vs Arc States

O-D Pair	Fastest Path	Arc States
1-10	1-3-7-10(1)	(1,3)(3)(.11),(2,3)(3)(.103),(3,8)(3)(.295), (4,8)(3)(.27), (5,8)(3)(.31)
	1-3-8-10(1)	(1,3)(1)(.13),(2,3)(1)(.13),(3,8)(1)(.37), (4,8)(1)(.34), (5,8)(1)(.37)
	1-3-8-10(1)	(4,8)(1)(.34),(5,8)(1)(.37),(3,8)(1)(.37)
	1-3-8-10(1)	(5,8)(1)(.37),(3,8)(1)(.37),(4,8)(1)(.34), (1,3)(1)(.13)
1-11	1-3-8-11(1)	Indifferent to arc states
2-10	2-3-7-10(1)	(3,8)(3)(.21),(4,8)(3)(.25),(4,9)(1)(.13), (5,8)(3)(.302)
	2-3-7-10(1)	(3,8)(3)(.21),(4,8)(3)(.25),(4,9)(3)(-.13), (5,8)(3)(.302)
	2-3-8-10(1)	(3,8)(1)(.19),(4,8)(1)(.19), (4,9)(3)(.11), (5,8)(1)(.22)
	2-3-8-10(1)	(3,8)(1)(.19),(4,8)(1)(.19), (4,9)(1)(-.13), (5,8)(1)(.22),(5,9)(3)(.23)
	2-4-8-10(1)	(1,4)(1)(.2107),(2,4)(1)(.1526)
	2-5-8-10(1)	(1,5)(1)(.28),(2,5)(1)(.25),(5,9)(1)(.19)
2-11	2-3-8-11(1)	(1,5)(3)(.203),(2,5)(3)(.174),(5,9)(3)(.23)
	2-3-8-11(1)	(1,5)(3)(.203),(5,9)(1)(-.23)
	2-4-8-11(1)	(1,4)(1)(.24),(2,4)(1)(.16)
	2-4-8-11(1)	(2,4)(1)(.16)
	2-5-8-11(1)	(5,9)(1)(.24),(1,5)(1)(.31),(2,5)(1)(.27)
	2-5-8-11(1)	(1,5)(1)(.31),(2,5)(1)(.27),(5,9)(1)(.24), (4,9)(1)(.13)

cluster was found to be greater than 0.56. A threshold value equal to 0.1 was set for testing the correlation between the traffic elements of each class. The correlation coefficient between the traffic elements i.e. input flow vs fastest path and arc state vs fastest path was checked against the threshold value for each class. Only those elements having correlation coefficient equal to or exceeding the threshold value were retained.

Table 5 presents the canonical correlation analysis results for the fastest paths obtained from the 7 clusters. It can be seen in Table 5 that the fastest paths are more correlated with arc states than the input flows. The reason being the weak correlation coefficient between the input flows and the paths of rank 1 which was found to be less than 0.1 in all the 7 clusters.

Thus, we can say that in this example the time required by a car to join an input node to an output node depends on the state of the arcs but not on the input flows. This does not mean that this condition always holds true because the impact of traffic flows on travel times also depends on the location of the input nodes inside the network.

We will use the information depicted in Table 5 for generating decision rules which will be used for making predictions about the fastest paths on the network. For instance, using Table 5 we can say that for O - D pair 1 - 10, if we observe that arcs (1,3), (2,3), (3,8), (4,8) and (5,8) have state ranking 3 i.e. their state lies between 0.9*maximum state and maximum state, then the fastest path to take between 1 and 10 is 1-3-7-10. It can also be seen in Table 5 that one path can be the fastest path under two different states of an arc. For instance, consider the path 2-3-8-10 for origin-destination pair 2-10. This path is the fastest path when arcs (3,8), (4,8) and (5,8) have rank 1 and arc (5,9) has rank 3. However for arc (4,9), we observe two ranks 1 and 3. It can be seen that positive correlation $(+.11)$ exists for rank 3 and negative correlation$(-.13)$ exists for rank 1. This means that the path 2-3-8-10 is chosen as the fastest path whenever rank 3 for arc(4,9) is observed and if rank 1 is observed then the path 2-3-8-10 should be avoided.

For the combination of other arc states not presented in Table 5, the correlations between the fastest paths and the arc states was found to be less than 0.1 indicating weak influence of other arc state ranks on the fastest paths of the network. Under these conditions we will consider that the shortest paths are also the fastest paths on the network.

Analysis of Table 5 leads us to the following rules for determining the fastest paths between various origin-destination pairs of the network given any combination of input flows and arc states. In the rules, the rank of arc (i,j) is denoted by $r(i,j)$ and the Boolean operators "equal to" and "not equal to" are denoted by "$=$" and "$\neq$" respectively. Each rule comprises of several combinations of arc states and/or input flows. These elements are arranged in the rules in the increasing order of their risk for deciding the fastest path.

For instance, consider the rule for origin node 2 and destination node 10.

IF $r(1,4) = 1$ **AND** $r(2,4) = 1$ **THEN** the shortest path is (2,4,8,10).

Using this rule we can say that if both the arcs (1,4) and (2,4) are of rank 1, then the path 2-4-8-10 is the fastest path between origin 2 and destination 10.

8.3 Results

The various rules used to make decisions about the fastest paths among the origins and the destinations of the network of Figure 4 are mentioned below. In this set of conditions, the two first conditions must apply. The greater the number of conditions that apply, the lesser is the risk attached to this decision.

ORIGIN 1 - DESTINATION 10

IF $\{r(5,8) = 3$ <u>AND</u> $r(3,8) = 3$ <u>AND</u> $r(4,8) = 3$ <u>AND</u> $r(1,3) = 3$ <u>AND</u> $r(2,3) = 3\}$ **THEN** the fastest path is (1,3,7,10).

IF $\{r(5,8) = 1$ <u>AND</u> $r(3,8) = 1$ <u>AND</u> $r(4,8) = 1$ <u>AND</u> $r(1,3) = 1$ <u>AND</u> $r(2,3) = 1\}$ **THEN** the fastest path is (1,3,8,10).

IF $\{r(5,8) = 1$ <u>AND</u> $r(3,8) = 1$ <u>AND</u> $r(4,8) = 1$ $\}$ **THEN** the fastest path is (1,3,8,10).

IF $\{r(5,8) = 1$ <u>AND</u> $r(3,8) = 1$ <u>AND</u> $r(4,8) = 1$ <u>AND</u> $r(1,3) = 1$ $\}$ **THEN** the fastest path is (1,3,8,10).

ORIGIN 1 - DESTINATION 11

Choose always the shortest path (1,3,8,11) as the fastest path.

ORIGIN 2 - DESTINATION 10

IF $\{r(5,8) = 3$ <u>AND</u> $r(4,8) = 3$ <u>AND</u> $r(3,8) = 3$ <u>AND</u> $r(4,9) \neq 3\}$ **THEN** the fastest path is (2,3,7,10).

IF $\{r(5,9) = 3$ <u>AND</u> $r(5,8) = 1$ <u>AND</u> $r(4,8) = 1$ <u>AND</u> $r(4,9) \neq 1$ <u>AND</u> $r(3,8) = 1$ $\}$ **THEN** the fastest path is (2,3,8,10).

IF $\{r(1,4) = 1$ <u>AND</u> $r(2,4) = 1\}$ **THEN** the fastest path is (2,4,8,10).

IF $\{r(1,5) = 1$ <u>AND</u> $r(2,5) = 1$ <u>AND</u> $r(5,9) = 1$ $\}$ **THEN** the fastest path is (2,5,8,10).

IF $\{$ $r(5,8) \neq 3$ <u>AND</u> $r(5,9) \neq 3$ <u>AND</u> $r(1,4) \neq 1$ <u>AND</u> $r(1,5) \neq 1$ $\}$ **THEN** the fastest path is (2,5,8,10).

ORIGIN 2 - DESTINATION 11

IF $\{r(5,9) \neq 1$ <u>AND</u> $r(1,5) = 3$ <u>AND</u> $r(2,5) = 3$ $\}$ **THEN** the fastest path is (2,3,8,11).

IF $\{r(1,4) = 1$ <u>AND</u> $r(2,4) = 1\}$ **THEN** the fastest path is (2,4,8,11).

IF $\{r(1,5) = 1$ <u>AND</u> $r(2,5) = 1$ <u>AND</u> $r(5,9) = 1$ <u>AND</u> $r(4,9) = 1$ $\}$ **THEN** the fastest path is (2,5,8,11).

IF $\{r(5,9) = 1$ <u>AND</u> $r(1,4) \neq 1$ <u>AND</u> $r(1,5) \neq 1$ $\}$ **THEN** choose (2,5,9,11) as the fastest path.

8.4 *Validation of the Results*

The results obtained from the data analysis were validated by comparison with simulation results. The input data (input flows, arc states) were generated at random for simulation. The value of input flow at node 1 was 70 and at node 2 was 60. The arc state of arc (1,3) was 50 (rank 2), arc (1,4) was 203 (rank 3), arc (1,5) was 80 (rank 1), arc (2,3) was 190 (rank 3), arc (2,4) was 152 (rank 3), arc (2,5) was 60 (rank 1), arc (2,6) was 174 (rank 3), arc (3,7) was 130 (rank 1), arc (3,8) was 170 (rank 3), arc (4,8) was 55 (rank 1), arc (4,9) was 121 (rank 3), arc (5,8) was 144 (rank 2), arc (5,9) was 135 (rank 2), arc (6,9) was 60 (rank 1), arc (7,10) was 112 (rank 3), arc (8,10) was 96 (rank 2), arc (8,11) was 50

Table 6 Simulation Results

Input		Output		Fastest path
1		10		1-3-8-10
1		11		1-3-8-11
2		10		2-5-8-10
2		11		2-5-8-11

Table 7 Results provided by the rules

Input Node	Output Node	Rule	Fastest path
1	10	**IF** {r(5,8) = 1} **THEN** the fastest path is (1,3,8,10).	1-3-8-10
1	11	Choose always the shortest path (1,3,8,11) as the fastest path.	1-3-8-11
2	10	**IF** {r(1,5) = 1 AND r(2,5) = 1 AND r(5,9) = 1 }**THEN** the fastest path is (2,5,8,10).	2-5-8-10
2	11	**IF** {r(1,5) = 1 AND r(2,5) = 1} AND r(5,9) = 1 AND r(4,9) = 1 } **THEN** the fastest path is (2,5,8,11).	2-5-8-11

(rank 1) and arc (9,11) was 121 (rank 3). The results of the simulation (fastest paths for each origin-destination pair) are presented in Table 6.

Table 7 presents the results of hybrid clustering i.e. rules for finding the fastest path between each origin-destination pair.

Let us now validate the data analysis results of Table 7 with the simulation results of Table 6. On verifying the input flow and arc state ranks used in the the rules of Table 7 with the input flow and arc state ranks presented above, we find that arc (5,8) has rank 2 when 1-3-8-10 is chosen as the fastest path between 1 and 10 whereas the rules yield 1-3-8-10 as the fastest path when arc (5,8) has rank 1. In this case, we can see that the simulation results do not match with decision rule results. Le us now consider the results for fastest path between origin 1 and destination 11. It can be seen in Table 6 that the fastest path between 1 and 11 is 1-3-8-11 which conforms with the table 12 results and remains uninfluenced by the arc's state ranks. The rules yield fastest paths between 2 and 10 and 2 and 11 as 2-5-8-10 and 2-5-8-11 using rank 1 of arcs (1,5) and (2,5), which also holds true for simulation results (Table 6) and the arc states. Therefore, we can say that the results obtained from the simulation experiment agree with the findings obtained from the hybrid clustering in 75% of the cases.

9 Conclusion

This paper presents a statistical approach for approximating fastest paths under stepwise constant input flows and initial states of the arcs on urban networks. Hybrid clustering and canonical correlation analysis have been used to find arc states and input flows that govern the fastest paths on the network. During the study it was found that once a car arrives at the entrance of the network then the input flows do not play a significant role and it is mainly the arc states that regulate the fastest paths on the network. However, the location of input nodes on the network also affects of the impact of traffic flow on travel times. There are certain arcs called critical arcs whose ranks decide the fastest paths on the network. Therefore, prediction of fastest paths would merely require testing the presence of these critical arc states with the ranks proposed by the rules. If agreement is found, then the next path to follow is the path extracted from the rule otherwise the shortest path is chosen as the fastest path.

Real networks are considerably huge in size and the present approach can be applied to networks of relatively small dimension. The next step of our study is to develop an algorithm for approximating fastest paths on real networks. This would be done by decomposing the real network into small sub-networks and computing the fastest path for the sub-networks using the statistical approach discussed in the paper. The fastest path between any origin destination pair of the real network will be obtained by joining the average fastest paths of the sub-networks.

References

1. Ambroise, C., Badran, F., Thiria, S., Séze, G.: Hierarchical Clustering of Self-Organizing Maps for Cloud Classification. Neurocomputing 30(1-4), 47–52 (2000)
2. Awasthi, A., Chauhan, S.S., Parent, M., Proth, J.M.: Partitioning algorithms for large scale urban networks (under submission) (2008)
3. CISIA, SPAD Reference Manuals, Centre International de Statistique et d'Informatique Appliquées, France (1997)
4. Cutting, D., Karger, D., Pedersen, J., Tukey, J.W.: Scatter/Gather: a cluster based approach to browsing large document collections. In: Proc. 15th Annual Int. ACM SIGIR Conf., Copenhagen (1992)
5. Diday, E., Simon, J.C.: Clustering Analysis. In: Fu, K.S. (ed.) Digital Pattern Recognition, pp. 47–94. Springer, Heidelberg (1976)
6. Diday, E., Govaert, G., Lechevallier, Y., Sidi, J.: Clustering in pattern recognition. In: 4th International Joint Conference on Pattern Recognition, Kyoto, Japan (1978)
7. Diday, E., Lemaire, J., Pouget, P., Testu, F.: Elements d'analyse des donnes, Dunod, Paris (1983)
8. Dougherty, M.S.: A review of neural networks applied to transport. Transportation Research 3(4), 247–260 (1995)
9. Everitt, B.S.: Cluster Analysis. Heinemann Educational Books (1974)
10. Hansen, P., Jaumard, B.: Cluster analysis and mathematical programming. Mathematical Programming 79, 191–215 (1997)
11. Jain, A.K., Murty, M.N., Flynn, P.J.: Data Clustering: A Review. ACM Computing Surveys 31(3), 264–323 (1999)
12. Kisgyorgy, L., Rillett, L.R.: Travel Time prediction by advanced neural network. Periodica Polytechnica Ser Civil Engineering 46(1), 15–32 (2002)
13. Kohonen, T.: Self organization and associative memory, 2nd edn. Springer, Berlin (1988)

14. Klein, G., Aronson, J.E.: Optimal Clustering: a model and method. Naval research Logistics 38, 447–461 (1991)
15. Legendre, P.: Program K-means (2001),
    http://www.fas.umontreal.ca/biol/legendre/
16. Marcotorchino, J.F., Proth, J.M., Janesen, J.: Data analysis in real life environments. North Holland, Amsterdam (1985)
17. Meila, M., Heckerman, D.: An experimental comparison of several clustering and initialization methods. In: Proc. Uncertainty in Artificial Intelligence, pp. 386–395 (1998)
18. Murtagh, F.: Interpreting the Kohonen self-organizing feature map using contiguity-constrained clustering. Pattern Recognition Letters 16, 399–408 (1995)
19. Park, D., Rillett, L.R.: Forecasting multiple period freeway link travel times using modular neural networks, Transportation Research Record, 1617, National Research Council., Washington D.C (1998)
20. Smith, R., Chou, J., Romeijn, E.: Approximating Shortest Paths in Large Scale Networks with Application to Intelligent Transportation Systems. INFORMS Journal on Computing 10(2), 163–179 (1998)
21. Tomassone, R., Danzart, M., Daudin, J.J., Masson, J.P.: Discrimination et classement, Masson, Paris (1988)
22. Roux, M.: Algorithmes de classification, Masson, Paris (1986)
23. Ward, J.H.: Hierarchical grouping to optimize an objective function. Journal of the American Statistical Association 58, 236–244 (1963)
24. Zhan Benjamin, F.: Three Fastest Shortest Path Algorithms on Real Road Networks: Data Structures and Procedures. Journal of Geographic Information and Decision Analysis 1(1), 70–82 (1997)

Linear Models for Visual Data Mining in Medical Images

Alexei Manso Corrêa Machado

Summary. Modern non-invasive imaging modalities such as Computer-Aided Tomography and Magnetic Resonance Imaging brought a new perspective to anatomic characterization, as they allowed detailed observation of in vivo structures. The amount of provided data became, however, progressively overwhelming. Data may be useless if we are not able to extract relevant information from them and to discard undesirable artifacts and redundancy. Ultimately, we aim to discover new knowledge from collected data, and to be able to statistically represent this knowledge in the form of prior distributions that may be used to validate new hypotheses, in addition to clinical and demographic information. In this chapter we propose an analysis of available methods for data mining in very high-dimensional sets of data obtained from medical imaging modalities. When applied to imaging studies, data reduction methods may be able to minimize data redundancy and reveal subtle or hidden patterns. Our analysis is concentrated on linear transformation models based on unsupervised learning that explores the relationships among morphologic variables, in order to find clusters with strong correlation. This clustering can potentially identify regions that have anatomic significance and thus lend insight to knowledge discovery and hypothesis testing. We illustrate this chapter with successful case studies related to the segmentation of neuroanatomic structures and the characterization of pathologies.

1 Introduction

An important problem related to medical image analysis is the efficient representation of the large amount of data provided by imaging modalities and

A.M. Corrêa Machado
Graduate Program on Electrical Engineering, Pontifical Catholic University of
Minas Gerais, Av. Dom Jose Gaspar, 500, Belo Horizonte, MG, Brazil, 30535-610
e-mail: alexei@pucminas.br

A. Abraham et al. (Eds.): Foundations of Comput. Intel. Vol. 6, SCI 206, pp. 321–344.
springerlink.com   © Springer-Verlag Berlin Heidelberg 2009

the extraction of relevant shape-related knowledge from the dataset. The data should not only be represented in a manageable way, but also facilitate hypothesis-driven explorations of regional shape differences and lend deeper insight to morphometric investigation. This chapter discusses the use of linear Gaussian models for data mining and knowledge discovery, based on unsupervised learning, that explores the correlation among morphometric variables and the possible anatomic significance of these relationships.

The traditional approach for medical diagnosis through images can be roughly defined as a comparative method, in which the physician visually identifies the anatomical structures in the patient's image. This process is quite robust, as the specialist makes use of his knowledge on human anatomy and his intrinsic visual recognition ability. This approach, however, can be extremely difficult if the three-dimensional shape of structures, which itself is reconstructed from a set of tomographic slices, is fundamental to the diagnostic assessment. It seems improbable that a human specialist may discover new knowledge regarding the relationship among thousands or millions of variables without the aid of computational methods.

In 1981, Bajcsy and Broit [1] presented the elastic matching method, a revolutionary approach to the modeling of anatomical structures. Based on continuum mechanics, the elastic matching method models a three-dimensional image as an elastic body, which can be deformed to match the shape of another image. When applied to the field of medical imaging recognition, this technique allowed for the representation of *a priori* knowledge on anatomic structures by constructing a deformable model that embedded the characteristics of a normal population. Then, given a set of images from a patient, it was possible to consider them as a deformable volume and verify the degree of deformation required to match a reference image to the patient's model. The obtained measurements are fundamental in assisting precocious diagnosis of several diseases that manifest themselves through morphological variation in the anatomy of structures.

The analysis of large datasets is a difficult task even for experienced researchers. Imaging modalities may provide too much data for manual pointwise registration, what motivates the development of automatic methods that implement machine learning algorithms. The result of registration may, nevertheless, increase even more the amount of data to be analyzed. In addition to the morphometric variables, clinical and demographic information can be considered in the analysis so as to allow for the investigation of the relationship between regions in the image and pathologies or features of special interest. In this scenario, even the computation of the covariance matrix itself becomes intractable.

The process of image registration and subsequent analysis is exemplified in Figure 1. In the acquisition step, the original images are obtained by a sensor (e.g. MRI, CAT or PET scanner) and stored in digital format as an array of two-dimensional slices, that together constitute the three-dimensional representation of the image volume. One of the images is used as a reference that

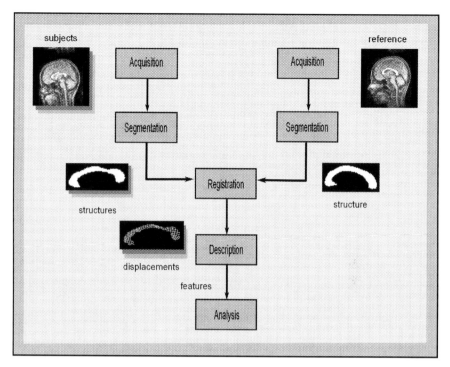

Fig. 1 The process of image registration and analysis. The subjects of the study have their images acquired by a sensor and the structure of interest is segmented. One of the images is taken as a reference and registered to all subjects in the sample. The result of registration is a displacement field that allows for detailed description of shape and volumetric variation. The extracted features are the input to exploratory analysis

will be deformed (or warped) to match each of the subjects in the study. If a specific structure is of interest, it may be segmented before registration to aid the warping process. Local registration step is responsible for the non-rigid deformation of the reference image onto the subject's image. The output of the local registration step is a set of displacement fields, one for each subject, that represents the point-wise deformation that was needed to perform registration.

Once the displacement fields for all subjects are determined, they can be used to describe the shape and volumetric variation embedded in the sample. However, the high dimensionality of these descriptions may require special representation in a less redundant variable space. The purpose of medical imaging is to extract information that will increase the knowledge about pathology and normal anatomical variability. The representation of the dataset in a variable space with lower dimension enables an efficient automatic analysis that will search for hidden associations among the original variables.

2   Image Registration and Description

Image registration aims to determine a correspondence between each pixel $\mathbf{q}$ in the reference image I_R to a pixel $\mathbf{p}$ in the subject or test image I_T. The problem of matching can be stated as finding two functions h and g such that

$$I_R(\mathbf{q}) = g(I_T(h(\mathbf{q}))). \tag{1}$$

The intensity transformation function g establishes a correspondence between the two image spectra. The spatial transformation function h maps corresponding pixels between the two images:

$$h(\mathbf{q}) = \mathbf{q} + \mathbf{u}(\mathbf{q}) = \mathbf{p}, \tag{2}$$

where $\mathbf{u}$ is a displacement field.

Image registration may be performed by first applying rigid transformations (translation and rotation), in order to approximately register corresponding features, and then warping the template to match the subject. Image volumes may be described as continuous media to which a constitutive model will be prescribed. The linear elasticity model [12], in which the image is deformed as an elastic body, guarantees smoothness to the deformation, so that neighboring structures in the reference image will be matched to neighboring structures in the subject's image, preserving the gross anatomy common to the majority of individuals in the population. Other models of spatial transformation include modal matching [28], active shape models [4, 9], fluid mechanics [3] and active contour models [6].

When the reference image is warped to match a subject image, some regions may get enlarged and some may be reduced. It is possible to determine the amount of scaling applied to an infinitesimal area around each point of the reference image, by computing the Jacobian determinant of the spatial transformation. In the case of two-dimensional images, the displacement vector field from the reference to subject i, $\mathbf{u}_i$, can be decomposed into its components u_i and v_i. Similarly, $\mathbf{q}$ can be expressed in terms of its coordinates (x, y). The Jacobian determinant $J_i(\mathbf{q})$ is defined as the determinant of the gradient of the mapping function $\mathbf{q} + \mathbf{u}_i(\mathbf{q})$:

$$J_i(\mathbf{q}) = |\nabla(\mathbf{q} + \mathbf{u}_i(\mathbf{q}))| = \begin{vmatrix} \frac{\partial u_i(\mathbf{q})}{\partial x} + 1 & \frac{\partial u_i(\mathbf{q})}{\partial y} \\ \frac{\partial v_i(\mathbf{q})}{\partial x} & \frac{\partial v_i(\mathbf{q})}{\partial y} + 1 \end{vmatrix}. \tag{3}$$

Since the result of image registration is a smooth displacement field, it is expected that the Jacobian determinants of a point be correlated to the determinants of its neighbors. Other features may be computed from the results of registration, including the pointwise curvature of the boundary of a structure or of its medial axis, the gross curvature of a structure (bending angle) or even the partial derivatives of the displacement fields, that yield a rotation-invariant description of shape.

3   Linear Gaussian Models

The purpose of linear models is to explore the redundancy of the data in the original variable space, so as to determine a new set of variables of lower cardinality. Although data reduction may be considered as the main objective of these transformations, it is also desirable that the new variables favor data interpretation. In this context, these models may become powerful data mining techniques, revealing patterns that were hidden in high-dimensional variable spaces. The most well known linear model is the Principal Component Analysis (PCA), in which the set of original variables is rotated in order to find the orthogonal axes that capture most of the variance embedded in the dataset. The subjects represented in the new basis are obtained by multiplying the original variables by an orthogonal matrix. Each new variable or *component* is therefore a linear combination of the original variables. It can be shown that the matrix that rotates the dataset, as to line up with the principal modes of the data variance, is composed of the transposed normalized eigenvectors of the covariance matrix.

The efficiency of PCA to summarize the variability in a data set is counterposed by its inadequacy to represent biological hypotheses. The fact that principal components are orthogonal linear combinations of variables may preclude the inference of relevant morphological information. A more useful method, although less explored, is Factor Analysis (FA) that aims at representing the covariance or correlation among the variables.

3.1   Factor Analysis

In factor analysis, a set of N subjects, each one denoted as a column vector of p variables, $\mathbf{y} = [y_1, \ldots, y_p]^T$ is represented as linear combinations of m hypothetical constructs called *factors*:

$$\mathbf{y} = \mathbf{A}\mathbf{f} + \boldsymbol{\epsilon}, \tag{4}$$

where $\mathbf{f} = [f_1, \ldots, f_m]^T$ is the vector of *common factors*, $\boldsymbol{\epsilon} = [\epsilon_1, \ldots, \epsilon_p]^T$ are the *unique factors* or residual terms which account for the portion of $\mathbf{y}$ that is not common to other variables, and $\mathbf{A} = [[a_{11}, \ldots, a_{1m}]^T, \ldots, [a_{p1}, \ldots, a_{pm}]^T]^T$ is the *loading matrix*. The coefficients a_{ij}, called *loadings*, express the correlation between variable y_i and factor f_j. Variables are standardized, so that their expected values and standard deviations in the dataset are respectively $\mathbf{0}$ and $\mathbf{1}$. Therefore, the sample covariance and correlation matrices are identical in this context. The factor analytic model assumes that common and unique factors are not correlated and have null expected values. The covariances among unique factors are represented by the diagonal matrix $\boldsymbol{\Psi} = diag(\psi_1, \ldots, \psi_p)$ and the covariance matrix for the common factor is the identity matrix. It should also be noticed that FA, as well as PCA, can be completely modeled from the information represented in the covariance matrix. In other words, FA is implemented in the context of the classical assumption of Gaussianity. Determining if the data fit a multivariate Gaussian distribution is an additional aspect of the problem, since the population parameters should be estimated from the sample. In order to justify the choice of a specific model, a test of Gaussianity should be made (e.g. forth-order cumulants [5, 21]), so that the results can be considered reliable.

Considering the assumptions in the factor analytic model, the variance σ_i^2 of a given variable y_i can be decomposed into components related to the m common factors, $a_{i1}^2 + \ldots + a_{im}^2$, called the *communality*, and a *specific variance* ψ_i:

$$\sigma_i^2 = \sum_{j=1}^{m} a_{ij}^2 + \psi_i. \tag{5}$$

Therefore, the population covariance matrix $\mathbf{\Sigma}$ can be represented in the new variable space. Since $\mathbf{Af}$ and ϵ are not correlated, the covariance matrix of their sum is the sum of the covariance matrix of each term. Also, since $cov(\mathbf{f}) = \mathbf{I}$, the relationship between $\mathbf{\Sigma}$, $\mathbf{A}$ and $\mathbf{\Psi}$ can be written as

$$\begin{aligned} \mathbf{\Sigma} &= cov(\mathbf{Af} + \epsilon) \\ &= cov(\mathbf{Af}) + cov(\epsilon) \\ &= \mathbf{A}cov(\mathbf{f})\mathbf{A}^T + \mathbf{\Psi} \\ &= \mathbf{AA}^T + \mathbf{\Psi}. \end{aligned} \tag{6}$$

3.2 *Rotation of Factor Loadings*

An important property of the loading matrix $\mathbf{A}$ is that it can be rotated and still be able to represent the covariance among factors and original variables. Since any orthogonal matrix $\mathbf{Q}$ multiplied by its transpose leads to the identity matrix, the basic model for factor analysis in (4) can be written as

$$\begin{aligned} \mathbf{y} &= \mathbf{AQQ}^T\mathbf{f} + \epsilon \\ &= \mathbf{A}^*\mathbf{f}^* + \epsilon, \end{aligned} \tag{7}$$

where $\mathbf{A}^* = \mathbf{AQ}$ and $\mathbf{f}^* = \mathbf{Q}^T\mathbf{f}$. If $\mathbf{\Sigma}$ in (6) is expressed in terms of $\mathbf{A}^*$, we have

$$\begin{aligned} \mathbf{\Sigma} &= \mathbf{A}^*\mathbf{A}^{*T} + \mathbf{\Psi} \\ &= \mathbf{AQ}(\mathbf{AQ})^T + \mathbf{\Psi} \\ &= \mathbf{AQQ}^T\mathbf{A}^T + \mathbf{\Psi} \\ &= \mathbf{AA}^T + \mathbf{\Psi}, \end{aligned} \tag{8}$$

showing that the original loading matrix $\mathbf{A}$ and the rotated matrix $\mathbf{A}^*$ yield the same representation of $\mathbf{\Sigma}$.

The rotation of loadings plays an important role in factor interpretation, as it is possible to obtain a matrix that assigns few high loading for each variable, keeping the other loadings small. The *quartimax* algorithm [26] is an orthogonal rotation method that maximizes the variance of the squared loadings in each column of the loading matrix, so that each variable presents high loading for fewer factors. Rotation may reveal hidden patterns and relationships in the data. A cluster of variables that are strongly correlated is related to a factor. In medical imaging studies, it is possible to plot these clusters and observe the main trends of shape and volumetric variation. It is also possible to determine how other variables, such as clinical and demographic data, correlate to shape variation, by looking at their loadings on each factor.

3.3 Evaluation of the Factor Analytic Model

Oftentimes, the choice of the number of factors to be considered in the analysis is not straightforward and is subject to controversy. In PCA, data reduction is obtained after the computation of the eigenvalues, by eliminating the components that do not contribute significantly to the representation of variance. The number of components to be kept can be determined based on the eigenvalues of the spectral decomposition. Each eigenvalue represents the amount of variance explained by the corresponding component. In FA, the number of factors to be considered in the analysis must also be previously determined.

The influence of the choice of the number of factors in the results can be evaluated based on the quantitative analysis of the statistical fit of the model, by computing two parameters:

1. The completeness of factor analysis

   If the original variables are naturally correlated so that clusters of variables related to independent factors can be found, the specific factor terms ϵ should be minimum. Since the variables are standardized, the variance for each variable y_i is 1. From (5) we have that

   $$\sum_{j=1}^{m} a_{ij}^2 = 1 - \psi_i \tag{9}$$

   for each variable y_i, meaning that the communality tends to unity, when the variance can be completely explained by the factors and consequently there is no significant specific variance ψ_i. The *completeness* of the FA can thus be defined as the average of all communalities:

   $$completeness = \frac{1}{p} \sum_{i=1}^{p} \sum_{j=1}^{m} \hat{a}_{ij}^2. \tag{10}$$

2. Analysis of the communality for specific features of interest

   For exploratory purpose, FA can be applied to a heterogeneous variable set composed of visual data and other measurements or features of

interest. The loadings related to these specific features of interest can be
separately investigated. If the number of factors is not properly chosen,
the specific variance may be too large, meaning that the model is not able
to correlate the feature of interest to the morphometric variables. Large
values for the specific variance may alternatively reveal that the feature is
actually not correlated to the visual data. The communality for feature i is
defined as

$$h_i^2 = \sum_{j=1}^{m} a_{ij}^2, \tag{11}$$

where i is the index of the feature of interest in the variable vector.

3.4   Estimation of Loadings

Many techniques have been proposed to determine $\mathbf{A}$. The simplest one,
called *principal factor method*, neglects $\mathbf{\Psi}$ and uses spectral decomposition
to represent the covariance matrix $\mathbf{S}$:

$$\begin{aligned}
\mathbf{S} &= cov(\mathbf{y}) \\
&= cov(\mathbf{Af} + \boldsymbol{\epsilon}) \\
&= \mathbf{A}cov(\mathbf{f})\mathbf{A}^T + \mathbf{\Psi} \\
&\approx \mathbf{A}\mathbf{I}\mathbf{A}^T \\
&= \mathbf{Q}\mathbf{\Lambda}\mathbf{Q}^T \\
&= (\mathbf{Q}\mathbf{\Lambda}^{1/2})(\mathbf{Q}\mathbf{\Lambda}^{1/2})^T,
\end{aligned} \tag{12}$$

where $\mathbf{\Lambda}^{1/2} = diag(\lambda_1^{1/2}, \ldots, \lambda_m^{1/2})$ is the diagonal matrix with the square root
of the eigenvalues of $\mathbf{S}$, and $\mathbf{Q} = [\mathbf{q}_1 \ldots \mathbf{q}_m]$ is the matrix of the corresponding
eigenvectors. Therefore, the loading matrix can be estimated based on the
sample covariance matrix as

$$\mathbf{A} \approx \mathbf{Q}\mathbf{\Lambda}^{1/2}. \tag{13}$$

The method described above for estimating the loading matrix $\mathbf{A}$ is very
similar to Principal Component Analysis and is known as *Principal Component Method* [33]. By neglecting the specific variance matrix, the factor
analysis of the covariance matrix is performed by placing communalities in
the diagonal elements. In this case, the recovered covariance matrix $\mathbf{A}\mathbf{A}^T$
have its off-diagonal elements affected.

A crucial issue to factor analysis is the definition of the number of factors to
be considered in the analysis. In this section we present an iterative algorithm
that determines the number of factors based exclusively on the characteristics of the data set. The first step in the process is the computation of the
eigenvalues $\mathbf{\Lambda}$ and eigenvectors $\mathbf{Q}$ of the correlation matrix $\mathbf{S}$. Fortunately, it

is possible to compute $\mathbf{Q}$ and $\mathbf{\Lambda}$ without having to compute $\mathbf{S}$, otherwise it would be intractable to deal with high-dimensional variable spaces. Singular value theory states that any $p \times N$ matrix, such as the sample matrix $\mathbf{Y}$, can be decomposed as:

$$\mathbf{Y} = \mathbf{QDW}^T,$$

where $\mathbf{Q}$ is a $p \times p$ matrix whose columns are the eigenvectors of $\mathbf{YY}^T/N$, $\mathbf{W}$ is a $N \times N$ matrix whose columns are the eigenvectors of $\mathbf{Y}^T\mathbf{Y}/N$ and $\mathbf{D} = \mathbf{\Lambda}^{1/2}$ is a $p \times N$ matrix whose values in the diagonal are the square roots of the eigenvalues of both $\mathbf{YY}^T/N$ and $\mathbf{Y}^T\mathbf{Y}/N$. Since both $\mathbf{Q}$ and $\mathbf{W}$ are orthogonal, we have that

$$
\begin{aligned}
\mathbf{Q\Lambda}^{1/2}\mathbf{W}^T &= \mathbf{Y} \\
\mathbf{Q\Lambda}^{1/2}\mathbf{W}^T\mathbf{W} &= \mathbf{YW} \\
\mathbf{Q\Lambda}^{1/2}\mathbf{\Lambda}^{-1/2} &= \mathbf{YW\Lambda}^{-1/2} \\
\mathbf{Q} &= \mathbf{YW\Lambda}^{-1/2}.
\end{aligned}
\tag{14}
$$

From (13) and (14), the estimate of the loading matrix is

$$\mathbf{A} \approx \mathbf{Q\Lambda}^{1/2} = \mathbf{YW}. \tag{15}$$

The initial number of factors (number of columns in $\mathbf{A}$) can be determined as the number of eigenvalues greater than 1, since they account for the variation of at least one variable. The computed loading matrix is then rotated so that each variable will present high loading for few factors. This can be achieved by finding a sequence of rotations that maximizes the variance of the squared loadings in each column of $\mathbf{A}$ (*quartimax* algorithm) [14]. The resulting loading values represent the correlation between variables and factors. A variable i is associated with a given factor j if it presents high absolute value for a_{ij}. The purpose of factor analysis is to represent the correlation among original variables by clustering them into factors. Based on this rationale, a factor that is not correlated to at least 2 variables may not be considered informative. The following algorithm reduces the initial number of factors by discarding factors that do not have high correlation to at least two variables. Since the absolute value for correlation goes from 0 to 1, we consider informative factors to have loadings with absolute value of at least 0.5. This threshold may vary according to specific conditions, such as the sample size, and the desired level of significance to reject the hypothesis of no correlation among variables and factors. At each iteration, $\mathbf{A}$ is computed based on the m_t columns of $\mathbf{Q}$, denoted as $\mathbf{Q}_t$, and the corresponding eigenvalues. Convergence is achieved when the number of factor m_t at iteration t equals the number of factors m_{t-1} computed in the previous iteration. The following algorithm summarizes the method:

330 A.M.C. Machado

Begin

    Decompose $\mathbf{Y}^T\mathbf{Y}/N$ *into its eigenvectors* $\mathbf{W}$ *and eigenvalues* $\mathbf{\Lambda}$

    Compute $\mathbf{Q} \leftarrow \mathbf{YW\Lambda}^{-1/2}$

    Set m_t *to the number of eigenvalues greater than 1*

    Repeat until $m_t = m_{t-1}$

        $m_{t-1} \leftarrow m_t$

        Estimate loadings as $\mathbf{A} \leftarrow \mathbf{Q}_t\mathbf{\Lambda}_t^{1/2}$

        Rotate loadings based on quartimax *algorithm*

        $m_t \leftarrow 0$

        For $j \leftarrow 1$ *to* m_{t-1} *do*

            $nvar \leftarrow 0$

            For $i \leftarrow 1$ *to* p *do*

                If $a_{ij} \geq 0.5$ *then* $nvar \leftarrow nvar + 1$

                If $nvar > 1$ *then* $m_t \leftarrow m_t + 1$

End

Fig. 2 Topology of the corpus callosum, adapted from Witelson [35]. The regions of the callosal structures are the rostrum(1), genu(2), rostral body(3), anterior midbody(4), posterior midbody(5), isthmus(6) and splenium(7)

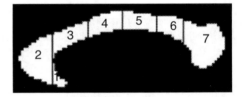

4 Examples

In this section, we present 3 case studies that illustrate the ability of factor analysis to act as a visual data mining tool. The examples involve different number of variables, from a dozen to few millions.

4.1 A Case Study on the 22q11.2 Chromosome Deletion Syndrome

The Chromosome 22q11.2 deletion syndrome (DS22q11.2) results from a 1.5 - 3Mb microdeletion on the long arm of chromosome 22 and encompasses a wide range of medical manifestations that include cardiac, palatal and immune disorders, as well as impairments in visuospatial and numerical cognition. In this example, consider the small dataset displayed at Table 1 that was part of the sample used in a study on the morphology of the corpus callosum and its relationship to cognitive skills in the DS22q11.2 [23]. The corpus callosum is a structure composed of axons that connects the two hemispheres

Table 1 Sample Data

CL	IQ	GM	VS	BA	RT	GE	RB	AM	PM	IS	SP
0	101	839.5	3.7	104.8	19.4	157.3	76.5	84.2	75.4	59.3	171.6
1	81	738.3	10.5	94.2	46.5	135.0	109.2	75.9	71.2	70.4	211.0
0	96	759.1	4.5	109.8	19.3	130.7	105.8	80.6	68.7	49.2	158.1
1	54	632.4	40.5	92.4	25.3	89.7	119.8	48.6	48.7	33.2	105.1
0	121	813.2	8.1	117.9	18.2	163.1	100.1	82.9	69.0	63.2	184.7
1	64	623.6	9.6	95.1	48.2	108.1	86.1	74.9	71.0	46.2	148.7
0	87	792.1	10.3	99.1	24.1	131.8	90.8	72.6	69.6	64.0	199.8
1	76	705.1	10.4	84.4	68.2	181.1	146.1	90.9	86.0	80.2	188.7
0	105	710.6	4.6	119.6	2.3	152.7	52.8	63.6	66.7	47.2	144.1
1	82	726.3	11.7	87.3	43.3	151.7	145.8	94.6	96.7	92.2	248.1

of the brain. The dataset is composed of 10 subjects, half of them drawn from a population of children with the syndrome and the remaining composed of typically developing controls. The variable space is composed of twelve gross morphometric measurements and clinical information obtained respectively from image registration procedures and clinical/psychometric tests: (a) class of diagnosis (CL), that is assigned a value of 0 for controls and 1 for children with DS22q11.2; (b) general intelligence quotient (IQ); (c) volume of gray matter (GM) in cubic centimeters; (d) ventricular size (VS) in cubic centimeters; (e) bending angle (BE) of the corpus callosum in degrees; (f) area of the rostrum (RS); (g) area of the genu (GE); (h) area of the rostral body (RB); (i) area of the anterior midbody (AM); (j) area of the posterior midbody (PM); (k) area of the isthmus (IS); and (l) area of the splenium. Variables (f) to (l) are given in square millimeters and are related to the topology of the corpus callosum proposed by Witelson [35] (see Fig. 2). All variables, except by CL, were corrected for age covariate.

The dataset was standardized and its correlation matrix, **S**, computed (Table 2). The eigendecomposition of **S** revealed 2 eigenvectors along which 86% of the total variance were represented. The 12 eigenvalues and the eigenvectors related to the 2 most significant modes of variation are displayed in Table 3. For this example, if the objective were only to reduce the dimensionality of the variable space, a PCA with 2 components would be sufficient to represent most of the variance in the dataset. The coordinates of the eigenvectors are nevertheless unable to provide further information about the relationship among the original variables.

Factor analysis was applied to the dataset. The initial number of factors considered in the analysis was 2, based on the number of eigenvalues greater than 1. The quartimax algorithm took 3 iterations to converge. The rotated loadings with corresponding communalities are displayed in Table 4. The completeness of the analysis was 0.81, showing a good statistical fit of the model with 2 factors. Five variables presented high absolute loadings with

Table 2 Correlation Matrix

	CL	IQ	GM	VS	BA	RT	GE	RB	AM	PM	IS	SP
CL	1.00											
IQ	-0.81	1.00										
GM	-0.73	0.80	1.00									
VS	0.51	-0.69	-0.57	1.00								
BA	-0.84	0.80	0.45	-0.42	1.00							
RT	0.80	-0.56	-0.38	0.09	-0.85	1.00						
GE	-0.27	0.64	0.58	-0.64	0.15	0.13	1.00					
RB	0.65	-0.43	-0.21	0.37	-0.75	0.72	0.12	1.00				
AM	0.01	0.36	0.46	-0.65	-0.20	0.45	0.73	0.41	1.00			
PM	0.20	0.15	0.24	-0.53	-0.38	0.52	0.67	0.45	0.91	1.00		
IS	0.24	0.17	0.35	-0.37	-0.43	0.54	0.68	0.58	0.83	0.90	1.00	
SP	0.11	0.25	0.44	-0.45	-0.30	0.39	0.55	0.43	0.78	0.83	0.94	1.00
	CL	IQ	GM	VS	BA	RT	GE	RB	AM	PM	IS	SP

Table 3 Eigenvalues and Eigenvectors

eigenvalues	5.31	4.82	0.70	0.44	0.28	0.21	0.16	0.07	0.02	0	0	0
eigenvector 1	0.04	0.13	0.19	-0.24	-0.13	0.22	0.34	0.21	0.41	0.41	0.42	0.39
eigenvector 2	-0.43	0.42	0.34	-0.29	0.41	-0.35	0.18	-0.32	0.04	-0.05	-0.07	-0.01

Table 4 Loadings and Communalities

variable	CL	IQ	GM	VS	BA	RT	GE	RB	AM	PM	IS	SP
factor 1	0.04	0.36	0.49	-0.60	-0.25	0.46	0.80	0.44	0.95	0.93	0.95	0.90
factor 2	-0.95	0.90	0.71	-0.61	0.93	-0.80	0.35	-0.73	0.03	-0.16	-0.20	-0.08
commun.	0.91	0.93	0.74	0.73	0.92	0.85	0.76	0.73	0.90	0.90	0.94	0.81

factor 1: GE, AM, PM, IS and SP. This factor seems to be related to the morphology of regions of the corpus callosum that are not associated with the deletion syndrome. Factor 2 was highly correlated to 7 variables, including diagnosis: CL, IQ, GM, VS, BA, RS and RB. This factor seems to capture the features that are associated with the syndrome, as already shown in other studies [29, 23]. The children with the DS22q11.2 have on average lower IQ, proportionally less gray matter and lager ventricles, a more arched corpus callosum with larger rostrum and rostral body. The larger anterior callosum is a recent finding that is under investigation. It seems to be related to alternative paths in the communication between left and right hemispheres, that is also associated with the syndrome. The results, in this case, are in accordance with other published studies based on classical hypothesis testing, showing the ability of factor analysis to act as a data mining tool.

4.2   A Case Study on the Morphology of the Corpus Callosum

In this example we investigate the exploratory aspect of FA, in a study on the substructural characterization of the corpus callosum. The MRI images used in the experiments, gently shared by the Mental Health Clinical Research Center of the University of Pennsylvania, are subjects recruited for a larger study on schizophrenia, composed of 42 male and 42 female right-handed controls. The age of the subjects is in the range of 19 to 68 years (mean±S.D.,30.4±11.8) for males and 18 to 68 years (26.5±9.0) for the females. The sample was chosen in order to provide an approximated distribution of age and race for both groups and to guarantee a minimum influence of these features in the analysis.

For each MRI volume, the middlemost sagittal slice was extracted and the corpus callosum structure segmented by manual delineation. Image registration was performed by warping one of the images, chosen as a template, to match each subject in the sample. The displacement fields obtained from registration were the basis for the computation of the Jacobian determinants at each of the 851 pixels in the callosal template. Since FA assumes that the variables fit Gaussian distributions, a test of normality was performed by estimating the skewness and kurtosis of the distribution for the populations. With a level of significance of 0.01, there were no evidences to reject the hypothesis of normality for 83.6 % of the variables, based on the skewness, and for 89.0% of the variables, based on the kurtosis of the distribution.

After determining the first estimate for the loading matrix, it was rotated in order to maximize the variance of the squared loadings in each column, so that each variable presented high loading for fewer factors. The algorithm used to determine the number of factors, described in Section 3.4, took 9 iterations to converge from 78 to 11 factors with correlation magnitude greater than 0.5 among at least 2 variables. With a level of significance of 0.01, a correlation coefficient magnitude of 0.5 computed for the sample gives an estimation that the population correlation coefficient, ρ, is in the confidence interval of $0.257 < \rho < 0.683$. The value of 0.5 is also sufficient to reject the hypothesis that $\rho = 0$ with level of significance $\alpha < 0.001$. Fig. 3 shows the factors that are correlated to the largest number of pixels in the callosal structure, after the algorithm's convergence. For each factor, the regions in the callosal structure that have loading values greater than 0.5 are shown in white and the regions that have loadings smaller than -0.5 are shown in light gray.

The results of the exploratory FA presented in this example are in accordance with the topology proposed by Witelson [35], regarding the subdivision of the callosal structure. Comparing Fig. 2 to the factors depicted in Fig. 3, it is possible to relate the rostrum with factor (a), the genu with factors (b) and (c), the rostral body with factor (d), the anterior midbody with factor (e), the posterior midbody with factor (f), the isthmus with factors number (g) and (h) and the splenium with factors (i) and (j). The results show that

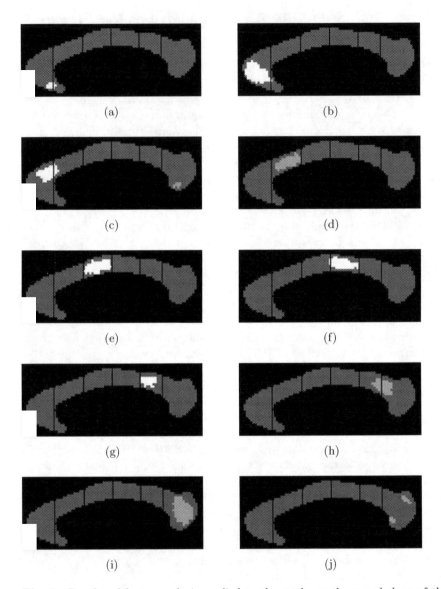

Fig. 3 Results of factor analysis applied to the study on the morphology of the corpus callosum. The 10 most significant factors are displayed, where the pixels with loading values greater than 0.5 are shown in white and the pixels with loadings smaller than -0.5 are shown in light gray. The topology proposed by Witelson is superimposed to the images

Fig. 4 Results of principal component analysis applied to the study on the morphology of the corpus callosum. The modes of variance are ordered left to right, top to bottom, according to the amount of variance they represent. Each variable is assigned to the mode for which it presents the greatest coefficient. Variables with positive coefficients are shown in white and the ones with negative values are shown in light gray. The topology proposed by Witelson is superimposed to the images

Table 5 Summary of the factor analysis results in the AD dataset. For each factor f_i, the table shows, from left to right, the number of voxels for which the factor presents the highest loading values, the loading of the factor to diagnosis and a description of the major regions associated to the factor, within the brain

f	volume	loading	description
1	1077759	0.32	overall cortical region at cerebrum and cerebellum
2	149742	-0.53	ventricles and regions in cerebellar white matter
3	54756	0.02	fiber tracts from the splenium
4	34317	0.02	anterior medial part of temporal lobes
5	26568	0.10	isolated regions of the cortex
6	28458	0.06	right superior frontal cortex
7	27135	-0.05	left insular gyrus
8	32292	-0.09	right frontal cingulated gyrus
9	13230	0.39	cortical regions in the frontal lobe and right insular gyrus
10	29106	0.05	left parietal cortex
11	15822	0.13	right anterior frontal cortex
12	18954	0.13	right posterior frontal cortex
13	22869	-0.15	left temporal white matter
14	15714	-0.07	left parietal cortex
15	23112	-0.04	left superior parietal cortex
16	21897	-0.25	right cerebellar white matter
17	6345	0.05	left inferior parietal white matter
18	13716	0.11	occipital cingulated gyrus
19	6993	-0.24	left temporal cortex
20	9666	-0.15	isolated regions of the cortex
21	9045	0.39	right occipital gyrus
22	4563	-0.03	isolated regions of the cortex
23	864	-0.24	isolated regions of the cortex

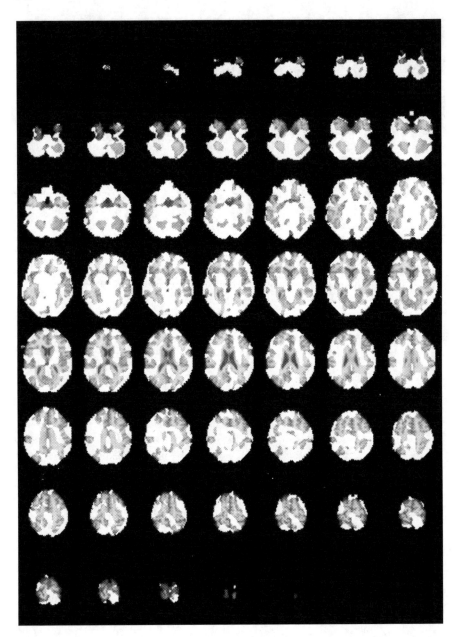

Fig. 5 Regions of the brain associated to factor 1 whose size variation present high correlation to Alzheimer's disease (in white). The analysis reveals that general cortical degeneration is a significant feature related to the pathology

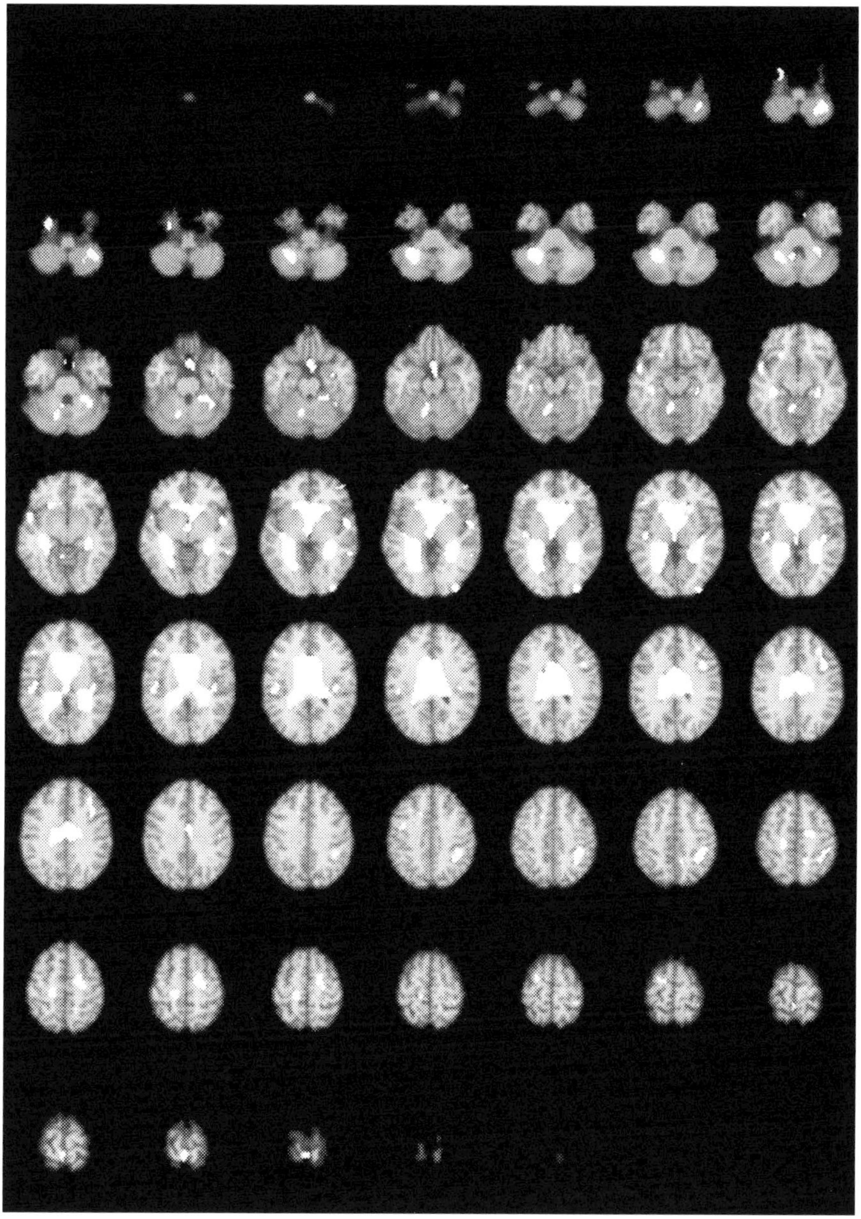

Fig. 6 Regions of the brain associated to factor 2 whose size variation present high correlation to Alzheimer's disease (in white). The analysis reveals that general ventricular enlargement is a significant feature related to the pathology

the subdivision of the corpus callosum proposed by Witelson, based on experimental procedures, is very similar to the one obtained by factor analysis, that were achieved by unsupervised learning.

A comparison with the results of PCA in the same dataset is appropriate. Fig. 4 shows the principal modes of variance, ordered from left to right, top to down, by the amount of variance they represent. Each variable is assigned to the mode for which it presents the greatest absolute coefficient, i.e., the coordinate in the corresponding eigenvector. Variables with positive coefficient are shown in white and the ones with negative values are shown in light gray. By comparing the results with the superimposed topology proposed by Witelson, it is clear that the PCA is unable to cluster the original variables into continuous regions that correspond to physiological substructures in the corpus callosum.

4.3 A Case Study on the Alzheimer's Disease

The last example illustrates the advantages of using FA in high-dimensional variable spaces, for which the computation of the correlation matrix is intractable. In this study, we investigate the Alzheimer's disease (AD) and its manifestations on the neuroanatomy of patients, compared to normal elderly controls. The dataset consists of 24 MRI image volumes from which 12 are diagnosed with AD (age = 70.8 ± 8.5) and the remaining are matched controls (age = 68.5 ± 9.4) [13].

The brain volumes were registered to an average image computed from 305 volumes [10], using SPM99 [11] with 12-parameter affine registration and non-linear registration with 12 iterations and $7 \times 8 \times 7$ basis function. The displacement fields obtained from image registration were the basis for the computation of the Jacobian determinants. The data matrix was formed by the determinant of the Jacobian matrix at each of the 2.1 million voxels of the stripped brain, together with the diagnosis, and used as input to FA. After each estimate of the loading matrix **A**, it was rotated in order to maximize the variance of the squared loadings in each column, so that each variable presented high loading for fewer factors.

The proposed algorithm for estimating the loadings took 5 iterations to converge to a solution in which the absolute difference between the loadings in 2 consecutive iterations was smaller than 0.01. The method revealed 23 factors associated to regional size variation, from which 4 was found to be correlated to the disease (absolute loadings greater than 0.30): factor 1 is associated with the overall cortical tissue at cerebrum and cerebellum; factor 2 is related to the ventricle enlargement; factor 9 encompasses cortical regions in the frontal lobe and right insular gyrus; and factor 21, related to the right occipital gyrus. The effects of cortical degeneration and ventricular enlargement are depicted in Fig. 5 and 6, respectively. A summary of the results is described in Table 5.

The use of FA as a statistical model may be questionable when the interpretation of the factors is not straightforward. Since factors are non-observed variables, they should have a natural meaning in order to provide new information about the data, otherwise the method would serve only as a data reduction tool. In the case of morphological studies such as this example, the factors are visually interpreted as regions in the organ being analyzed. An advantageous feature of FA over other linear models such PCA is that it allows for the observation of substructural regions that present a coherent behavior regarding dilation during image registration. The findings in this study are in accordance to the anatomical differences between patients and controls, reported in the literature [32], that reveal cortical degeneration and proportional enlargement of the ventricles associated to the disease.

5 Concluding Remarks

The use of factor analysis as a data mining tool has been considerably restricted because of its similarity with PCA, so a brief discussion on the fundamental differences between these two models may be helpful. Although the main objective of PCA and FA is data reduction, they differ fundamentally on two aspects: the algebraic model of the transformation and how data reduction is achieved. In FA, the original variables are represented as a linear combination of new variables, while in PCA, the new variables (principal components) are linear combinations of the original variables. In PCA, data reduction is achieved by changing the basis of the variable space, so that the new orthogonal axes represent most of the variance embedded in the dataset. The objective of PCA can be defined as maximizing the variance of a linear combination of the original variables. Data reduction is obtained (with possible loss of information) by ignoring the axes in which the data present small variance. In contrast, FA aims to find a new low-dimensional set of non-observed variables that maximally represents the covariance (or correlation) among the original variables.

Another difference to be highlighted is that PCA is closely related to the dataset behavior, while FA aims to understand the relationship among the many variables of the problem. In this sense, FA can be considered a more useful exploratory tool, rather than just a data reduction technique. The use of factor analysis as a statistical method to explore causality deserves further research effort, as correlation and causation are closely related phenomena. Despite of the fact that correlation does not imply causation, it is known that causation *does* imply correlation (although not necessarily linear), so the discovery of common factors in a dataset is of great interest.

When data mining is to be performed in a dataset where the number of variables is many orders of magnitude larger than the number of subjects, it is of fundamental importance that the variable space be initially reduced. In PCA, data reduction is obtained after the computation of the eigenvectors of the correlation matrix by eliminating the components that do not contribute significantly to the representation of variance. In principle, the dimension of the new basis is the same of the original variable space, so data reduction is achieved by discarding components, with possible loss of information. In fact, since PCA is directed to the representation of variance, the covariance information is guaranteed to be preserved only if all components are kept. Furthermore, data reduction is possible only when the original variables are correlated. In the case of independent variables, all the eigenvalues will have similar values. This contrasts with FA, in which dimension reduction is accomplished during the computation of the loading matrix. The number of factors to be considered must be chosen in order to compute the loading matrix. As a consequence, its elements change as the number of factors m considered in the analysis varies.

The differences in the way data is reduced impact particularly in the interpretation of the results. Since PCA aims to maximally represent the variance of the data, the resulting components may be linear combinations of original variables that do not share strong relationship. On the other hand, the goal of factors analysis is to group correlated variables. The variables are partitioned into groups that are associated to different factors, and can be visually observed as regions in the image. For exploratory studies, this may be essentially helpful, as it reveals regions of interest in which shape variability behaves in a correlated way.

6 Bibliographic Notes

The origin of factor analysis is usually ascribed to Spearman [30], although early works on genetics, in the nineteenth century, already tried to unveil hidden patterns of inherited characters. Spearman proposed a psychological theory composed of a single common factor and a number of specific factors that would represent measurements of general intelligence. The development of factor analysis in the following decades was impelled by psychological theories on the human behavior. These theories aimed at determining underlying variables that could explain patterns of behavior associated to clusters of observed variables quantified by psychological tests. The works of Thurstone [34] and Hotelling [15] were fundamental to the popularization and full development of the factor analytic theory.

Despite of the advantages of factor analysis in the analysis of high-dimensional morphometric studies, the most used linear model to describe general shape variation has been PCA. Marcus [24] used PCA to study the variation in the skull measurements of rodent and bird species. The resulting principal modes of variation were subjectively interpreted as size and gross shape components. Marcus concluded that no specific interpretation should be expected from the method, since it did not embed a biological model.

Cootes et al. [4] applied the theory of PCA to build a statistical shape model of organs based on manually chosen landmarks. The organs were represented by a set of labeled points located at particular regions in order to outline their characteristic shape. The model provided the average positions of the points and the principal modes of variation computed from the dataset.

The ability of the method to locate structures in medical images was demonstrated in a set of experiments with echocardiograms, brain ventricle tracking and prostate segmentation.

Generalizing the use of PCA to high-dimensional sets of variables, Le Briquer and Gee [20] applied the method to analyze the displacement fields obtained from registering a reference image volume of the brain to a set of subjects, based on the elastic matching framework [2]. The analysis provided the inference of morphological variability within a population and was the basis for the construction of a statistical model for the brain shape, which could be used as prior information to guide the registration process.

Davatzikos *et al.* [7] showed how the results obtained from matching boundaries of structures could be interpolated to determine an estimate for the complete displacement field. The method was useful in the registration of structures such as the corpus callosum, whose contour was of more interest than its inner texture. Further analysis on the gradient determinant of the resulting displacement fields showed the amount of area enlargement/reduction while deforming the reference image to match the images in the study. The method was applied to a small set of images of the human corpus callosum, revealing gender-related morphological differences. Using the same dataset, Machado and Gee [22] performed elastic matching to both the boundary and the interior of the structure. Based on the displacements fields obtained from image registration, the method was able to reproduce Davatzikos' results and additionally determine the principal modes of callosal shape variation between sexes.

Martin *et al.* [25] also applied PCA to the shape characterization of the brain, in a study of schizophrenia and Alzheimer's disease. The putamen and ventricles were modeled as a linear elastic material and warped to match the same structures of a normal brain volume. The principal modes of variation computed from the results of image registration were fed to a Gaussian quadratic classifier. The experiments showed that using principal components instead of gross volume as the features for the classifier increased the rate of correct classification from 60% to 72% while discriminating the putamen of normal and schizophrenic patients.

The use of FA in exploratory morphometry is much more recent [27] and has been restricted to the representation of gross measurements and landmarks, in most studies, regardless of its potential to explore the relationship between pointwise shape-related variables, as the ones obtained from image registration. Marcus [24] compared the application of PCA and FA on a set of length measurements for several hundreds skeletons of birds. The extracted factors were interpreted as general features related to the overall size of the subjects. Reyment and Jöreskog [26] presented a thorough discussion on the factor analysis of shape-related landmarks. Scalar features such as the distances between landmarks in the carapace of ostracod species were considered in the analysis. Some of the resulting factors were interpreted as shape-changes in specific regions of the shell, location of eye tubercles and valves. Other factors, however, were related to global features such as the dimensions and curvature of the shell.

Stievenart *et al.* [31] applied FA to study the correlation among parts of the corpus callosum, whose boundary curvature was measured at 11 different positions. The results revealed 3 factors that explained 69% of the variation of the original curvature values. The first and second values were clearly related to the curvature of the isthmus and posterior region of the splenium, respectively.

Another relevant work on the factor analysis of the corpus callosum was presented by Deneberg *et al.* [8], in which the structure was divided into 100 segments taken along equally spaced intervals of the longitudinal axis. The width of each interval, together with other scalar variables such as the callosal area, perimeter and axis length, was examined. Although the structure partitioning criteria was deliberately chosen to result on transversal segments, the study was able to identify regions in the corpus callosum, particularly the isthmus, which presented morphological differences related to gender and handiness. Factor analysis was based on the principal factor method with oblique rotation and the criterion for retaining factors was based on the eigenvalues and the loading magnitudes.

The most common model used to estimate the loading matrix is the principal component method, reason why FA has been frequently mistaken as PCA. As an attempt to provide more robust and efficient solutions to the computation of loadings, Harman [14] proposed the *Minres* method (*min*immum *res*iduals) in which a solution for the loading matrix is searched so as to minimize only the off-diagonal residuals of the difference between the observed correlation matrix and the recovered correlations in the solution.

Also of special interest is the *Maximum Likelihood* method, proposed by Lawley [19] and enhanced by Jöreskog and collaborators [16, 18, 17]. In this method, the loadings are computed and the goodness of fit of the model assessed simultaneously. Methods based on minimization, such as Minres and Maximum Likelihood compare the observed correlations in the original variablke space to the solution and are therefore limited by the dimensionality of the problem.

Acknowledgements. This work was partially supported by CNPq-Brazil, grants 471518/2007-7 and 306193/2007-8, and by FAPEMIG, grants PPM 347-08. The author is grateful to the University of Pennsylvania for sharing the corpus callosum and AD datasets.

References

1. Bajcsy, R., Broit, C.: Matching of deformed images. In: VI International Conference on Pattern Recognition, pp. 351–353 (1982)
2. Bajcsy, R., Kovacic, S.: Multiresolution elastic matching. Computer Vision, Graphics and Image Processing 46, 1–21 (1989)

3. Christensen, G., Rabbitt, R., Miller, M.: Deformable templates using large deformation kinematics. IEEE Transactions on Image Processing 5(10), 1435–1447 (1996)
4. Cootes, T., Hill, A., Taylor, C.J., Haslam, J.: Use of active shape models for locating structures in medical images. Image and Vision Computing 12(6), 355–365 (1994)
5. D'Agostino, R.B.: Small sample probability points for the D test of normality. Biometrika 59, 219–221 (1972)
6. Davatzikos, C., Prince, J.: An active contour model for mapping the cortex. IEEE Transactions on Medical Imaging 14, 65–80 (1995)
7. Davatzikos, C., Vaillant, M., Resnick, S., Prince, J., Letovsky, S., Bryan, R.: A computerized approach for morphological analysis of the corpus callosum. Journal of Computer Assisted Tomography 20(1), 88–97 (1996)
8. Denenberg, V.H., Kertesz, A., Cowell, P.E.: A factor analysis of the human's corpus callosum. Brain Research 548, 126–132 (1991)
9. Duta, N., Sonka, M.: Segmentation and interpretation of mr brain images: An improved active shape model. IEEE Transactions on Medical Imaging 17(6), 1049–1062 (1995)
10. Evans, A., Collins, D., Brown, E., Kelly, R., Petters, T.: A 3D statistical neuroanatomical models for 305 MRI volumes. In: IEEE Nuclear Science Symposium/Medical Imaging Conference, pp. 1813–1817 (1993)
11. Frackowiak, R., Friston, K., Frith, C., Dolan, R., Mazziotta, J.: Human Brain Function. Academic Press, San Diego (1997)
12. Gee, J.C.: On matching brain volumes. Pattern Recognition 32, 99–111 (1999)
13. Grossman, M., McMillan, C., Moore, P., Ding, L., Glosser, G., Work, M., Gee, J.: What's in a name: voxel-based morphometric analyses of MRI and naming difficulty in alzheimer's disease, frontotemporal dementia and corticobasal degeneration. Brain 127, 1–22 (2004)
14. Harman, H.: Modern Factor Analysis. University of Chicago Press, Chicago (1976)
15. Hotelling, H.: The relations of the new multivariate statistical methods to factor analysis. British Journal of Statistical Psychology 10, 69–79 (1957)
16. Jöreskog, K.: Some contributions to maximum likelihood factor analysis. Psychometrika 32, 443–482 (1967)
17. Jöreskog, K., Goldberger, A.S.: Factor analysis by generalized least squares. Psychometrika 37, 243–260 (1968)
18. Jöreskog, K., Lawley, D.N.: New metrods in maximum likelihood factor analysis. British Journal of Mathematical and Statistical Psychology 21, 85–96 (1968)
19. lawley, D.N.: The estimation of factor loadings by the method of maximum likelihood. Proceedings of the Royal Society of Edinburgh 60, 64–82 (1940)
20. Le Briquer, L., Gee, J.C.: Design of a Statistical Model of Brain Shape. In: Duncan, J.S., Gindi, G. (eds.) IPMI 1997. LNCS, vol. 1230, pp. 477–482. Springer, Heidelberg (1997)
21. Lin, C.C., Muldholkar, G.S.: A simple test for normality against asymetric alternatives. Biometrika 67, 455–461 (1980)
22. Machado, A., Gee, J.C.: Atlas warping for brain morphometry. In: Hanson, K.M. (ed.) Proceedings of the SPIE Medical Imaging 1998: Image Processing, pp. 642–651. Bellingham, San Diego (1998)
23. Machado, A., Simon, T., Nguyen, V., McDonald-McGinn, D., Zackai, E., Gee, J.: Corpus callosum morphology and ventricular size in chromosome 22q11.2 deletion syndrome. Brain Research 1131, 197–210 (2007)

24. Marcus, L.: Traditional Morphometrics. In: Rohlf, F.J., Bookstein, F.L. (eds.) Proceedings of the Michigan Morphometrics Workshop, Ann Arbor, The University of Michigan Museum of Zoology, pp. 77–122 (1990)
25. Martin, J., Pentland, A., Sclaroff, S., Kikinis, R.: Characterization of neuropathological shape deformations. IEEE Transactions on Pattern Analysis and Machine Intelligence 20(2), 97–112 (1998)
26. Reyment, R., Jöreskog, K.: Applied Factor Analysis in the Natural Sciences. Cambridge University Press, Cambridge (1996)
27. Roweis, S., Ghahramani, Z.: A unifying review of linear gaussian models. Neural Computation 11(2), 305–345 (1999)
28. Sclaroff, S., Pentland, A.: Modal matching for correspondence and recognition. IEEE Transactions on Pattern Analysis and Machine Intelligence 17(6), 545–561 (1995)
29. Simon, T.J., Bearden, C.E., McDonald-McGinn, D.M., Zackai: Visuospatial and numerical cognitive deficits in children with chromosome 22q11.2 deletion syndrome. Cortex 41, 145–155 (2005)
30. Spearman, C.: General intelligence, objectively determined and measured. American Journal of Psychology 15, 201–293 (1904)
31. Stievenart, J.L., Iba-Zizen, M.T., Tourbah, A., Lopez, A., Thibierge, M., Abanou, A., Canabis, A.: Minimal surface: A useful paradigm to describe the deeper part of the corpus callosum? Brain Research Bulletin 44(2), 117–124 (1997)
32. Thompson, P.M., Moussai, J., Zohoori, S., Goldkorn, A., Khan, A., Mega, M.S., Small, G., Cummings, J., Toga, A.W.: Cortical variability and assymetry in normal aging and Alzheimer's disease. Cerebral Cortex 8, 492–509 (1998)
33. Thomson, G.H.: Hotelling's method modified to give Spearman's g. Journal of Educational Psychology 25, 366–374 (1934)
34. Thurstone, L.L.: Multiple factor analysis. Psychological Reviews 38, 406–427 (1931)
35. Witelson, S.F.: Hand and sex differences in the isthmus and genu of the human corpus callosum: a postmortem morphological study. Brain 112, 799–835 (1989)

A Framework for Composing Knowledge Discovery Workflows in Grids

Marco Lackovic, Domenico Talia, and Paolo Trunfio

Summary. Grid computing platforms provide middleware and services for coordinating the use of data and computational resources available throughout the network. Grids are used to implement a wide range of distributed applications and systems, including frameworks for distributed data mining and knowledge discovery. This chapter presents a framework we developed to support the execution of knowledge discovery workflows in Grid environments by executing data mining and computation intelligence algorithms on a set of Grid nodes. Our framework is an extension of Weka, an open-source toolkit for data mining and knowledge discovery, and makes use of Web Service technologies to access Grid resources and distribute the computation. We present the implementation of the framework and show through some applications how it supports the design of knowledge discovery workflows and their execution on a Grid.

1 Introduction

Data mining often requires huge amounts of resources in terms of storage space and computation time. Distributed data mining explores techniques of how to apply data mining in a non-centralized way, therefore distributed computing, providing mechanisms that distribute the work load among several sites, can play a significant role for data mining.

The Grid [1] offers support for both parallel and distributed computing by integrating heterogeneous computers and resources. Therefore Grids can provide an effective data management and computational infrastructure support for distributed data mining.

Marco Lackovic, Domenico Talia, and Paolo Trunfio
University of Calabria, Italy
e-mail: {mlackovic,talia,trunfio}@deis.unical.it

A. Abraham et al. (Eds.): Foundations of Comput. Intel. Vol. 6, SCI 206, pp. 345–369.
springerlink.com   © Springer-Verlag Berlin Heidelberg 2009

Rather than building a distributed data mining system from scratch, we took an already existing data mining software, which works on a single machine, and we extended it to support distributed data mining applications in a Grid environment. The system we chose for this purpose is Weka [2]: being it a well established and wide-spread project, cross-platform (written in Java), and open source (available under the GNU General Public License), made it the most suitable candidate.

This extended version of Weka has been called *Weka4WS*, which stands for Weka for Web Services, meaning that the data mining algorithms are executed remotely through Web Services. Weka4WS uses the *Web Services Resource Framework* (WSRF) [3] technology for the implementation of the Grid services and Globus Toolkit [4] for the remote execution of the data mining algorithms and for managing the remote computations.

Distributed data mining applications can be built in Weka4WS using either *Explorer* or *Knowledge Flow*, two of the four Weka front ends which have been extended to allow to run single or multiple data mining tasks on remote hosts of the Grid.

In Explorer a data mining application is set up using menu selection and form filling: at the end of the set up a drop down menu shows a list of locations where the algorithm may be executed. Knowledge Flow allows to build directed graphs representing data mining applications workflows: clicking on the vertices corresponding to the data mining algorithms the user can choose the locations where they will be executed. The locations, both in Explorer and in Knowledge Flow, may be either specified by the user or automatically chosen by the system.

The work presented in this chapter is an extended version of the work presented in [5]. This new version describes an improved implementation of the framework, includes new experimental results, and outlines system extensions on which we are currently working.

The rest of this chapter is organized as follows. Section 2 introduces the goals of the system. Section 3 contains a structural description of the various Weka4WS components and their implementation. Section 4 explains the modifications made to Weka Explorer and Knowledge Flow interfaces to add the remote invocation of the data mining tasks. Section 5 presents a detailed description of the interactions among the Weka4WS components. Section 6 provides some examples of data mining workflows designed and executed using Weka4WS. Finally, Section 7 concludes the chapter and outlines future work.

2 System Goals

The objective that guided us in the design of Weka4WS is allowing users to perform distributed data mining on the Grid in a easy and effective way. In particular, the goals of the system are supporting both:

- the execution of a single data mining task on a remote Grid node; and
- the execution of multiple data mining tasks, defined as a workflow, on multiple nodes of a Grid.

Supporting remote execution of single or multiple data mining tasks allows users to exploit the computational power and the distribution of data and algorithms of a Grid.

To make as easier as possible the use of our framework, we decided to start from a well established data mining environment (the Weka toolkit) and to extend it with remote execution features. In this way, domain experts can focus on designing their data mining applications, without worrying about learning complex tools or languages for Grid submission and management. Indeed, the Weka4WS visual interface allows users to set up their data mining tasks or workflows as in Weka, with the additional capability of specifying the Grid nodes where to execute the data mining algorithms.

To achieve integration with standard Grid environments, Weka4WS uses the *Web Services Resource Framework* (*WSRF*) as enabling technology. WSRF is a family of technical specifications concerned with the creation, addressing, inspection and lifetime management of *stateful resources* using Web Services. To enable remote invocation, Weka4WS exposes all the data mining algorithms originally provided by Weka through a WSRF-compliant Web Service, which can be easily deployed on the available Grid nodes.

Other Grid systems share with Weka4WS the use of a service-oriented approach to support distributed data mining, such as *Discovery Net* [6], *Grid Miner* [7], and the *Knowledge Grid* [8]. However, the design approach of Weka4WS is different from that of those systems. In fact, Weka4WS aims at extending a widely-used framework to minimize the efforts needed by users to learn to use it, while the objective of the systems mentioned above is implementing complete frameworks providing ad hoc services and/or languages to perform data mining on the Grid.

3   System Architecture

Weka4WS has been developed using the Java *Web Services Resource Framework* (WSRF) libraries provided by *Globus Toolkit* (GT) and uses its services for the standard Grid functionalities such as security and data transfer. The application is made up of two separate parts:

- **User node:** it is the part where the client side of the application runs. It is made by an extension of the Weka Graphical User Interface, a *Client Module* and the Weka library. It requires the presence of the *Globus Java WS Core* (a component of Globus Toolkit);
- **Computing node:** the server side of the application, it allows the execution of data mining tasks through Web Services. A Globus Toolkit full installation is required.

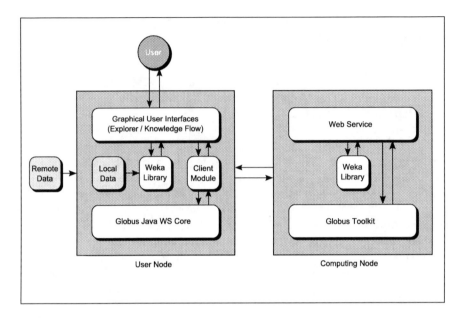

Fig. 1 Weka4WS user node and computing node

Data to be mined may be located either at the user node, or at the computing node, or at some other remote resource (for example some shared repositories). When data are not available at the computing node they are transferred by means of GridFTP [9], a high-performance, secure and reliable data transfer protocol which is part of Globus Toolkit.

Figure 1 shows the Weka4WS software components of the user node and the computing node, and the interactions among them. A user operates on the *Graphical User Interface* to prepare and control the data mining tasks: those to be executed locally will be managed through the local *Weka Library*, whereas those to be executed on a remote host will be handled by the *Client Module* which will interact with the computing node using the services offered by the Globus Java WS Core installed on the user node machine.

In the best network scenario the communication between the client module and the Web Service is based on the "push-style" [10] mode of the *NotificationMessage* delivery mechanism, where the client module is the *NotificationConsumer* and the Web Service is the *NotificationProducer*. The client module invokes the service and waits to be notified of the required task completion.

There are certain circumstances in which the basic "push-style" of NotificationMessage delivery is not appropriate, for example when the client resides behind a a NAT Router/Firewall as shown in Figure 2, because the messages sent to the client will be blocked unless they are over a connection initiated by the client itself.

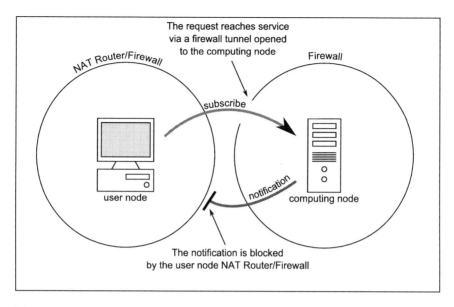

Fig. 2 Notifications are blocked when the client runs on a machine behind a NAT Router/Firewall

When the client subscribes to notification of the completion of the required task it also passes to the service a port number, randomly generated, to which the client will be listening to receive the notification: from that moment the client will act as server, awaiting for notifications to the given port. When the results are ready at the computing node a connection is initiated by the computing node but the attempt to send the notification fails because it is blocked by the user node NAT Router/Firewall. In the scenario just depicted the only way for the client module to get the results is to work in "pull-style" mode, that is to continuously try to *pull* (retrieve), at certain given intervals, the results from the computing node until they will be available.

In order to make the system to automatically adapt to all possible network scenarios in a way transparent to the user, the user node starts in pull-mode by default. At the moment of the notification subscription it also asks for the immediate send of a dummy notification whose solely purpose is to check whether the user node may receive notifications: when and if the user node will receive this dummy notification it will switch to the "push-style" mode, otherwise it will persist in the "pull-style" one.

At the computing node side, the *Web Service* uses the *Globus Toolkit* services to interact with the user node and answers to its requests by invoking the requested algorithm in the underlying *Weka Library*. The relation between Web Services and Globus Toolkit is threefold: Web Services are built using some libraries provided by Globus Toolkit, they are deployed and hosted over the Globus Toolkit container, and they use some Globus Toolkit services like security and the notification mechanism [11] to access data and interact with the user node.

Let's now examine the two components in more details.

3.1 User Node

The user node is formed by three components: the *Graphical User Interface* (GUI), the *Client Module* and the Weka library. The GUI is made by the Weka *Explorer* and *Knowledge Flow* components, extended to support the execution of remote data mining tasks. Explorer is a tool for "exploring" data through data preprocessing, mining and visualization. Knowledge Flow has essentially the same features of the Explorer but with a drag-and-drop interface which allows to build data mining workflows.

The local tasks are executed invoking the local *Weka library*, while the remote ones are performed through the *Client Module* which acts as intermediary between the GUI and the remote Web Services. Each task is carried out in a thread of its own thus allowing to run multiple tasks in parallel, either using Explorer or Knowledge Flow.

The remote hosts addresses are loaded from a text file located inside the application directory. This text file is read in background when the application is launched and for each remote host is checked that:

• the Globus container and GridFTP are running;
• the Weka4WSService is deployed;
• the versions of the client and the service are the same.

Only those hosts which pass all the checks are made available to the user in the GUI. In order to take into account possible alteration of the Grid network configuration without having to restart the application, the remote hosts addresses may be reloaded at any time simply by pressing a given button provided for the purpose.

A static list of addresses stored in a text file on the client machine is not actually the ideal solution, as Globus Toolkit already provides a point of inquiry (the *Globus Index Service*) regarding the characteristics of a physical system in a Grid and allows clients or agents to discover services that come and go dynamically on the Grid. The choice of using that file has been taken as temporary solution in order to speed up the development of a Weka4WS prototype and focus on more demanding and crucial aspects of the application. The use of the Globus Index Service in place of the text file is planned to be introduced in the near future versions of Weka4WS.

3.2 Computing Node

A computing node includes two components: a *Web Service* and the *Weka Library*. The *Web Service* answers the user node query by invoking the requested algorithm in the underlying *Weka Library*. The invocation of the

Table 1 Operations provided by each Web Service in the Weka4WS framework

Operation	Description
classification	Submits the execution of a classification task.
clustering	Submits the execution of a clustering task.
associationRules	Submits the execution of an association rules task.
stopTask	Explicitly requests the termination of a given task.
getVersion	Returns the version of the service.
notifCheck	Checks whether the client is able to receive notifications.
createResource	Creates a new stateful resource.
subscribe	Subscribes to notifications about resource properties changes.
getResourceProperty	Retrieves the resource property values.
destroy	Explicitly requests destruction of a resource.

algorithms is performed in an asynchronous way, that is, the client submits the task in a non-blocking mode and results are either notified to it whenever they are computed (push-style mode) or they are repeatedly checked for readiness by the client (pull-style mode) depending on the network configuration, as described in Section 3.

Table 1 lists the operations provided by each Web Service in the Weka4WS framework. The first three operations are used to request the execution of a particular data mining task; operations in the middle row of the table are some useful extra operations, while the last four operations are related to WSRF-specific invocation mechanisms.

The getVersion operation is invoked at the moment of the hosts check, described in Section 3.1, and is used to check whether the client and the service versions are the same. The notifCheck operation is invoked just after the subscribe operation to check whether the client is able to receive notifications, as described in Section 3. All the other operations are described in Table 1.

The parameters required for each operation are shown in Table 2; the operations getVersion and notifCheck do not require any parameter.

The taskID field is required solely for the purpose of stopping a task. The algorithm field is of a complex type, shown in Table 3, which contains two subfields: name, a string containing the Java class in the Weka library to be invoked (e.g., weka.classifiers.trees.J48), and parameters, a string containing the sequence of arguments to be passed to the algorithm (e.g., -C 0.25 -M 2). The dataset and testset are other two fields, of a complex type which contains four subfields: fileName, a string containing the name of the file of the dataset (or test set), filePath, a string containing the full path (file name included) of the dataset , dirPath, a string containing the path (file name excluded) of the dataset, and crc which specifies the checksum of the dataset (or test set) file.

Table 2 Weka4WS Web Services input parameters

Parameters type	Field name	Field type
classificationParameters	taskID	long
	algorithm	algorithmType
	dataset	datasetType
	testset	datasetType
	classIndex	int
	testOptions	testOptionsType
	evalOptions	evalOptionsType
clusteringParameters	taskID	long
	algorithm	algorithmType
	dataset	datasetType
	testset	datasetType
	classIndex	int
	testOptions	testOptionsType
	selectedAttributes	array of int
associationRulesParameters	taskID	long
	algorithm	algorithmType
	dataset	datasetType
stopParameters	taskID	long

The `classIndex` field is an integer designating which attribute of the dataset is to be considered the class attribute when invoking a classifying or clustering algorithm. The `testOptions` field is of a complex type containing three subfields: `testMode`, an integer representing the test mode to be applied to the classifying or clustering algorithm (1 for *Cross-validation*, 2 for *Percentage split*, 3 to use the training set and 4 to use a separate test set), `numFolds` and `percent` are two optional fields used when applying a *Cross-validation* or a *Percentage split* test mode respectively.

The `evalOptions` field is used specifically for the classification algorithms and contains several subfields like `costMatrix` (to evaluate errors with respect to a cost matrix), `outputModel` (to output the model obtained from the full training set), and others shown in Table 3. The field `selectedAttributes` is used specifically for the clustering algorithms and contains those attributes in the data which are to be ignored when clustering.

4   Graphical User Interface

The application starts with the three windows shown in Figure 3:

- the *Gui Chooser* (left side in Figure 3), used to launch Weka's four graphical environments;

Table 3 Weka4WS Web Services input parameters fields types

Fields type	Subfield name	Subfield type
algorithmType	name	string
	parameters	string
datasetType	fileName	string
	filePath	string
	dirPath	string
	crc	long
testOptionsType	testMode	int
	numFolds	int
	percent	int
evalOptionsType	costMatrix	string
	outputModel	boolean
	outputConfusion	boolean
	outputPerClasss	boolean
	outputSummary	boolean
	outputEntropy	boolean
	rnd	int

- the remote hosts list checking window (top right side in Figure 3), used to give a visual confirmation of the hosts checking described in Section 3.1;
- the Grid Proxy Initialization window (middle right side in Figure 3), automatically loaded at startup only if the user credentials are not available or have expired.

We will now examine the two Weka components which have been extended in Weka4WS: *Explorer* and *Knowledge Flow*.

4.1 Explorer

Explorer, Weka's main graphical user interface, is a comprehensive tool with six tabbed panes, each one dedicated to a specific Weka facility like data preprocessing (loading from file, URL or database, filtering, saving, etc.), data mining (classification, clustering, association rules discovery) and data visualization. A more detailed description of Explorer may be found in [2].

In Weka4WS the Explorer component is essentially the same as the Weka one with the exception of the three tabbed panes associated to classification, clustering and association rules discovery: in those panes the two buttons for starting and stopping the algorithms have been replaced with a *Control Panel*, and a button named *Proxy* has been added in the lower left corner of the window. Modifications are highlighted in Figure 4.

The drop down menu in the Control Panel allows to choose either the exact Grid location where we want the current algorithm to be executed (where

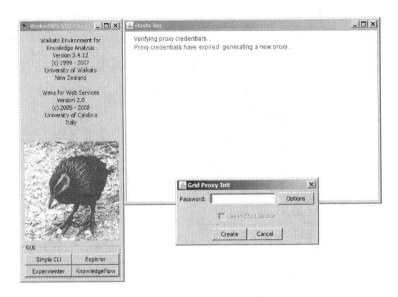

Fig. 3 Weka4WS start up

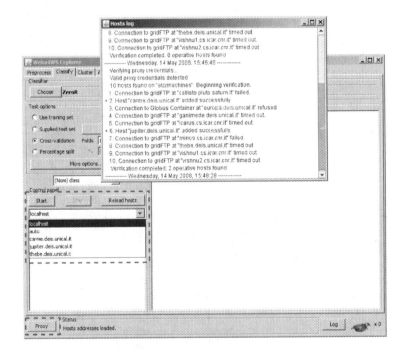

Fig. 4 Weka4WS Explorer: Control Panel and hosts reloading

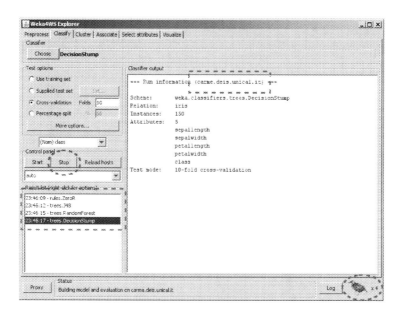

Fig. 5 Weka4WS Explorer: multiple tasks execution

localhost will make the algorithm be computed on the local machine) or to let the system automatically choose one by selecting the *auto* entry. The currently used strategy in the *auto* mode is round robin: on each invocation the host in the list next to the previously used one is chosen.

The *Reload hosts* button, when pressed, brings up the hosts list checking window and starts the hosts checking procedure described in Section 3.1. The *Proxy* button, when pressed, brings up the Grid Proxy Initialization window described earlier.

Once the Grid node is chosen, be that local or remote, the task may be started by pressing the *Start* button and stopped by pressing the *Stop* button. As mentioned in Section 3.1 in Weka4WS, unlike in Weka, a task is carried out in a thread of its own thus allowing to run multiple tasks in parallel. In Figure 5, in the lower right corner of the window, the number of running tasks is displayed. The list of started tasks is displayed in the *Result list* pane, just below the Control panel.

The *Output panel*, at the right side of the window, shows the run information and results (as soon as they are known) of the task currently selected in the Result list; at the top of the Output Panel, as highlighted in Figure 5, is shown the host name where the task is being computed.

It is possible to follow the remote computations in their very single steps as well as to know their execution times through the log window, which is shown by pressing the *Log* button in the lower right corner of the Window, as highlighted in Figure 6.

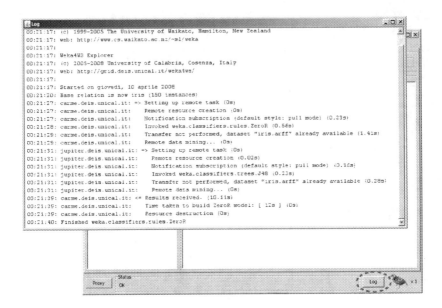

Fig. 6 Weka4WS Explorer: detailed log

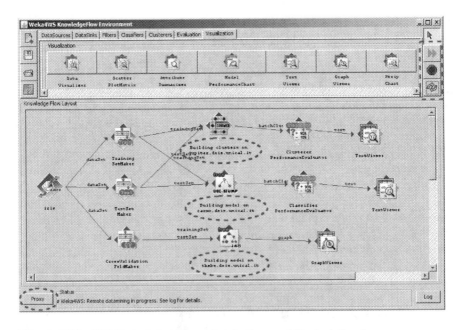

Fig. 7 Weka4WS Knowledge Flow: simple flow and Control Panel

4.2   Knowledge Flow

Knowledge Flow is the component of Weka which allows to compose work-flows for processing and analyzing data. A workflow can be done by selecting components from a tool bar, placing them on a layout canvas and connecting them together: each component of the workflow is demanded to carry out a specific step of the data mining process. A more detailed description of Knowledge Flow may be found in [2].

We extended the Weka Knoledge Flow to allow the execution of distributed data mining workflow on Grids by adding annotations into the Knowledge Flow. Through annotations a user can specify how the workflow nodes can be mapped onto Grid nodes.

In Figure 7 some of the changes introduced in Weka4WS are highlighted: in the upper right corner three buttons have been added whose purpose is, from top to bottom, to start all the tasks at the same time, to stop all the tasks and to reload the hosts list, as seen in Section 3.1. A button named *Proxy* has been added in the lower left corner of the window which when pressed, just like in the Explorer component, brings up the Grid Proxy Initialization window described earlier. The labels under each component associated to an algorithm indicate, during the computation, the hosts where that algorithm is being computed.

The choice of the location where to run a certain algorithm is made into the configuration panel of each algorithm, accessible right clicking on the given algorithm and choosing *Configure*, as shown in Figure 8: within the highlighted area it can be seen the part added in Weka4WS which, as previously seen for the Control Panel of the Explorer component, consists of a drop down menu containing the available locations where the selected algorithm can be executed.

Although the algorithms and their performance evaluators are represented by two separate nodes, the model building and its evaluation are actually performed in conjunction at the computing node when the chosen location is not local.

For complex configurations of workflows the sub-flow grouping feature of Knowledge Flow turns out to be useful in order to easily and quickly set the remote hosts for the execution of the algorithms. Through this feature it is possible to group together a set of components of the flow which will then be represented graphically by only one component, the black-to-gray faded one shown in Figure 9: right clicking on this component it is possible to either set to *auto* all the computing locations of the algorithms belonging to the group, or choosing the specific location of each algorithm by accessing the relative configuration listed in the menu.

The computations may be started, as shown in Figure 10, either by selecting the *Start loading* entry in the right-click context menu of each loader component of the flow (just like usually done in the conventional Weka Knowledge Flow) or

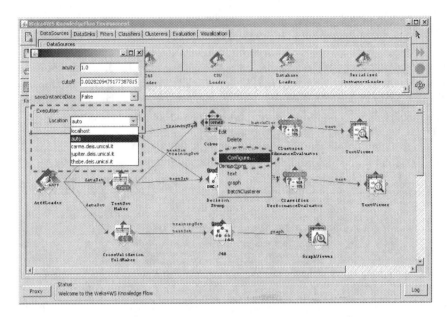

Fig. 8 Weka4WS Knowledge Flow: selection of the remote host

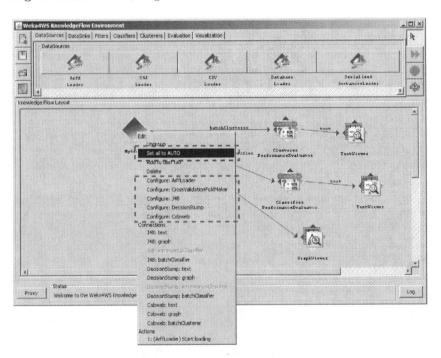

Fig. 9 Weka4WS Knowledge Flow: computing locations selection

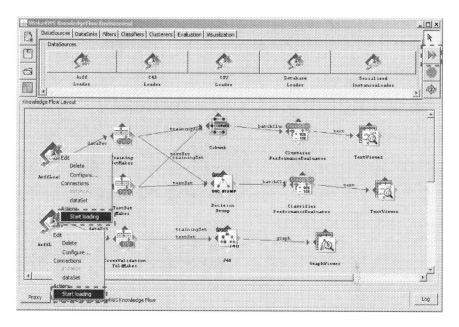

Fig. 10 Weka4WS Knowledge Flow: computations start

by pressing the *Start all executions* button in the right-top corner of the window (which is more convenient in flows with multiple loader components).

As for the Explorer component, pressing the Log button in the lower-right corner it is possible to follow the computations in their very single steps as well as to know their execution times.

5 How the System Works

In this section we are going to see in details, through an invocation example, all the steps and mechanisms involved in the execution of one single data mining task on the Grid; these steps are the same regardless on whether the task is invoked from the Explorer or the Knowledge Flow interface.

In this example we are assuming that the client is requesting the execution of a classification task on a dataset which is present on the user node, but not on the computing node. This is to be considered a worst scenario because in many cases the dataset to be mined is already available (or, more specifically, replicated) on the Grid node where the task is to be submitted.

When a remote data mining task is started an unambiguous identification number, called `taskID`, is generated: this number is associated to that particular task and is used when a `stopTask` operation is invoked to identify the task among all the others running at the computing node.

The whole invocation process may be divided in the 8 steps shown in Figure 11:

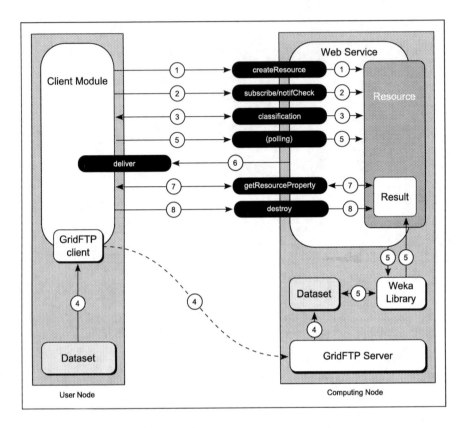

Fig. 11 Weka4WS: task invocation

1. **Resource creation:** the *createResource* operation is invoked to create a new resource that will maintain its state throughout all the subsequent invocations of the Web Service until its destruction. The state is stored as properties of the resource; more specifically a `Result` property, detailed in Table 4, is used to store the results of the data mining task.

   The first three fields of the property stores the inferred model and/or models, the evaluation outcomes and additional information about visualization and prediction. The last three fields, `exception`, `stopped` and `ready` are used only when certain circumstances arise:

   - if during the computation phase something goes wrong and an exception is thrown then the field `exception` is set accordingly: its boolean parameter `thrown` is set to true and in its string parameter, `message`, is stored the generated exception message;

Table 4 Result resource property composition

Field name	Type	Subfield names
model	ModelResult	model, models
evaluation	EvalResult	summary, classDetails, matrix
visualization	VisResult	predInstances, predictions, plotShape, plotSize
exception	Weka4WSException	thrown, message
stopped	boolean	
ready	boolean	

- if during the computation a request of termination is received through the stopTask operation then the boolean field stopped is set to true;
- after the end of the computation the results are put into the Result property and the field ready is set to true: this field is used by the client when it's unable to receive notifications, like in the scenario depicted in Figure 2, to periodically check whether the results have been computed;

After the resource has been created the Web Service returns the end-point reference (EPR) of the created resource. The EPR is unique within the Web Service, and differentiates this resource from all the other resources in that service. Subsequent requests from the Client Module will be directed to the resource identified by that EPR;

2. **Notification subscription and notifications check:** the *subscribe* operation is invoked in order to be notified about changes that will occur to the Result resource property. Upon this property value change (that is upon the conclusion of the data mining task) the Client Module will be notified of it. Just after the *subscribe* operation the *notifCheck* operation is invoked to request the immediate send of a dummy notification to check whether the client is able to receive notifications, as described in Section 3: when and if the user node will receive this dummy notification it will switch to the "push-style" mode, otherwise will persist in the "pull-style" one;

3. **Task submission:** the *classification* operation is invoked in order to ask for the execution of the classification task. This operation requires the 7

Table 5 Response composition

Field name	Type
datasetFound	boolean
testsetFound	boolean
dirPath	string
exception	Weka4WSException

parameters shown in Table 2, among which the `taskID` previously mentioned. The operation returns the `Response` object, detailed in Table 5. If a copy of the dataset is not already available at the computing node, then the field `datasetFound` is set to false and the `dirPath` field is set to the URL where the dataset has to be uploaded; similarly, when a validation is required on a test set which is different from the dataset and the test set is not already available at the computing node, the `testsetFound` field is set to false. The URL where the test set has to be uploaded is the same as for the dataset. If during the invocation phase something goes wrong and an exception is thrown then the field `exception` is set accordingly: its boolean parameter `thrown` is set to true and in its string parameter, `message`, is stored the generated exception message;

4. **File transfer:** since in this example we assumed that the dataset was not already available at the computing node, the Client Module needs to transfer it to the computing node. To that end, a Java GridFTP client [12] is instantiated and initialized to interoperate with the computing node GridFTP server: the dataset (or test set) is then transferred to the computing node machine and saved in the directory whose path was specified in the `dirPath` field contained in the `Response` object returned by the classification operation;

5. **Data mining:** the classification analysis is started by the Web Service through the invocation of the appropriate Java class in the Weka library. The results of the computation are stored in the `Result` property of the resource created on Step 1;

6. **Notification reception:** as soon as the `Result` property is changed a notification is sent to the Client Module by invoking its implicit `deliver` operation. This mechanism allows the asynchronous delivery of the execution results whenever they are generated. In those cases where the client is unable to receive notifications the client will be periodically checking the results for readiness through the value of the Result's field `ready` (see Table 4);

7. **Results retrieving:** the Client Module invokes the operation `getResourceProperty` in order to retrieve the `Result` property containing the results of the computation;

8. **Resource destruction:** The Client Module invokes the `destroy` operation, which eliminates the resource created on Step 1.

6   Examples and Performance

In this section we present two examples of distributed data mining workflows designed and executed on a Grid using the Knowledge Flow component of Weka4WS. The first workflow defines a classification application while the second one defines a clustering application. Both of these workflows have been executed on a Grid environment composed of five machines, to evaluate performance and scalability of the system. In addition, we present the execution of a workflow on a multi-core machine to show how Weka4WS can obtain lower execution times compared to Weka even when executed on a single computer.

6.1 Classification Workflow

Data mining applications that easily exploit the Weka4WS approach are those where a dataset is analyzed in parallel on multiple Grid nodes using different data mining algorithms. For example, a given dataset can be concurrently classified using different classification algorithms with the aim of finding the "best" classifier on the basis of some evaluation criteria (e.g., error rate, confusion matrix, etc.). We used the Knowledge Flow component of Weka4WS to build such kind of application, in which a dataset is analyzed in parallel using four different classification algorithms: *Decision Stump*, *Naive Bayes*, *J48* and *Random Forest*.

The dataset analyzed is *kddcup99*, publicly available at the UCI KDD archive [13]. This dataset contains a wide set of data produced during seven weeks of monitoring in a military network environment subject to simulated intrusions. From the original dataset we removed all but 9 attributes and the class attribute, using a selection filter provided by the Preprocess panel of the Explorer component. Then, from the resulting dataset, we used another filter to extract three datasets with a number of instances of 215000, 430000 and 860000, and a file size of about 7.5 MB, 15 MB and 30 MB respectively. The same classification workflow has been executed for each of those datasets.

Figure 12 shows the workflow designed to build the application. The workflow begins (on the left side) with an *ArffLoader* node, used to load the *kddcup99* dataset from file, which is connected to a *CrossValidation FoldMaker* node (set to 5 folds), used for splitting the dataset into training and test sets according to a cross validation. The *CrossValidation FoldMaker* node is connected to four nodes, each one performing the four algorithms mentioned earlier. These are in turn connected to a *Classifier PerformanceEvaluator* node for the model validation, and then to a *TextViewer* node for results visualization.

When the application is started by the user, the four branches of the workflow are executed in parallel. For each of the three dataset sizes (215, 430 and 860 thousands instances), the workflow has been executed using 1 to 4 Grid nodes in order to evaluate the speedup of the system. The machines used for the experiments had Intel Pentium processors ranging from 2.8 GHz to 3.2 GHz, RAM ranging from 1 GB to 2 GB, and belong to two local area networks. The results of the experiments are shown in Figure 13.

For the largest dataset (860k instances) the total execution time decreases from 2456 sec (about 41 min) using 1 node, to 1132 sec (about 19 min) using 4 nodes, achieving a speedup of 2.17. For the smallest dataset (215k instances)

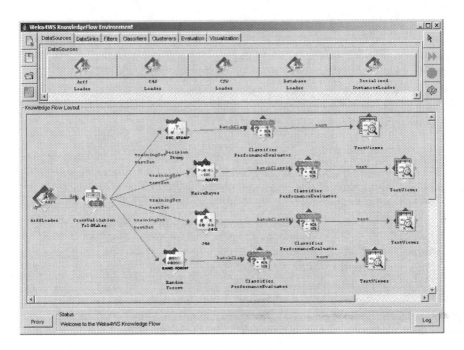

Fig. 12 Workflow of the classification application

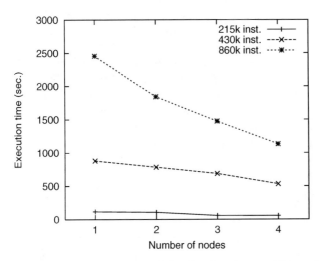

Fig. 13 Execution times of the classification workflow using 1 to 4 Grid nodes for datasets with 215, 430 and 860 thousands instances

the total execution time passes from 117 seconds with 1 node, to 55 seconds using 4 nodes, with a speedup of 2.13. We can observe that the speedup value is not so high as one might expect. The reason of that is the diversity of the classification algorithms used in the example, coupled with the heterogeneity of computing nodes.

Since to execute the model evaluation all the four classifiers must have completed their execution, the total classification time is given by the slower classifier. However this experiment shows that the workflow execution time decreases significantly when large datasets are analyzed on Grids.

It is worth noticing that the total execution time is made by the sum of three contributions: file transfer time, WSRF overhead, and data mining time. In all cases, the file transfer took just a few seconds to complete (about 6 seconds in the worst case). Also the WSRF overhead, that is the overall time needed to invoke the service, subscribe to notification, and receive the results as explained in Section 5, took a small time (less than 4 seconds). The sum of file transfer time and WSRF overhead is therefore negligible compared to the data mining time which is the most relevant part of the total execution time, as already detailed in [5].

6.2 Clustering Workflow

The data mining workflow described in this section implements a parameter sweeping application in which a given dataset is analyzed using multiple instances of the same algorithm with different parameters. In particular, we used the Knowledge Flow to compose an application in which a dataset is analyzed by running multiple instances of the same clustering algorithm, with the goal of obtaining multiple clustering models from the same data source.

The dataset *covertype*, from the UCI KDD archive, has been used as data source. The dataset contains information about forest cover type for 581012 sites in the United States. Each dataset instance, corresponding to a site observation, is described by 54 attributes that give information about the main features of a site (e.g., elevation, aspect, slope, etc.). The 55th attribute contains the cover type, represented as an integer in the range 1 to 7. From this dataset we extracted three datasets with about 72, 145 and 290 thousands instances and a file size of about 9 MB, 18 MB and 36 MB respectively. Then we used Knowledge Flow to perform a clustering analysis on each of those datasets. The workflow corresponding to the application is shown in Figure 14.

This workflow is similar to the one shown in Figure 12: it includes an *Arf-fLoader* node connected to a *Training SetMaker* node, used for accepting a dataset and producing a training set. The *Training SetMaker* node is connected to 5 nodes, each one performing the *KMeans* clustering algorithm, and each one set to group data into a different number of clusters (3 to 7), based on all the attributes but the last one (the cover type). These nodes are in turn connected to a *TextViewer* node for results visualization.

The workflow has been executed using a number of computing nodes ranging from 1 to 5 for each of the three datasets (72k, 145k and 290k instances). The execution times are shown in Figure 15.

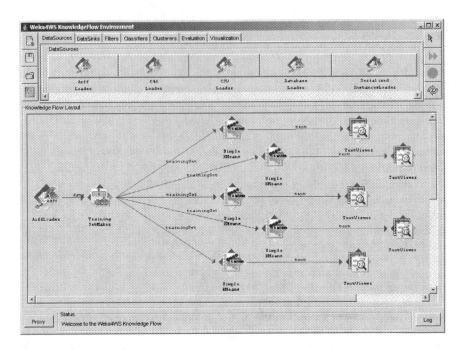

Fig. 14 Workflow of the clustering application

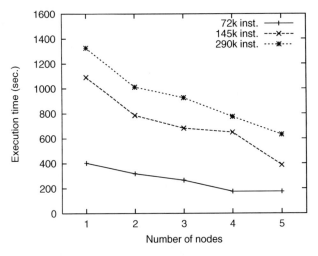

Fig. 15 Execution times of the clustering workflow using 1 to 5 Grid nodes for datasets with 72, 145 and 290 thousands instances

With the largest dataset as input, the total execution time decreases from 1326 seconds obtained using 1 computing node, to 632 seconds obtained on 5 nodes. For the smallest dataset, the execution time passed from 403 to 175 seconds using 1 to 5 nodes. The execution speedup with 5 nodes ranged between 2.10 to 2.82. In this case the speedup is mainly limited by the different amounts of time taken by the various clustering tasks included in the workflow. Finally, as for the classification workflow, the file transfer time and the WSRF overhead took a negligible percentage of the total execution time.

Also in this case the slower algorithm determines the total execution time. However, the speedup values for large datasets are good.

6.3 Execution on a Multi-core Machine

In this section we present the execution times of a classification workflow when it is executed locally on a multi-core machine. Since Weka4WS executes an independent thread for each branch of the workflow, we obtain lower execution times compared to Weka even when the workflow is run on a single multi-processor and/or multi-core machine.

The workflow considered here is a variant of the parameter sweeping workflow presented in the previous section. In this case, a data source is analyzed in parallel using 4 instances of the J48 classification algorithm, configured to use a confidence factor of 0.20, 0.30, 0.40 and 0.50.

As data source, we used 6 datasets extracted from the *covertype* dataset introduced earlier. Those datasets have a number of instances ranging from 39 to 237 thousands, with a size ranging from 5 MB to 30 MB. For each of those 6 datasets as input, we executed the same workflow with Weka and Weka4WS. The machine used for this experiment has two Intel Xeon dual-core processors with a clock frequency of 3 GHz and 2 GB of RAM. The execution times are reported in Figure 15.

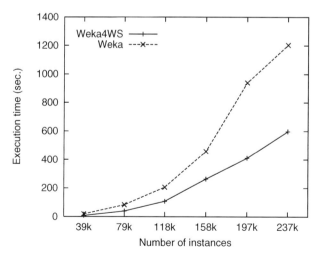

Fig. 16 Execution times of the classification workflow using Weka and Weka4WS on a two-processor dual-core machine, for a dataset having 39k to 237k instances

With Weka, the execution time ranges from 19 seconds for the dataset with 39k instances, to 1204 seconds for the dataset with 237k instances. Using Weka4WS the execution time passes from 8 to 598 seconds, saving an amount of time ranging from a minimum of 42% to a maximum of 58%. These results confirm that the multi-threaded approach of Weka4WS is well suited also to fully exploit the computational power of multi-processor and/or multi-core machines.

7   Conclusions

We described the design and implementation of Weka4WS, an extension of Weka which, adopting both WSRF technologies and the services offered by Globus Toolkit, provides support to the designing of distributed applications which coordinate the execution of multiple data mining tasks on a set of Grid nodes.

Our framework extends, in particular, the functionalities of the Knowledge Flow component of Weka which allow users to compose knowledge discovery workflows made by several algorithms and data analysis processes. Thanks to the extensions implemented, the Knowledge Flow of Weka4WS allows the parallel execution of the data mining algorithms which are part of the workflow on several Grid nodes, hence allowing to reduce the execution time, as ascertained by the performance tests results presented above. Weka4WS can help reducing the execution time of multiple data mining algorithms also when used on a single multi-processor and/or multi-core machine.

Future developments of Weka4WS we are considering are:

- to make use of the dynamic information on the resources to make an efficient scheduling of the task on the Grid nodes, in order to reduce the execution times of particularly complex applications which operate on big amount of data. To that end, it will be implemented a higher integration with the Globus services, in particular with the resources monitoring and discovery system;
- to support *data parallelism*, that is the distribution of the data across different parallel computing nodes, besides the currently employed *task parallelism* which focuses on distributing execution tasks across different computing nodes.

The Weka4WS code is available for research and application purposes. It can be downloaded from http://grid.deis.unical.it/weka4ws.

Acknowledgements. This research work is carried out under the FP6 Network of Excellence CoreGRID funded by the European Commission (Contract IST-2002-004265).

References

1. Foster, I., Kesselman, C., Nick, J., Tuecke, S.: The Physiology of the Grid. In: Berman, F., Fox, G., Hey, A. (eds.) Grid Computing: Making the Global Infrastructure a Reality, pp. 217–249. Wiley, New York (2003)
2. Witten, H., Frank, E.: Data Mining: Practical machine learning tools with Java implementations. Morgan Kaufmann, San Francisco (2000)
3. Czajkowski, K., et al.: The WS-Resource Framework Version 1.0. (2006), http://www-106.ibm.com/developerworks/library/ws-resource/ws-wsrf.pdf (visited May 21, 2008)
4. Foster, I.: Globus Toolkit Version 4: Software for service-oriented systems. In: Jin, H., Reed, D., Jiang, W. (eds.) NPC 2005. LNCS, vol. 3779, pp. 2–13. Springer, Heidelberg (2005)
5. Talia, D., Trunfio, P., Verta, O.: Weka4WS: a WSRF-enabled Weka Toolkit for distributed data mining on Grids. In: Jorge, A.M., Torgo, L., Brazdil, P.B., Camacho, R., Gama, J. (eds.) PKDD 2005. LNCS, vol. 3721, pp. 309–320. Springer, Heidelberg (2005)
6. Al Sairafi, S., Emmanouil, F.-S., Ghanem, M., Giannadakis, N., Guo, Y., Kalaitzopoulos, D., Osmond, M., Rowe, A., Syed, J., Wendel, P.: The Design of Discovery Net: Towards Open Grid Services for Knowledge Discovery. Int. Journal of High Performance Computing Applications 17(3), 297–315 (2003)
7. Brezany, P., Hofer, J., Min Tjoa, A., Woehrer, A.: GridMiner: An Infrastructure for Data Mining on Computational Grids. In: APAC Conference and Exhibition on Advanced Computing, Grid Applications and eResearch, Queensland, Australia (2003)
8. Congiusta, A., Talia, D., Trunfio, P.: Distributed data mining services leveraging WSRF. Future Generation Computer Systems 23(1), 34–41 (2007)
9. Allcock, W., Bresnahan, J., Kettimuthu, R., Link, M., Dumitrescu, C., Raicu, I., Foster, I.: The Globus striped GridFTP framework and server. In: Supercomputing Conf. (2005)
10. Web Services Base Notification 1.3, OASIS Standard (2006), http://docs.oasis-open.org/wsn/wsn-ws_base_notification-1.3-spec-os.pdf (visited May 21, 2008)
11. Graham, S., et al.: Publish-Subscribe Notification for Web services (2004), http://www.oasis-open.org/committees/download.php/6661/WSNpubsub-1-0.pdf (visited May 21, 2008)
12. Java GridFTP client, http://www.globus.org/cog/jftp/ (visited May 21, 2008)
13. Hettich, S., Bay, S.D.: The UCI KDD Archive, University of California, Department of Information and Computer Science, http://kdd.ics.uci.edu (visited March 19, 2007)

Distributed Data Clustering: A Comparative Analysis

N. Karthikeyani Visalakshi and K. Thangavel

Abstract. Due to explosion in the number of autonomous data sources, there is a growing need for effective approaches to distributed clustering. This paper compares the performance of two distributed clustering algorithms namely, Improved Distributed Combining Algorithm and Distributed K-Means algorithm against traditional Centralized Clustering Algorithm. Both algorithms use cluster centroid to form a cluster ensemble, which is required to perform global clustering. The centroid based partitioned clustering algorithms K-Means, Fuzzy K-Means and Rough K-Means are used with each distributed clustering algorithm, in order to analyze the performance of both hard and soft clusters in distributed environment. The experiments are carried out for an artificial dataset and four benchmark datasets of UCI machine learning data repository.

Keywords: Centroid, Cluster Ensemble, Distributed Clustering, Fuzzy K-Means, K-Means, Rough K-Means.

1 Introduction

Clustering is an unsupervised learning method that constitutes a cornerstone of an intelligent data mining process. It is the process of grouping objects into clusters such that objects within a cluster have high similarity in comparison to one another, but are very dissimilar to objects in other clusters [36]. It groups data objects on information found in the data that describes the objects and their relationships. Clustering algorithms can be applied to a wide range of problems, including customer segmentation, image segmentation, information retrieval and scientific data analysis. There are different clustering algorithms, each has its own characteristics. There is no clustering algorithm performing best for all datasets. Each dataset requires both expertise and insight to choose a single best clustering algorithm, and it depends on the nature of application and patterns to be extracted.

N. Karthikeyani Visalakshi
Department of Computer Science, Vellalar College for Women, Erode-12,
Tamil Nadu, India
karthichitru@yahoo.co.in

K. Thangavel
Department of Computer Science, Periyar University, Salem-11, Tamil Nadu, India
drktvelu@yahoo.com

A. Abraham et al. (Eds.): Foundations of Comput. Intel. Vol. 6, SCI 206, pp. 371–397.
springerlink.com   © Springer-Verlag Berlin Heidelberg 2009

Over the years, dataset sizes have grown rapidly with the advances in technology, the ever-increasing computing power and computer storage capacity, the permeation of internet into daily life and the increasingly automated business, manufacturing and scientific processes. Moreover, many of these datasets are, in nature, geographically distributed across multiple sites. For example, the huge number of sales records of hundreds of chain stores is stored at different locations [21]. Traditional clustering methods require that all data have to be located at the site, where they are analyzed and cannot be applied in the case of multiple distributed datasets, unless all data are transferred in a single location and then clustered. Due to technical, economical or security reasons, it is not always possible to transmit all the data from different local sites to single location and then perform global clustering. To cluster such large and distributed datasets, it is important to investigate efficient distributed clustering algorithms in order to reduce the communication overhead, central storage requirements, and computation times [28].

The main objective of distributed clustering algorithms is to cluster the distributed datasets without necessarily downloading all the data to the single site. It assumes that the objects to be clustered reside on different sites. This process is carried out on two different levels: local level and global level. On the local level, all sites carry out clustering process independently from each other. After having completed the clustering, a local model or local representative is determined, which should reflect an optimum trade-off between complexity and accuracy. Next, the local model is transferred to a central site, where the local models are merged in order to form a global model. The global model is again transmitted to local sites to update local models [11]. The key idea of distributed clustering is to achieve a global clustering that is as good as the best Centralized Clustering Algorithm (CCA), limited communication required to collect the local models in a single location, regardless of the crucial choice of any clustering technique in local site.

This study compares the performance of two centralized ensemble based distributed clustering algorithms, namely, Improved Distributed Combining Algorithm (IDCA) [16] and Distributed K-Means (DKMA) [14] algorithm against CCA. Both the algorithms use centroid as local model for communication during distributed clustering, which can be obtained through any centroid based conventional clustering technique. These two distributed clustering algorithms are generalized to work based on three centroid based algorithms K-Means (KM), Fuzzy K-Means (FKM) and Rough K-Means (RKM), to retrieve hard and soft clusters in distributed environment. The experiments are carried out for an artificial dataset and four benchmark datasets of UCI machine learning data repository [26] in order to compare the efficiency of the algorithms. The comparative analysis assumes that all datasets are having uniform schema with same underlying distribution and shows the importance of fuzzy and interval set representation of clusters as like crisp clusters in distributed environment.

The rest of this chapter is organized as follows: Section 2 details the motivation for this chapter. Section 3 provides some background discussion on clustering and distributed clustering and related works in the domain of distributed clustering. The centroid based distributed clustering is described in Section 4. In Section 5, the experimental analysis are performed with synthetic and benchmark datasets. Finally, Section 6 concludes the chapter.

2 Motivation

Clusters can be hard or soft in nature. In conventional clustering, objects that are similar are allocated to the same cluster while objects that differ significantly are put in different clusters. These clusters are disjoint and are called hard clusters. In soft clustering, an object may be a member of two or more clusters. Soft clusters may have fuzzy or rough boundaries [24].

The conventional hard K-Means method [10] always leads to hard clusters, which is found to be too restrictive in many data mining applications. In practice, an object may display characteristics of different clusters. In such cases, an object should belong to more than one cluster, and as a result, cluster boundaries necessarily overlap. Fuzzy set representation of clusters, using algorithm like fuzzy C-means [35], make it possible for an object to belong to multiple clusters with a degree of membership between 0 and 1. In some cases, the fuzzy degree of membership may be too descriptive for interpreting clustering results. Rough set based clustering provides a solution that is less restrictive than conventional clustering and less descriptive than fuzzy clustering [29, 35, 36]. The rough sets have been incorporated in the K-Means framework to develop Rough K-Means [22, 23, 24] algorithm to deal with uncertainty, vagueness and incompleteness in data.

Distributed clustering faces several additional challenges, compared to conventional clustering, such as communication overhead, security and integrity. It is required that an efficient distributed clustering algorithm needs to exchange a few data and avoids synchronization as much as possible. Despite the success of data clustering in centralized environments, only a few approaches to the problem in a distributed environment are available to date. Most of the existing distributed clustering algorithms available in the literature [33] aim to provide crisp clusters using K-Means algorithm or density based algorithm [12]. Hence, it is essential to integrate FKM and RKM in existing distributed clustering to obtain fuzzy and interval set clusters respectively.

3 Background and Related Works

In this section, the overview of three centroid based clustering algorithms is presented. The types of distributed clustering algorithms and the use of cluster ensemble are also discussed. Related works on distributed clustering are presented at the end of this section.

3.1 Centroid Based Partitioned Clustering

Generally, clustering algorithms can be categorized into partitioning methods, hierarchical methods, density-based methods, grid-based methods, and model-based methods. An excellent survey of clustering techniques can be found in [10, 29, 35, 36]. Partitioned clustering procedures typically start with the patterns of partitioning into a number of clusters and divide the data patterns by increasing the number of partitions. The partitioning methods are divided into two major subcategories, the centroid and the mediods algorithms [36]. In centroid based clustering, each object is assigned to the cluster whose centroid is closer to it. The centroid is average of all points in the clusters, where as mediod is most representative point of a cluster.

Almost all centroid based clustering algorithms follow the following four steps to perform the process of clustering [31]:

Step 1 – Initialization

A set of objects to be partitioned, the number of clusters and a centroid for each cluster are defined. Some implementations of the standard algorithm determine randomly the initial centroids; while some others preprocess the data and determine the centroids through calculations.

Step 2 - Classification

For each database object its distance to each of the centroids is calculated, the closest centroid is determined, and the object is incorporated to the cluster related to this centroid.

Step 3 - Centroid calculation

For each cluster generated in the previous step its centroid is recalculated.

Step 4 - Convergence condition

Several convergence conditions have been used from which the most utilized are the following: stopping when reaching a given number of iterations, stopping when there is no exchange of objects among groups, or stopping when the difference among centroids at two consecutive iterations is smaller than a given threshold. If the convergence condition is not satisfied, steps two, three and four of the algorithm are repeated.

Table 1 Summary of Symbols and Definitions

Symbols	Description
C	**Cluster Centroid Matrix**
d	**Data Dimension**
d_{ij}	$\|x_i - C_j\|$ **Distance from data object x_i and centroid C_j**
K	**Number of Clusters**
m	**Fuzzy Parameter**
n	**Number of Data Objects (Patterns)**
n_z	**Number of Objects in Cluster Z**
p	**Number of Distributed Datasets**
U	**Membership Matrix**
$\underline{U}(X)$	**Lower Approximation**
$\overline{U}(X)$	**Upper Approximation**
x	**Data Object Vector**
X	**Dataset Matrix**
Z	**Cluster**

The most well-known centroid algorithm is K-Means [10, 29]. The K-Means algorithm results in hard clusters, where each object belongs to one and only one cluster. The Fuzzy K-Means and Rough K-Means are variants of K-Means, but produce soft clusters, by incorporating fuzzy set and rough set respectively.

To facilitate subsequent discussion the main symbols used throughout the chapter and their definitions are summarized in Table 1.

3.1.1 K-Means Clustering

K-Means algorithm is a prototype-based, partitioned clustering technique that attempts to find user-specified number of clusters, which are represented by their centroids. This algorithm relies on finding cluster centers by trying to minimize a cost function or an objective function of dissimilarity or distance measure. In most cases this dissimilarity measure is chosen as the Euclidean distance.

A set of n vectors $x_i, i = 1, 2, \ldots, n$, are to be partitioned into K groups, $C_j, j = 1, 2, \ldots, K$. The cost function, based on the Euclidean distance between a vector x in group j and the corresponding cluster centroid C_j, can be defined by:

$$J = \sum_{j=1}^{K} \sum_{i=1}^{n} \left\| x_i - C_j \right\|^2 \tag{1}$$

The formal description of K-Means is given in Figure 1. Generally, the K-Means algorithm has the following important properties: (i) it is efficient in processing large datasets, (ii) it often terminates at a local optimum, (iii) the clusters have spherical shapes, (iv) it is sensitive to noise. Choosing the proper initial centroids is the key step of the basic K-Means procedure [29, 36].

3.1.2 Fuzzy K-Means Clustering

Fuzzy K-Means or Fuzzy C-Means [6, 29] is a method developed by Dunn in 1973 and improved by Bezdek in 1981 mainly to smooth the hard nature of K-Means algorithm. The FKM employs fuzzy partitioning such that a data point can belong to all clusters with different membership grades between 0 and 1. Larger membership values indicate higher confidence in the assignment of the object to the cluster. It is based on minimization of the following objective function:

$$J = \sum_{j=1}^{k} \sum_{i=1}^{n} U_{ij}^{m} \left\| x_i - C_j \right\|^2, \quad 1 \le m \le \infty \tag{2}$$

where m is any real number greater than 1, U_{ij} is the degree of membership of x_i in the cluster j, x_i is the ith of d dimensional measured data, C_j is the d dimension center of the cluster, and $\|*\|$ is Euclidean distance between any measured data and the centroid. The steps involved in Fuzzy K-Means algorithm is described in Figure 2.

Algorithm: Hard K-Means Algorithm

Input : Dataset of n objects with d features and the value of K
Output: Partition of the input data into K clusters
Procedure:

Step 1: Declare a membership matrix U of size n x K.
Step 2: Generate K cluster centroids randomly within the range of the data or select K objects randomly as initial cluster centroids. Let the centroids be $C_1, C_2, \ldots, C_K$.
Step 3: Calculate the distance measure $d_{ij} = \|X_i - C_j\|$ using Euclidean distance, for all cluster centroids $j = 1, 2, \ldots, K$ and data objects $x_i, i = 1, 2, \ldots, n$.
Step 4: Compute the U membership matrix

$$U_{ij} = \begin{cases} 1; & d_{ij} \leq d_{il}, \ j \neq l \\ 0; & otherwise \end{cases} \begin{array}{l} i=1,2,\ldots,n \\ j=1,2,\ldots,K \end{array}$$

Step 5: Compute new cluster centroids C_j

$$C_j = \frac{\sum_{i=1}^{n} (U_{ij}) \, x_i}{\sum_{i=1}^{n} (U_{ij})} \quad for \ j=1,2,\ldots,K$$

Step 6: Repeat steps 3 to 5 until convergence.

Fig. 1 Hard K-Means Algorithm

3.1.3 Rough K-Means Clustering

Rough set or Interval set is a mathematical tool used to deal with uncertainty. It is used, when there is insufficient knowledge to precisely define clusters as sets. The properties of rough sets are defined by Pawlak [30]. In rough clustering each cluster has two approximations, a lower and an upper approximation. The lower approximation is the set of objects definitely belonging to the vague concept, whereas the upper approximation is a set of objects possibly belonging to the same [23].

Rough set theory has made substantial progress as a classification tool in data minin. The basic concept of representing a set as lower and upper approximation can be used in broader context such as clustering. Clustering in relation to rough set theory is attracting increasing interest among researchers [22]. Lingras et al. [22, 23, 24] proposed Rough K-Means (RKM) algorithm, which does not verify all the properties of rough set theory, but they used following basic properties for their algorithm:

Property 1: a data object can be a member of one lower approximation at most.

Property 2: a data object that is a member of the lower approximation of a cluster is also member of the upper approximation of the same cluster.

Property 3: a data object that does not belong to any lower approximation is member of at least two upper approximations.

According to the above three properties, the lower approximation is a subset of the upper approximation. The difference between upper and lower approximation is called boundary region, which contains objects in multiple clusters. The algorithm is outlined in Figure 3.

Algorithm: Fuzzy K-Means Algorithm

Input : Dataset of n objects with d features, value of K and fuzzification value $m > 1$

Output: Membership matrix U_{ij} for n objects and K clusters

Procedure:

Step 1: Declare a membership matrix U of size n x K.

Step 2: Generate K cluster centroids randomly within the range of the data or select K objects randomly as initial cluster centroids. Let the centroids be $C_1, C_2, ..., C_K$.

Step 3: Calculate the distance measure $d_{ij} = \|X_i - C_j\|$ using Euclidean distance, for all cluster centroids $1, 2, \cdots, K$ and data objects $i = 1, 2, \cdots, n$.

Step 4: Compute the Fuzzy membership matrix U_{ij}

$$U_{ij} = \begin{cases} \left[\sum_{l=1}^{n} \left(\frac{d_{ij}}{d_{il}} \right)^{\frac{2}{m-1}} \right]^{-1} & \text{if } d_{ij} > 0 \\ = 1 & \text{if } d_{ij} = 0 \end{cases} \quad \begin{array}{l} i = 1, 2, ..., n \\ j = 1, 2, ..., K \end{array}$$

Step 5: Compute new cluster centroids C_j

$$C_j = \frac{\sum_{i=1}^{n} (U_{ij})_i}{\sum_{i=1}^{n} (U_{ij})} \quad for \ j = 1, 2, ..., K$$

Step 6: Repeat steps 3 to 5 until convergence

Fig. 2 Fuzzy K-Means Algorithm

Algorithm: Rough K-Means Algorithm

Input : Dataset of n objects with d features, value of K and values of W_{lower}, W_{upper} and *epsilon*

Output: Lower approximation $\underline{U}(K)$ and upper approximation $\overline{U}(K)$ of K clusters

Procedure:

Step 1: Randomly assign each data object to exactly one lower approximation $\underline{U}(K)$. By property 2, the data object also belongs to upper approximation $\overline{U}(K)$ of the same cluster.

Step 2: Compute cluster centroids C_j

If $\underline{U}(K) \neq \phi$ and $\overline{U}(K) - \underline{U}(K) = \phi$

$$C_j = \frac{\sum_{x \in \underline{U}(K)} x_j}{\left|\underline{U}(K)\right|}$$

Else

If $\underline{U}(K) = \phi$ and $\overline{U}(K) - \underline{U}(K) \neq \phi$

$$C_j = \frac{\sum_{x \in (\overline{U}(K) - \underline{U}(K))} x_j}{\left|\overline{U}(K) - \underline{U}(K)\right|}$$

Else

$$C_j = W_{lower} \times \frac{\sum_{x \in \underline{U}(K)} x_j}{\left|\underline{U}(K)\right|} + W_{upper} \times \frac{\sum_{x \in (\overline{U}(K) - \underline{U}(K))} x_j}{\left|\overline{U}(K) - \underline{U}(K)\right|}$$

$$where \ \ j = 1, 2, \cdots, K$$

Step 3: Assign each object to the lower approximation $\underline{U}(K)$ or upper approximation $\overline{U}(K)$ of cluster / clusters respectively. For each object vector x, let $d(x, C_j)$ be the distance between itself and the centroid of cluster C_j. Let $d(x, C_i)$ be $\min_{1 \leq j \leq K} d(x, C_j)$. The ratio $d(x, C_i)/d(x, C_j)$, $1 \leq i, j \leq k$ is used to determine the membership of x as follows:

If $d(x, C_i) / d(x, C_j) \leq$ *epsilon*, for any pair (i, j), the $x \in \overline{U}(C_i)$ and $x \in \overline{U}(C_j)$ and x will not be a part of any lower approximation. Otherwise, $x \in \underline{U}(C_i)$, such that $d(x, C_i)$ is the minimum for $1 \leq i \leq K$. In addition, $x \in \overline{U}(C_i)$.

Step 4: Repeat steps 2 and 3, until convergence.

Fig. 3 Rough K-Means Algorithm

In RKM algorithm, the concept of K-Means is extended by viewing each cluster as an interval or rough set. The membership of each object in lower approximation and upper approximation is determined by three parameters W_{lower}, W_{upper} and $epsilon$. The parameters W_{lower} and W_{upper} correspond to the relative importance of lower and upper bounds, and $W_{lower}+ W_{upper} = 1$. The $epsilon$ is a threshold parameter used to control the size of boundary region. Experimentation with various values of the parameters is necessary to develop a reasonable rough set clustering.

These three methods belong to iterative clustering algorithms. In each iteration, the object is reassigned and centroid is updated. The algorithm is repeated, until the criterion function converges. Both FKM and RKM use parameter values in membership identification and centroid calculation, to identify the uncertainty in clusters. In FKM, the parameter m as set to 2 gives consistent solutions in most cases [7]. But, RKM requires fine tuning in parameters to reach the required solution.

3.2 Distributed Clustering Algorithm

Distributed clustering assumes that the objects to be clustered reside on different sites. Instead of transmitting all objects to a central site where any standard clustering procedure can be applied to analyze the data, the data are clustered independently on the local sites. In a subsequent step, the central site tries to establish a global clustering based on the local models. Distributed clustering algorithms [33] can be classified along two independent dimensions: classification based on data distribution and classification based on data communication.

3.2.1 Classification Based on Data Distribution

A common classification based on data distribution in the literature [1, 25] is those, which apply to homogeneously distributed or heterogeneously distributed data. Homogeneous datasets contain the same set of attributes across distributed data sites. Examples include local weather databases at different geographical locations and market-basket data collected at different locations of a grocery chain. Heterogeneous data model supports different data sites with different schemata. For example, a disease emergence detection problem may require collective information from a disease database, a demographic database, and biological surveillance databases.

3.2.2 Classification Based on Data Communication

According to the type of data communication, distributed clustering algorithms are classified into two categories: multiple communications round algorithms and centralized ensemble-based algorithms. The first group consists of methods requiring multiple rounds of message passing. These methods require a significant amount of synchronization, whereas the second groups work asynchronously. Many of the distributed clustering algorithms work in an asynchronous manner by first generating the local clusters and then combining those at the central site. These approaches potentially offer two nice properties in addition to lower synchronization requirements. If the local models are much smaller than the local data, their transmission will result in excellent message complexity. Moreover, sharing only the local models may be a reasonable solution to privacy constraints in some situations. Both the ensemble approach and the multiple communication round-based clustering algorithms usually work a lot better than their centralized counterparts in a distributed environment [33].

3.3 Cluster Ensemble

Cluster ensemble [4] is the problem of combining multiple partitioning of a set of objects into a single consolidated clustering without accessing the features or algorithms that determined these partitioning. It emerged as a powerful method of improving both the robustness as well as the stability of unsupervised learning solutions. Cluster ensemble can be formed in a number of different ways such as

- Using different clustering algorithms to produce partitions for combination
- Changing initialization or other parameters of a clustering algorithm
- Using different features via feature extraction for subsequent clustering
- Partitioning different subsets of the original data

All the above four mechanisms try to produce more diversity by considering data from different aspects. The major hardship in cluster ensemble is consensus function and partitions combination algorithm to produce final partition, or in the other words finding a consensus partition from the output partitions of various clustering algorithms. The combination of multiple clustering can also be viewed as finding a median partition with respect to the given partitions, which is proven to be NP-complete.

Cluster ensemble usually is two stage processes. At the first, they store the results of some independent runs of different clustering algorithms or same clustering algorithm with different parameters. Then, they use the specific consensus function to find a final partition from stored results. There are many types of consensus function such as: hyper graph partitioning, voting approach, cluster-based similarity partitioning and meta clustering [34].

3.4 Related Works

There are various distributed clustering solutions proposed in the literature and their comprehensive survey can be obtained from [33]. Some of the recent research works on distributed clustering are discussed in this section.

Jongil Jeong et al. [13] proposed a distributed clustering scenario and modified K-Means algorithm for clustering huge quantities of biological data. By this algorithm, after clustering local datasets using K-Means, local centroids are transferred to central site and average centroid is calculated. Then local datasets are again clustered using averaged centroid as initial central set in K-Means algorithm.

Khanuja J. and Kamlakar K. [17] presented CLOUD, a cohesive framework for the identification of clusters and outliers in the distributed environment. CLOUD used a parameter free clustering algorithm SRA (Stability based RECORD Algorithm) for the clustering of data at local sites. It also proposed an efficient way to compute the K-Nearest Neighbour.

Jin R. et al. [15] presented distributed version of Fast and Exact K-Means (FEKM) algorithm, which collected sample data from each data source, and communicated it to the central node. The main data structure of FEKM, the cluster abstract table is computed and sent to all data sources to get global clusters. The boundary point and other sufficient statistics are communicated to central node.

The P2P K-Means algorithm is proposed in [33] for distributed clustering of data streams in a peer-to-peer sensor network environment. In the P2P K-Means algorithm, computation is performed locally, and communication of the local data models, represented by the corresponding centroids and the cluster counts is restricted only within a limited neighborhood.

Lamine M. Aouad et al. [19] proposed a lightweight distributed clustering technique based on a merging of independent local sub clusters according to a increasing variance constraint. The key idea of this algorithm is to choose a relatively high number of clusters locally, or an optimal local number using an approximation technique, and to merge them at the global level according to an increasing variance criterion which require a very limited communication overhead.

Cormode G. et al. [2] have introduced the problem of continuous, distributed clustering, and given a selection of algorithms, based on the paradigms of local vs. global computations, and furthest point or parallel guessing clustering. In their experimental evaluation, the combination of local and parallel guessing gave the least communication cost.

In [37], Zhou A. et al. proposed an EM-based (Expectation Maximization) framework to effectively cluster the distributed data streams. In the presence of noisy or incomplete data records, their algorithms learn the distribution of underlying data streams by maximizing the likelihood of the data clusters. A test-and-cluster strategy is proposed to reduce the average processing cost, which is especially effective for online clustering over large data streams.

Le-Khac N. et al. [20] presented an approach for distributed density-based clustering. The local models are created by DBSCAN at each node of the system and these local models are aggregated by using tree based topologies to construct global models. In [3], P-SPARROW algorithm is proposed for distributed clustering of data in peer-to-peer environments. The algorithm combined a smart exploratory strategy based on a flock of birds with a density-based strategy to discover clusters of arbitrary shape, size in spatial data.

In [8] Prodip Hore and L. Hall proposed Distributed Combining Algorithm to cluster large scale datasets without clustering all the data at a time. Data is randomly divided into almost equal size disjoint subsets. Each subset is clustered using the hard K-Means or Fuzzy K-Means algorithm. The centroids of subsets form an ensemble. A centroid correspondence algorithm transitively solves the correspondence problem among the ensemble of centroids. When the number of clusters in each subset is large, the complexity increases in centroid mapping due to collision. Moreover, when the number of clusters in each dataset is different, this type of centroid mapping is found not suitable. Prodip Hore extended this algorithm in [9], to avoid collision and filter bad centroids, but restricted to same number of clusters in each data source.

A series of research papers [6, 7, 24, 36] about comparison of various clustering techniques seems to be the best effort to identify the best technique for specific application, but concentrated on centralized environment. This proposed study compares the results of two distributed clustering algorithms against centralized clustering algorithm based on three centroid based clustering techniques.

4   Centroid Based Distributed Clustering

The framework for Centroid Based Distributed Clustering (CBDC) used for this comparative study is shown in Figure 4. The local subsets are clustered using any one of the Centroid Based Partitioned Clustering (CBPC) algorithms discussed in section 4. The local centroids are merged using distributed clustering algorithm to compute global centroid. Here, two types of distributed clustering algorithms namely Improved Distributed Clustering Algorithm (IDCA) and Distributed K-Means Algorithm (DKMA) are used for comparison. The global centroid is passed to local subsets, where the respective centroid based clustering algorithm is executed with single iteration named as Single Pass CBPC (SPCBPC), to obtain global cluster results.

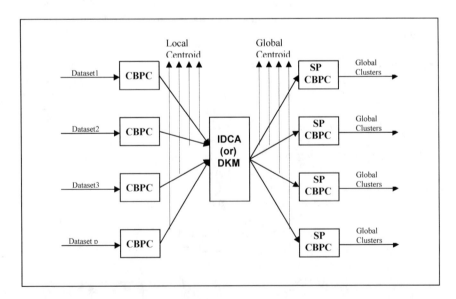

Fig. 4 Framework for Centroid Based Distributed Clustering

Algorithm: Generalized Improved Distributed Combining Algorithm

---

Input : p datasets with same underlying distribution
Output: Global partitions of p datasets
Procedure:

Step 1: Cluster each datasets by KM / FKM / RKM algorithm and obtain centroid matrix for each dataset. (The value of K need not be same for every dataset)
Step 2: For every pair of datasets, Euclidean distance among centroid vectors and store it in a distance matrix.
Step 3: Introduce dummy rows if necessary in the distance matrices, when the K value differs in each dataset.
Step 4: Find the correspondence relation between the centroid vectors using Hungarian method of assignment problem.
Step 5: Obtain global consensus chain by applying transitivity mapping
Step 6: Compute the weighted arithmetic mean of centroids in consensus chain to represent global set of centroids.
Step 7: Re-cluster local dataset using single pass KM / FKM / RKM algorithm, using global centroid.

Fig. 5 Generalized Improved Distributed Combining Algorithm

4.1 Improved Distributed Combining Algorithm

The IDCA [16] is refined version of Distributed Combining Algorithm [8], designed for distributed clustering. The improved algorithm avoids collision during centroid mapping process, by replacing greedy algorithm with hungarian method [18] of assignment problem. Moreover, when each dataset produces different number of clusters, the process of centroid mapping is performed effectively, with the support of hungarian method of unbalanced assignment problem. In this study, the IDCA is generalized to support K-Means, Fuzzy K-Means and Rough K-Means clustering techniques for local clustering. According to the type of clustering techniques used, this algorithm can produce either hard or soft clusters. The step by step procedure of the generalized IDCA is described in Figure 5.

4.2 Distributed K-Means Algorithm

The Distributed K-Means Algorithm (DKMA) is an ensemble learning based distributed clustering algorithm introduced in [14]. The DKMA first does clustering in local site using K-Means, then sent all centroid values to central site, finally global centroid values of underlying global clustering are obtained by using

K-Means again. This algorithm is generalized to support FKM and RKM and used for comparative analysis. The generalized version of Distributed K-Means Algorithm is described in Figure 6.

Algorithm: Generalized Distributed K-Means Algorithm

Input : p datasets with same underlying distribution
Output: Global partitions of p datasets
Procedure:

Step 1: Cluster each datasets by KM / FKM / RKM algorithm and obtain centroid matrix for each dataset. The value of K need not be same for every dataset.
Step 2: Combine centroid vectors of local datasets and cluster using K-Means to obtain global centroid.
Step 3: Re cluster local dataset using single pass KM / FKM / RKM algorithm, using global centroid.

Fig. 6 Distributed K-Means Algorithm

5   Experimental Analysis

In this section, the efficiency of Improved Distributed Combining Algorithm and Distributed K-Means Algorithm are compared with Centralized Clustering Algorithm based on three centroid based clustering algorithms, K-Means, Fuzzy K-Means and Rough K-Means separately. All experiments are conducted with the assumption of three distributed subsets and all are having non overlapping objects with same set of features. The results of synthetic dataset are illustrated with uniform and non uniform data distribution. In the case of benchmark datasets, only uniform data distribution is considered for the evaluation of performance.

Table 2 Details of Datasets

S. No.	Dataset	No. of Attributes	No. of Classes	No. of Instances
1	Synthetic	2	2	69
2	Iris	4	3	150
3	Breast Cancer	10	2	699
4	Pima Indian Diabetes	8	2	768
5	Page Block	10	5	5473

5.1 Datasets

All the algorithms discussed in section 4 have been implemented and tested with one small size synthetic dataset [22] and four bench mark datasets available in the UCI machine learning data repository [26]. Synthetic dataset is a two dimensional dataset with 69 objects. The detailed information of the datasets is shown in Table 2.

5.2 Evaluation Methodology

In order to evaluate the results of clustering algorithms, the most commonly used validity index - Davies Bouldin (DB) index [5] is used. It is a function of the ratio of the sum of within-cluster distance to between-cluster separation. As clusters have to be compact and separated, the lower DB index means better cluster configuration. The DB index [27] is defined as

$$DB = \frac{1}{n_Z} \left[\sum_{i=1}^{n_Z} \max_{\substack{j=1..n_Z \\ i=1..n_{zi}, i \neq j}} \left(\frac{S(Z_i) + S(Z_j)}{d(C_i, C_j)} \right) \right] \tag{3}$$

The formula to find the average distance between objects within the cluster Z_i for crisp clusters is expressed as

$$S(Z_i) = \frac{1}{n_Z} \sum_{x \in Z_i} d(x, C_i) \tag{4}$$

The between-cluster separation is defined as $d(C_i, C_j)$. The Fuzzy DB index uses fuzzy within-cluster distance, which is formulated as

$$S(Z_i) = \frac{\sum_{j=1}^{n} U_{ij}^{m} d(x_j, C_i)}{\sum_{j=1}^{n} U_{ij}^{m}} \qquad \begin{aligned} & i = 1, 2, \ldots, K \\ & m \text{ is fuzzification value (it is set as } 2) \end{aligned} \tag{5}$$

Similarly, the rough within-cluster distance is formulated based on W_{lower} and W_{upper} to find Rough DB index:

$$S(Z_i) = \begin{cases} w_{lower} \dfrac{\sum_{x \in \underline{U}(K)} \left\| x_j - C_i \right\|^2}{\left| \underline{U}(K) \right|} + w_{upper} \dfrac{\sum_{x \in (\overline{U}(K) - \underline{U}(K))} \left\| x_j - C_i \right\|^2}{\left| \overline{U}(K) - \underline{U}(K) \right|}, & \text{if } \underline{U}(K) \neq \phi \text{ and } \overline{U}(K) - \underline{U}(K) \neq \phi \\[4mm] \dfrac{\sum_{x \in (\overline{U}(K) - \underline{U}(K))} \left\| x_j - C_i \right\|^2}{\left| \overline{U}(K) - \underline{U}(K) \right|}, & \text{if } \underline{U}(K) = \phi \text{ and } \overline{U}(K) - \underline{U}(K) \neq \phi \\[4mm] \dfrac{\sum_{x \in \underline{U}(K)} \left\| x_j - C_i \right\|^2}{\left| \underline{U}(K) \right|}, & \text{otherwise} \end{cases}$$

$$\text{where } 1 \leq i \leq K \tag{6}$$

5.3 Parameter Tuning

Parameter tuning is concerned with selecting appropriate parameter values for Rough K-Means, by considering Rough DB index as evaluation criteria. The selection of initial parameters in Rough K-Means for every dataset is still a challenging problem in research [32]. Several experiments have been conducted on each dataset, to choose the optimal values from given range of parameters W_{lower} and *epsilon*, where $0.7 \leq W_{lower} \leq 0.9$ and $1.0 \leq epsilon \leq 1.5$. The same values of parameters are assigned for local clustering as well as global clustering.

During the tuning of parameter, it is observed that minimizing DB index will slowly move the clusters to crisp. In other words, increasing the size of boundary area may increase the DB index. Great care has been taken to compromise the size of boundary area, between the values of parameters and DB index.

5.4 Comparative Analysis of Synthetic Dataset

The distributed clustering process based on K-Means clustering is illustrated with synthetic dataset. This is followed by comparative analysis of IDCA, DKMA and CCA based on both Fuzzy K-Means and Rough K-Means clustering.

5.4.1 Uniformly Distributed Data

Initially, all objects of synthetic dataset are uniformly distributed to three subsets, such that each having 23 objects, 3 clusters of uniform size and 3 outliers as can be seen in Figures 7, 8 and 9. Next, each local subset is clustered through K-Means algorithm. The set of centroids obtained from local subsets, after clustering are shown in Table 2 and the same are highlighted in Figures 1, 2 and 3. These three sets of centroids are applied to K-Means version of both IDCA and DKMA for performance analysis.

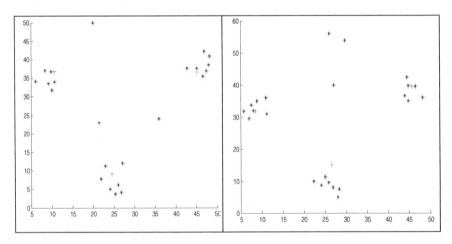

Fig. 7 Clusters of subset-1 (uniform type)   **Fig. 8** Clusters of subset-2 (uniform type)

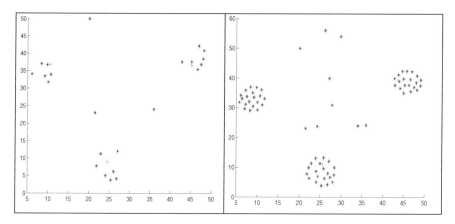

Fig. 9 Clusters of subset-3 (uniform type)   **Fig. 10** Global Clusters (uniform type)

Table 3 Centroids of three subsets

Centroid	Subset-1(S1)	Subset-2 (S2)	Subset-3 (S3)
C1	24.4000 9.1750	26.4667 14.9667	25.7429 8.5714
C2	10.5857 36.7429	8.5429 31.6571	39.5333 42.1556
C3	45.1500 36.6750	45.7286 39.5429	8.5000 32.7429

When they are applied to IDCA, the centroids of each local dataset forms an ensemble of centroids at the central place with the process called centroid mapping using step 2 to step 5 as in Figure 5. For every pair of subsets, the Euclidean distance between centroid vectors is computed and stored in a distance matrix, as shown in Table 4. The rows of the distance matrix represent the clusters of the first subset and the columns represent the clusters of the second subset. From the first row of matrix D1, it is observed that 6.1494 is minimum value and so first cluster of subset-1 is related with first cluster of subset-2. To identify the optimal relationship among the centroids of subsets, the two distance matrices D1 and D2 are applied to Hungarian method of assignment problem. The results of centroid mapping are obtained as shown in Figure 11.

Table 4 Distance Matrices

Distance Matrix-1 (D1) (S1 × S2)			Distance Matrix-2 (D2) (S2 × S3)		
6.1494	27.5117	37.1095	6.4361	30.1658	25.2744
26.9519	5.4807	35.2542	28.7887	32.7204	1.0866
28.6412	36.9494	2.9256	36.8600	6.7236	37.8445

388   N.K. Visalakshi and K. Thangavel

Fig. 11 Centroid Mapping

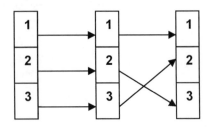

The centroid-1 of subset-1 is mapped with centroid-1 of subset-2. The centroid-2 of subset-1 is mapped with centroid-1 of subset-3. This will produce consensus chain of 1→1→1. Similarly other rows of consensus chain are generated by transitivity relation (2→2→3 and 3→3→2). The consensus chains ensemble similar types of cluster centroids. After grouping the centroids into consensus chains, the weighted arithmetic mean of centroids in a consensus chain is computed to represent a global cluster centroid. The results of global centroid calculation on this example are shown in first column of Table 5. Finally, the local clusters are updated using single pass K-Means algorithm and considering the global centroid as initial centroid. In other words, the Euclidean distance between the object and every centroid is computed and the object is moved to the one of the cluster centroid which yields minimum distance.

As per the Distributed K-Means algorithm, the same three set of centroids obtained from local subsets are combined and clustered using K-Means algorithm. The centroid of this intermediate clustering process will be considered as global centroid of distributed clustering, which is shown in second column of the Table 5. The same procedure is followed to obtain global clusters, as above.

The global centroid of both IDCA and DKMA are same for the synthetic dataset. After making several experiments, it is observed that DKMA may sometimes produce non optimum results, because the results of K-Means are purely dependant on the initial centroid. Although it is computationally complex, IDCA will always produce consistent global centroid. The last column of Table 5 shows the centroid obtained through centralized global K-Means algorithm. The comparison of DB index values of three types of distributed clustering algorithm based on three centroid based algorithms is shown in Table 6. The values of parameters W_{lower}, W_{upper} and epsilon are set as 0.72, 0.28 and 1.03 respectively.

Table 5 Comparison of Global Centroids

Algorithm	IDCA		DKMA		CCA	
Global Centroid	25.5365	10.9044	9.2095	33.7143	42.5960	39.2240
	9.2095	33.7143	25.5365	10.9044	25.4696	10.3087
	43.4706	39.4578	43.4706	39.4578	9.2095	33.7143

Table 6 Comparison of DB index for Synthetic Dataset

Validity Index	IDCA	DKMA	CCA
DB Index	0.3471	0.3471	0.3436
Fuzzy DB Index	0.1306	0.1306	0.1306
Rough DB Index	0.4927	0.4927	0.1627

5.4.2 *Non Uniformly Distributed Data*

The same synthetic dataset is non uniformly distributed, in such a way to produce different number of clusters in each local subset, as an example shown in Figures 12, 13 and 14. In case of IDCA, when the number of clusters is different, hungarian method of unbalanced assignment problem is applied to distance matrix for seeking centroid correspondence relationship. But, DKMA can simply provide required number of global clusters, independent of the number of local clusters. In IDCA, the number of global clusters should be equal to the maximum value among the number of local clusters, where as, DKMA allows the global number may be greater than or less than the local cluster numbers. The experiments are conducted for various types of distribution. The performance of IDCA, DKMA and CCA based on K-Means is shown in Table 7. The numbers provided in braces indicate the number of objects in corresponding subset.

The performance of IDCA, DKMA and CCA based on Fuzzy K-Means and Rough K-Means are tabulated in Table 8 and Table 9 respectively. It is obvious that fuzzy global clustering will provide overlapping clusters and rough global clustering will produce interval set clusters. While using RKM, though the dataset is non uniformly distributed, the values of parameter for RKM is not

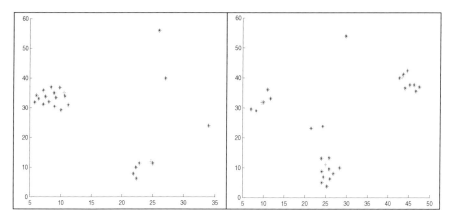

Fig. 12 Clusters of subset-1 (non uniform type)

Fig. 13 Clusters of subset-2 (non uniform type)

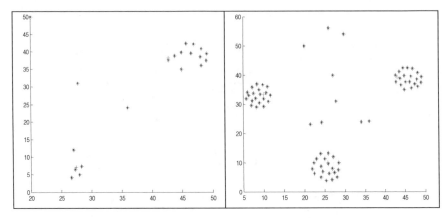

Fig. 14 Clusters of subset-3 (non uniform type)   **Fig. 15** Global Clusters (non uniform type)

Table 7 Comparison Based on K-Means for Non Uniform Type of Distribution

No. of Clusters			No. of Global	DB Index		
S1	S2	S3	Clusters	IDCA	DKMA	CCA
2(23)	3(26)	2(20)	3	1.393	0.379	0.344
3(28)	3(23)	2(18)	3	0.820	0.378	0.344
2(23)	3(23)	3(23)	3	0.938	0.531	0.344

Table 8 Comparison Based on FKM for Non Uniform Type of Distribution

No. of Clusters			No. of Global	Fuzzy DB Index		
S1	S2	S3	Clusters	IDCA	DKMA	CCA
2(23)	3(26)	2(20)	3	2.880	0.402	0.262
3(28)	3(23)	2(18)	3	1.093	0.275	0.262
2(23)	3(23)	3(23)	3	1.571	0.485	0.262

Table 9 Comparison Based on RKM for Non Uniform Type of Distribution

No. of Clusters			No. of Global	Rough DB Index		
S1	S2	S3	Clusters	IDCA	DKMA	CCA
2(23)	3(26)	2(20)	3	1.621	0.973	0.163
3(28)	3(23)	2(18)	3	2.013	0.193	0.163
2(23)	3(23)	3(23)	3	0.682	0.484	0.163

correspondingly tuned. The same values as in uniform data distribution are assigned to all local datasets. It reflects that there is much variation among the values of rough DB index.

5.5   Comparative Analysis of Benchmark Datasets

The centroid based partitioning clustering algorithms start with initial centroid. The choice of initial centroid may lead to better or worst cluster results. Hence, the experiment on each dataset runs 25 times and the best and average results are considered for analysis. The results of   IDCA, DKMA and CCA on four benchmark datasets based on K-Means clustering are represented in Table 10. From this table, it is observed that the average DB index value of IDCA is closer to Centralized Clustering except for breast cancer database. Similarly, the average DB index value of DKMA is almost consistent with CCA, except for iris dataset. While considering the best DB Index value, it is noted that DKMA has achieved minimum index value than the other two algorithms except for iris dataset. The average performance of IDCA and DKMA against CCA based on K-Means is compared in Figure 16.

The results of IDCA, DKMA and CCA on four benchmark datasets based on Fuzzy K-Means clustering are tabulated in Table 11. It is observed that IDCA has

Table 10 Comparison Based on KM for Benchmark Datasets

Dataset	IDCA		DKMA		CCA	
	Best	**Average**	**Best**	**Average**	**Best**	**Average**
Iris	0.664	0.665	0.664	0.962	0.645	0.692
Breast Cancer	0.468	0.591	0.231	0.455	0.290	0.461
Diabetes	0.674	0.722	0.674	0.722	0.713	0.713
Page Block	0.631	0.633	0.542	0.572	0.627	0.627

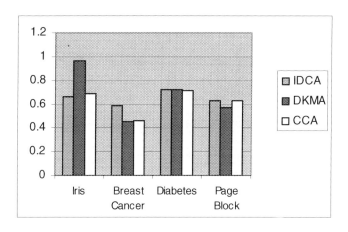

Fig. 16 Comparison based on KM

achieved consistent performance as like CCA for all datasets. But, DKMA has reached the same performance as like CCA, only for Breast Cancer and Diabetes datasets. When the best index values are considered, the performance of both DKMA and IDCA are closer to CCA. The index value of DKMA on page block data set is better than other two algorithms. The average performance of three algorithms based on FKM and RKM is depicted in Figure 17.

Table 11 Comparison Based on FKM for Benchmark Datasets

Dataset	IDCA		DKMA		CCA	
	Best	Average	Best	Average	Best	Average
Iris	0.648	0.649	0.648	1.873	0.616	0.616
Breast Cancer	0.467	0.467	0.466	0.467	0.469	0.469
Diabetes	0.690	0.690	0.682	0.690	0.693	0.693
Page Block	0.524	0.524	0.444	1.070	0.511	0.511

The Table 12 shows the results of IDCA, DKMA and CCA on four benchmark datasets based on Rough K-Means clustering along with corresponding parameter values. It is observed from the table, both IDCA and DKMA have achieved better results than CCA for diabetes dataset. The result of IDCA on breast cancer database is quite discouraging compared to the results achieved in DKMA and CCA. The performance of IDCA and DCA are same on page block dataset, but it is lesser than the performance of CCA. For iris dataset, the performance of IDCA is better than DKMA, but it is little bit behind the result of CCA. The variance in average performance of three algorithms based on RKM is depicted in Figure 18.

In general, it is observed from all the above three tables, there is a much variation in DKMA between the average index value and the best index value. This is due to the fact that the result of global centroid purely dependant on the initial seed of K-Means algorithm used in the intermediate stage of Distributed K-Means algorithm. Moreover, all our experiments are only for three distributed subsets, which automatically influence limited number of local centroid entries to K-Means algorithm for finding out global centroid.

Table 12 Comparison Based on RKM for Benchmark Datasets

Dataset	Parameters		IDCA		DKMA		CCA	
	W_{lower}	Epsilon	Best	Avg.	Best	Avg.	Best	Avg.
Iris	0.72	1.06	0.778	0.778	0.688	1.205	0.686	0.686
Breast Cancer	0.71	1.46	2.098	2.098	1.215	1.215	0.840	0.840
Diabetes	0.90	1.20	0.673	0.673	0.673	0.673	0.761	0.761
Page Block	0.71	1.46	1.001	1.001	0.564	1.047	0.735	0.735

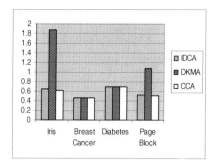

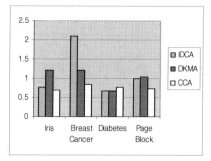

Fig. 17 Comparison based on FKM   **Fig. 18** Comparison based on RKM

5.6 Comparison of Hard and Soft Clusters

This part of the section compares the results of hard and soft clusters obtained by two distributed clustering algorithms each of which is integrated with three centroids based clustering algorithms. Hard clusters are generated using K-Means and Fuzzy K-Means, whereas soft clusters are retrieved using Fuzzy K-Means and Rough K-Means. While using FKM, objects with membership values higher than 0.6 are moved to hard clusters. But, objects with membership value higher than 0.4 obviously belong to soft clusters. By default, the outer boundaries of each cluster are considered as soft clusters. These four types of cluster results are evaluated based on crisp version of DB index and the best results are shown in Table 13.

Table 13 Comparison of Hard and Soft Clusters based on DB Index

Dataset	Hard Clusters				Soft Clusters			
	KM		FKM		FKM		RKM	
	IDCA	DKMA	IDCA	DKMA	IDCA	DKMA	IDCA	DKMA
Iris	0.664	0.664	0.678	0.678	0.699	0.699	0.778	0.819
Breast Cancer	0.468	0.231	0.489	0.489	0.674	0.674	1.869	0.955
Diabetes	0.674	0.674	0.728	0.728	0.746	0.746	0.758	0.758
Page Block	0.631	0.542	0.605	0.510	0.581	0.514	0.699	0.699

The variance of DB index for hard clusters based on KM and FKM is compared in Figure 19. Similarly, the variance of DB index for soft clusters based on FKM and RKM is compared in Figure 20. It is noticed that the DB index values of soft clusters are almost greater than the DB index values of hard clusters because of overlapping in soft clusters.

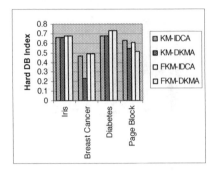

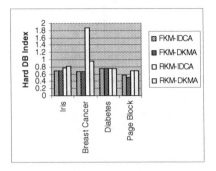

Fig. 19 Comparison of Hard Clusters   **Fig. 20** Comparison of Soft Clusters

6 Conclusion

Two distributed clustering algorithms IDCA and DKMA have been reviewed and
generalized to support K-Means, Fuzzy K-Means and Rough K-Means. The key
idea of both IDCA and DKMA is to regard distributed clustering as cluster
ensembles. The problem of obtaining global clusters in distributed environment is
solved by combining local centroids based on ensemble approach. The
performance of two distributed clustering algorithms has been compared with
centralized clustering and it is measured with popular validity measure DB index.
After analysis, it is shown that without compromising much with the accuracy, the
time taken by distributed clustering is much less than centralized clustering.

From this experimental analysis the following observations are made:

- The results presented in this paper show that in general the best index
  values of DKMA based on KM, FKM and RKM are always closer to or
  sometimes better than CCA. So, by applying any heuristics, DKMA can
  still be optimized to have optimum consistent results.
- The performance of IDCA in uniform distribution is stable almost in all
  situations achieving closer performance as Centralized Clustering. But its
  performance in non uniform distribution is fair compared to DKMA and
  CCA.
- It is observed from the results of synthetic dataset, DKMA is more
  suitable than IDCA for non uniform distribution. Anyway, its
  performance is still to be improved towards Centralized Clustering
  Algorithm.
- In case of non uniform distribution, the proper value of parameters to
  RKM needs to be set, separately for each dataset in order to improve the
  performance of distributed clustering.
- Both DKMA and IDCA can be applied for obtaining hard and soft
  clusters in distributed environment. Hard clusters can be obtained by
  integrating K-Means or Fuzzy K-Means, whereas Rough or Fuzzy K-
  Means algorithm is to be used with distributed clustering algorithms to
  obtain soft clusters.

- Since both DKMA and IDCA are centroid based algorithms, they can be used with any other centroid based clustering algorithm like Expectation Maximization, K-Harmonic and other variants of K-Means and Fuzzy K-Means algorithms.

References

[1] Chen, R., Sivakumar, K., Kargupta, H.: Collective Mining of Bayesian Networks from Distributed Heterogeneous Data. Knowledge and Information Systems Journal 6, 164–187 (2004)

[2] Cormode, G., Muthukrishnan, S., Zhuang, W.: Conquering the Divide: Continuous Clustering of Distributed Data Streams. In: IEEE 23rd International Conference on Data Engineering, pp. 1036–1045 (2007)

[3] Folino, G., Forestiero, A., Spezzano, G.: Swarm-Based Distributed Clustering in Peer-to-Peer Systems. In: Talbi, E.-G., Liardet, P., Collet, P., Lutton, E., Schoenauer, M. (eds.) EA 2005. LNCS, vol. 3871, pp. 37–48. Springer, Heidelberg (2006)

[4] Ghosh, J., Merugu, S.: Distributed Clustering with Limited Knowledge Sharing. In: Proceedings of the 5th International Conference on Advances in Pattern Recognition, pp. 48–53 (2003)

[5] Halkidi, M., Batistakis, Y., Vazirgiannis, M.: Cluster validity methods: part II. ACM SIGMOD Record 31(3), 19–27 (2002)

[6] Hamerly, G., Elkan, C.: Alternatives to the K-Means algorithm that find better clusterings. In: Proceedings of the Eleventh International Conference on Information and Knowledge Management, pp. 600–607 (2002)

[7] Hammouda, K.: A Comparative Study of Data Clustering Techniques. In: Tools of Intelligent Systems Design. Course Project SYDE 625 (2000)

[8] Hore, P., Hall Lawrence, O.: Scalable Clustering: A Distributed Approach. In: IEEE International Conference on Fuzzy Systems, pp. 25–29 (2004)

[9] Hore, P., Hall Lawrence, O., Goldgofz, D.: A Cluster Ensemble Framework for Large Datasets. In: Proceedings of IEEE Conference on Systems, Man Cybernetics B (2006)

[10] Jain, A.K., Murthy, M.N., Flynn, P.J.: Data Clustering: A Review. ACM Computing Surveys 31(3), 265–323 (1999)

[11] Januzaj, E., Kriegel Hans, P., Pfeifle, M.: Towards Effective and Efficient Distributed Clustering. In: Proceedings of International Workshop on Clustering Large Datasets, 3rd IEEE International Conference on Data Mining, pp. 49–58 (2003)

[12] Januzaj, E., Kriegel Hans, P., Pfeifle, M.: DBDC: Density Based Distributed Clustering. In: Bertino, E., Christodoulakis, S., Plexousakis, D., Christophides, V., Koubarakis, M., Böhm, K., Ferrari, E. (eds.) EDBT 2004. LNCS, vol. 2992, pp. 88–105. Springer, Heidelberg (2004)

[13] Jeong, J., Ryu, B., Shin, D., Shin, D.: Integration of Distributed Biological Data using modified K-means algorithm. In: Washio, T., Zhou, Z.-H., Huang, J.Z., Hu, X., Li, J., Xie, C., He, J., Zou, D., Li, K.-C., Freire, M.M. (eds.) PAKDD 2007. LNCS, vol. 4819, pp. 469–475. Springer, Heidelberg (2007)

[14] Genlin, J., Xiaohan, L.: Ensemble learning based distributed clustering. In: Washio, T., Zhou, Z.-H., Huang, J.Z., Hu, X., Li, J., Xie, C., He, J., Zou, D., Li, K.-C., Freire, M.M. (eds.) PAKDD 2007. LNCS, vol. 4819, pp. 312–321. Springer, Heidelberg (2007)

[15] Jin, R., Goswami, A., Agarwal, G.: Fast and exact out-of-core and distributed K-means clustering. Knowledge and Information Systems 10(1), 17–40 (2006)

[16] Karthikeyani, N.V., Thangavel, K., Alagambigai, P.: Ensemble Approach to Distributed Clustering. In: Natarajan (ed.) Mathematical and Computational Model, pp. 252–261. Narosa Publishing House, New Delhi (2007)

[17] Khanuja, J., Karlapalem, K.: CLOUD: Cluster Identification and Outlier Detection for Distributed Data. Technical report (2007)

[18] Kuhn, H.W.: The Hungarian Method for the Assignment Problem. Naval. Res. Logist. Quart 2, 83–97 (1995)

[19] Lamine, M.A., Le-Khac, N., Tahar, M.K.: Lightweight Clustering Technique for Distributed Data Mining Applications. In: Perner, P. (ed.) ICDM 2007. LNCS (LNAI), vol. 4597, pp. 120–134. Springer, Heidelberg (2007)

[20] Le-Khac, N., Lamine, M.A., Tahar, M.K.: A New Approach for Distributed Density Based Clustering on Grid Platform. In: Cooper, R., Kennedy, J. (eds.) BNCOD 2007. LNCS, vol. 4587, pp. 247–258. Springer, Heidelberg (2007)

[21] Li, T., Zhu, S., Ogihara, M.: A New distributed data mining model based on similarity. In: Proceedings of the 2003 ACM symposium on Applied Computing, pp. 432–436 (2003)

[22] Lingras, P., Chen, M., Miao, D.: Precision of Rough Set Clustering. In: The Sixth International Conference on Rough Sets and Current Trends in Computing Akron, Ohio, USA (submitted, 2008)

[23] Lingras, P., West, C.: Interval set clustering of web users with rough k-means. Journal of Intelligent Information Systems 23(1), 5–16 (2004)

[24] Lingras, P., Yan, R., Jain, A.: Web usage mining: Comparison of conventional, fuzzy, and rough set clustering. In: Zhang, Y., Liu, J., Yao, Y. (eds.) Computational Web Intelligence: Intelligent Technology for Web Applications, ch. 7, pp. 133–148. Springer, Heidelberg (2004)

[25] Merugu, S., Ghosh, J.: A Distributed Learning Framework for Heterogeneous Data Sources. In: Proceedings of the 11th International Conference on Knowledge Discovery and Data Mining (KDD 2005) (2005)

[26] Merz, C.J., Murphy, P.M.: UCI Repository of Machine Learning Databases. Irvine, University of California (1998), http://www.ics.uci.eedu/~mlearn/

[27] Mitra, S., Banka, H., Pedrycz, W.: Rough-Fuzzy Collaborative Clustering. IEEE Transactions on Systems, Man, and Cybernetics –Part B: Cybernetics 36(4), 795–805 (2006)

[28] Park, B., Kargupta, H.: Distributed Data Mining. In: Ye, N. (ed.) The Hand Book of Data Mining. Lawrence Erlabum Associates, Publishers, Mahwah (2003)

[29] Tan, P.-N., Steinbach, M., Kumar, V.: Cluster Analysis: Basic Concepts and Algorithms. In: Introduction to Data Mining. Pearson Addison Wesley, Boston (2006)

[30] Pawlak, Z.: Rough sets. Internationl Journal of Information and Computer Sciences 11, 145–172 (1982)

[31] Perez, J.O., Pazos, R.R., Cruz, L.R., et al.: Improving the Efficiency and Efficacy of the K-Means Clustering Algorithm through a new convergence condition. In: Gervasi, O., Gavrilova, M.L. (eds.) ICCSA 2007, Part III. LNCS, vol. 4707, pp. 674–682. Springer, Heidelberg (2007)

[32] Peters, G.: Some Refinements of Rough K-Means clustering. Pattern Recognition 39(8), 1481–1491 (2006)

[33] Sanghamitra, B., Giannella, C., Maulik, U., et al.: Clustering Distributed Data Streams in Peer-to-Peer Environments. Information Science 176(4), 1952–1985 (2006)

[34] Strehl, A., Ghosh, J.: Cluster Ensembles – A Knowledge Reuse Framework for Combining Multiple Partitions. Journal of Machine Learning Research 3, 583–617 (2002)

[35] Xiong, X., Lee, K.T.: Similarity-Driven Cluster Merging method for Unsupervised fuzzy clustering. In: Proceedings of the 20th conference on Uncertainty in Artificial Intelligence, pp. 611–618 (2004)

[36] Xu, R., Wunsch II, D.: Survey of clustering algorithms. IEEE Transaction on Neural Networks 16(3), 645–678 (2005)

[37] Zhou, A., Cao, F., Yan, Y., Sha, C., He, X.: Distributed Data Stream Clustering: A Fast EM-based Approach. In: ICDE 2007, IEEE 23rd International Conference on Data Engineering, pp. 736–745 (2007)

Author Index